Tourism and Climate Change

CLIMATE CHANGE, ECONOMIES AND SOCIETY: LEADERSHIP & INNOVATION
Series Editors: Susanne Becken, Lincoln University, New Zealand and John E. Hay, JEH Ltd, New Zealand and Ibaraki University, Japan.

Aims of the Series
This book series seeks to stimulate further development and dissemination of knowledge and understanding in relation to climate change and society, including a perspective of economic implications and measures. We encourage contributions from a wide range of fields including Geopolitics, Social Sciences, Human Geography, Anthropology, Resource Management, Development Studies, Ecological/Environmental Economics, and Sustainability Sciences. The aim of the series is to provide high-quality interdisciplinary assessments of critical aspects of the Climate Change and Society interface that can only come from climate and social scientists and economists with a commitment to undertaking and synthesising research in ways that deliver policy-relevant findings.

Other books of Interest:
Tourism, Recreation and Climate Change
C. Michael Hall and James Higham (eds)
Wildlife Tourism
David Newsome, Ross Dowling and Susan Moore
Rural Tourism and Sustainable Business
Derek Hall, Irene Kirkpatrick and Morag Mitchell (eds)
Tourism Ethics
David A. Fennell

For more details of these or any other of our publications, please contact:
Channel View Publications, Frankfurt Lodge, Clevedon Hall,
Victoria Road, Clevedon, BS21 7HH, England
http://www.channelviewpublications.com

CLIMATE CHANGE, ECONOMIES AND SOCIETY
Series Editors: Susanne Becken and John E. Hay

Tourism and Climate Change

Risks and Opportunities

Susanne Becken and John E. Hay

CHANNEL VIEW PUBLICATIONS
Clevedon • Buffalo • Toronto

Library of Congress Cataloging in Publication Data
Becken, Susanne
Tourism and Climate Change: Risks and Opportunities/Susanne Becken and John Hay.
Includes bibliographical references and index.
1. Tourism–Environmental aspects. 2. Climatic changes–Environmental aspects.
I. Hay, John E. II. Title.
G155.A1B3819 2007
338.4'791–dc22 2007004441

British Library Cataloguing in Publication Data
A catalogue entry for this book is available from the British Library.

ISBN-13: 978-1-84541-067-4 (hbk)
ISBN-13: 978-1-84541-066-7 (pbk)

Channel View Publications
An imprint of Multilingual Matters Ltd

UK: Frankfurt Lodge, Clevedon Hall, Victoria Road, Clevedon BS21 7HH.
USA: 2250 Military Road, Tonawanda, NY 14150, USA.
Canada: 5201 Dufferin Street, North York, Ontario, Canada M3H 5T8.

The policy of Multilingual Matters/Channel View Publications is to use papers that are natural, renewable and recyclable products, made from wood grown in sustainable forests. In the manufacturing process of our books, and to further support our policy, preference is given to printers that have FSC and PEFC Chain of Custody certification. The FSC and/or PEFC logos will appear on those books where full certification has been granted to the printer concerned.

Typeset by Datapage International Ltd.
Printed and bound in Great Britain by MPG Books Ltd.

Contents

Illustrative material . . . vii
Abbreviations . . . xiii
Acknowledgements . . . xv
Preface . . . xvii

1 Introduction . . . 1
2 The Tourism–Climate System . . . 6
Introduction . . . 7
The Tourism System . . . 10
The Climate System . . . 16
Examples of Tourism–Climate Interactions . . . 20
Conclusions . . . 34
3 Case Studies of the Tourism–Climate System . . . 35
Introduction . . . 36
Climate Change and Tourism in Alpine Europe . . . 36
Climate Change and Tourism in Small Island Countries . . . 45
Insurance, Climate Change and Tourism . . . 59
International Aviation . . . 71
Conclusions . . . 81
4 An Overview of Tourism . . . 83
Introduction . . . 84
Who Is a Tourist and What Is Tourism? . . . 86
Tourist Destinations – Main Global Tourist Flows . . . 88
Economic Importance of Tourism . . . 90
Observed Tourism Trends . . . 97
Conclusions . . . 114
5 Global and Regional Climate Change . . . 116
Introduction . . . 117
The Causes of Global Climate Change . . . 117
Climate Variability and Extremes . . . 119
Atmospheric Ozone and Climate Change . . . 125
Futures for Climate and Tourism . . . 126
Uncertainties, Abrupt Climate Change and 'Surprises' . . . 140
Conclusions . . . 143
6 Methodologies for Greenhouse Gas Accounting . . . 144
Introduction . . . 145
Energy Use and Greenhouse Gas Emissions . . . 145

Accounting Framework 150
Bottom-up Analysis of Tourism Industries 152
Bottom-up Analysis of Tourist Travel Behaviour 161
Top-down Analysis 165
Conclusions 171
7 Climate Change Mitigation Measures 173
Introduction 174
Transport 176
Tourism Businesses 203
Carbon Compensation Projects for Tourism 217
Conclusions 221
8 Climate Change-related Risks and Adaptation 223
Introduction 224
Overview of Adaptation 225
A Risk-based Approach to Adaptation by the Tourism Sector 228
Characterisation and Management of Climate Change-related Risks to Tourism 229
Conclusions 260
9 Climate Change Policies and Practices for Tourism 261
Introduction 262
Global Context 263
Underlying Policy Principles 272
Business Practices and Managing Climate Change 278
National Policy Making 283
Conclusions 299
10 Conclusion 301

References 306

Index 328

Illustrative Material

Figures

2.1 Framework for assessing the relationship between tourism and climate change . . . 8
2.2 A model of the tourism system . . . 12
2.3 Contributions of tourism to global climate change . . . 17
2.4 The greenhouse effect . . . 18
2.5 The Earth system: switch and choke elements . . . 19
2.6 Climate–tourism hotspots . . . 26
2.7 Chaos model of tourism . . . 33
3.1 The tourism–climate system: Alpine Europe . . . 37
3.2 Adaptation strategies for ski resorts . . . 40
3.3 Artificial snow making using snow guns . . . 41
3.4 The tourism–climate system: small island destinations . . . 46
3.5 Direct and indirect consequences of climate change for tourism and adaptation measures . . . 48
3.6 Backpacker accommodation with a solar hot water system on the roof . . . 57
3.7 Climate change impact on the insurance industry and tourism . . . 60
3.8 Number of great natural catastrophes per year . . . 61
3.9 Economic losses and insured losses . . . 63
3.10 Costs of damage-causing weather events . . . 66
3.11 Complete devastation of Niue after cyclone Heta . . . 68
3.12 GHG emissions from aviation . . . 72
3.13 International tourist arrivals worldwide: Long-haul versus intra-regional trips 1995–2020 . . . 73
3.14 Impacts of aviation on the atmosphere . . . 74
3.15 Estimates of the globally and annually averaged radiative forcing (W_m^{-2}) from subsonic aircraft emissions in 1992 . . . 75
3.16 Example of an emission charge . . . 81

4.1 Domestic departures by country of origin. 93
4.2 International departures by country of origin 94
4.3 Aiding planning of tourism operations by providing information on the weather/climate 106
4.4 A definition of sustainable tourism provided by the UNWTO 110
4.5 Keep winter cool campaign 113
5.1 Atmospheric concentrations for three GHGs, CO_2, CH_4 and NO_x 118
5.2 Storylines for the four IPCC GHG Emissions Scenarios . . . 128
5.3 Scenarios of CO_2 gas emissions and consequential atmospheric concentrations of CO_2 128
5.4 Multi-model ensemble annual mean change of temperature for the period 2071–2100, relative to the period 1961–1990, for the B2 emissions scenario 135
5.5 Multi-model ensemble annual mean change of precipitation for the period 2071–2100, relative to the period 1961–1990, for the B2 emissions scenario 136
5.6 Relative risks of hospital admissions on consecutive days with high daily maximum Physiological Equivalent Temperature assessed during summer 2003 139
5.7 Estimates of changes in temperature and precipitation. . . . 141
6.1 Example of energy-related questions for accommodation businesses. 159
6.2 Deriving energy use and CO_2 emissions for a tourist attraction 160
6.3 Comparison of travel distances for different transport options 164
7.1 Global CO_2-equivalent emissions from tourism 175
7.2 Passengers at major airports in Great Britain 179
7.3 Airbus 380 182
7.4 Comparison of the energy use and CO_2 emissions for a journey from Frankfurt to Munich by rail and car 194
7.5 Sightseeing of Angkor Watt (Cambodia) by bike. 202
7.6 Annual energy demand and capacity for different accommodation categories in New Zealand; and CO_2 emissions per visitor night 204

7.7 Breakdown of energy use into fuel sources for accommodation businesses in Viti Levu, Fiji 207

7.8 Example of a solar water system 214

7.9 Solar panels at the Mauna Lani Bay Resort, Hawaii 215

7.10 Expanded scheme for an autarkic CO_2-emission free energy supply for holiday facilities 217

7.11 Carbon cycle and sequestration of CO_2 through forest sinks 220

8.1 The likelihood of a given extreme event occurring in Maldives 225

8.2 Risk-based approaches to identifying and assessing options for managing the adverse consequences of climate change 228

8.3 The principal components of tourism–climate futures 231

8.4 Change in international arrivals and departures due to a global mean temperature increase of 1.03 °C by 2050 232

8.5 Components of the tourism transportation system sensitive to changes in climate 234

8.6 Tourism dependencies on coral reefs 239

8.7 Reforestation of mangrove forest at a tourist resort in Fiji 243

8.8 The resort development path 256

9.1 Timeline of major events relating to climate change and tourism 264

9.2 Steps of the Adaptation Policy Framework 270

9.3 Guidelines for adaptation mainstreaming 272

9.4 Framework for assessing company climate change exposure 278

9.5 Decisions made during the construction process and climate change-related factors that need to be considered . . . 281

9.6 National policies that have an impact on climate-change risks 285

9.7 Policy framework for climate change mitigation and adaptation 286

9.8 Seawall to protect beach and tourist accommodation 290

9.9 A normative model of participatory tourism planning 292

9.10 The roles of knowledge, skills and motivation in responding to climate change 294

9.11 The technology transfer process 296

Tables

2.1 Facets of climate and impact on tourists 21
2.2 Importance of destination and climate attributes 23
2.3 Climatic information on the three climate–tourism hot spot regions 24
3.1 Projected increases in (a) air temperature and (b) changes in precipitation (%) for small island regions 47
3.2 Possible adaptation measures for tourism in small island countries and barriers to implementation 51
3.3 Adaptation measures taken by accommodation providers in Fiji 54
3.4 Adaptation measures for tourism and their positive or negative ancillary effects 58
3.5 Damage to livelihoods by hurricane Ivan 62
3.6 Actions to reduce climate change-related loss to the insurance industry 69
3.7 Potential climate change challenges and opportunities 70
3.8 Possible schemes for a levy on air travel emissions within the EU 80
4.1 Top destinations for international tourism 89
4.2 Total tourist activity: top 10 countries in the world 90
4.3 World's top tourism earners 95
4.4 Tourism trends 98
4.5 Examples of tourism niche markets 100
4.6 World arrivals by mode of transport 102
4.7 Exemplified positions for light and dark green variants of sustainable tourism 110
5.1 Historical and present concentrations of the main GHGs, their atmospheric lifetimes and their GWPs 120
5.2 Forecast international tourist arrivals for 2010 and 2020 132
6.1 Conversion from one form of energy into another and typical efficiencies 146
6.2 Estimated GHG emission factors for European cars 149
6.3 Fuel use and average sector distance for representative types of aircraft 154

6.4 Fleet and fuel consumption data for New Zealand campervan companies . . . 156
6.5 Emission factors for different transport modes . . . 157
6.6 A simplified environmental accounting matrix . . . 167
6.7 A tourism sector distinguished in an input–output model. . . .170
7.1 Framework for considering transport GHG reductions 177
7.2 Tourist arrivals for main countries of origin, average flying distance, energy use and CO_2 emissions . . . 180
7.3 Energy intensities for various airlines . . . 182
7.4 Breakdown of transport modes for tourism travel in different countries . . . 187
7.5 Transport energy efficiencies of transport modes used by tourists . . . 191
7.6 Energy consumption and emissions per passenger-kilometre (g/pkm) for passenger trains in Denmark . . . 193
7.7 Comparison of alternative fuel and vehicle technologies for vehicles . . . 197
7.8 Examples of various transport management initiatives 199
7.9 Examples of energy intensities for different accommodation categories . . . 205
7.10 Electricity use in Vietnamese hotels . . . 206
7.11 Energy intensities for different leisure activities . . . 208
7.12 Practices for energy saving in accommodation businesses . . .210
8.1 Climate change adaptation portfolio for protected area agencies . . . 244
8.2 Ideal climate-related requirements for water- and land-based recreation activities . . . 249
9.1 GHG emissions and reduction targets for Annex I Parties 266
9.2 Adaptive responses in the case of tourism . . . 276
9.3 Including climate change risk into financial models for businesses . . . 280
9.4 Policy instruments for environmental improvements 288

Text Boxes

1 Aspen Skiing Company . . . 44
2 Hurricane Katrina, 29 August 2005 . . . 65
3 China – major emerging player in global tourism . . . 91

4 Hypermobility – high-frequency, long-distance travellers . . . 108
5 Are tourists sensitive to GHG emissions? 111
6 Recent changes in the observed climate. 121
7 Extreme climatic events in the European Alps 123
8 Storylines for tourism, based on the IPCC Scenarios 129
9 El Niño and tourism 137
10 Basic information needed for energy and GHG accounting. . . . 147
11 Auckland International Airport 186
12 European MUSST Project 188
13 East Japan Railway Company 192
14 Bad Hofgastein/Werfenweng, Austria. 201
15 Hong Kong Disneyland. 212
16 Tortoise Head Guest House, Australia. 216
17 Carbon offsets and air travel 219
18 Adaptation for coastal tourism 230
19 Climate change and wine tourism. 240
20 Using climate models to assess tourism destinations' competitiveness. 247
21 Climate change and the ski industry 258
22 Kyoto mechanisms for mitigation 269
23 Integrating adaptation and mitigation 284
24 Methods to support participatory planning. 290

Abbreviations

ADB	Asian Development Bank
CAEP	Committee on Aviation Environment Protection
CDM	Clean Development Mechanism
CER	Certified Emission Reduction
CNG	compressed natural gas
COP	Conference of the Parties
ENSO	El Niño – Southern Oscillation
ERU	Emission Reduction Unit
ET	emissions trading
ETC	European Travel Commission
EU	European Union
Eurostat	Statistical Office of the European Communities
FIT	free and independent tourists
GDP	gross domestic product
GEF	Global Environment Facility
GHG	greenhouse gas
GWP	global warming potential
IATA	International Air Transport Association
ICAO	International Civil Aviation Organisation
IHEI	Hotels Environment Initiative
IO	input–output analysis
IPCC	Intergovernmental Panel on Climate Change
IVS	International Visitor Survey
JI	Joint Implementation
LPG	liquefied petroleum gas
LTO	landing and take-off cycle
MICE	Meeting, Incentive, Convention, Event
NGOs	non-governmental organisations
NOAI	North Atlantic Oscillation Index
NTS	National Travel Survey

OECD	Organisation for Economic Cooperation and Development
PATA	Pacific Asia Tourism Association
PV	photovoltaic
SBSTA	Subsidiary Body for Scientific and Technical Advice
SEEA	System of Economic and Environmental Accounts
SME	small and medium-sized enterprise
SNA	System of National Accounts
SPTO	South Pacific Tourism Organization
TOI	Tour Operators Initiative
TPR	tourism product ratio
TSA	Tourism Satellite Account
UN	United Nations
UNDP	United Nations Development Programme
UNEP	United Nations Environment Programme
UNFCCC	United Nations Framework Convention on Climate Change
UNICEF	United Nations Children's Fund
UNWTO	United Nations World Tourism Organization
WMO	World Meteorological Organization
WTTC	World Travel & Tourism Council

Acknowledgements

Contributions to this book have been made by a number of experts in the field of tourism and climate change:

- Martin Beniston, Climate Change and Climate Impact, University of Geneva, Switzerland. Email: Martin. Beniston@unige.ch
- Jean Paul Ceron, CRIDEAU, Université de Limoges, France. Email: ceron@chello.fr
- Stefan Gössling, Department of Service Management, Lund University, Sweden. Email: stefan.gossling@msm.hbg.lu.se
- James Lennox, Sustainable Business and Government Group, Landcare Research, New Zealand. Email: lennoxj@landcareresearch.co.nz
- Tom Morton, Climate Care, United Kingdom. Email: Tom.Morton@climatecare.org
- Paul Peeters, NHTV, Breda University, The Netherlands. Email: pmpeeters@peetersadvies.nl

We would also like to thank the following people for their valuable input and contribution at different stages of producing this book: Phil Hart, David Simmons, David Viner, Maurice Marquart, James Lennox and Axel Reiser. We also wish to thank Christine Bezar for editing the chapters, and Anouk Wanrooy and David Hunt for assisting with the graphics. We are grateful for financial and institutional support from Landcare Research, New Zealand.

Preface

The rapid and sustained rise of tourism over the past 50 years is one of the most remarkable phenomena of our time. The number of international tourist arrivals has increased from 25 million in 1950 to 808 million in 2005. International tourism represents approximately 7% of the worldwide export of goods and services. For over 80% of countries tourism is one of the top five export categories, and for around 40% of countries it is the main source of foreign exchange earnings. The 11% increase in global revenues generated annually by international tourist arrivals is well above the rate of growth of the world economy.

Favourable climatic conditions at destinations are key attractions for tourists. Weather can ruin a vacation, while climate can devastate a holiday destination. Climate is especially important for the success of beach destinations and conventional sun-and-sea tourism, the dominant form of tourism. Tourists are attracted to coastal areas and islands by ample sunshine, warm temperatures and little precipitation, escaping from harsher weather conditions and seasons in their home countries. Other forms of tourism, such as mountain tourism and winter sports, are also highly dependent on favourable climate and weather conditions, such as adequate precipitation and snow cover.

Climate change will not only impact on tourism directly by changes in temperature, extreme weather events and other climatic factors, but also indirectly as it will transform the natural environment that attracts tourists in the first place – for example, by accelerating coastal erosion, damaging coral reefs and other sensitive ecosystems and by reducing snowfall and snow cover in mountainous regions. It will also affect the basic services that are so critical for tourism, such as water supplies, especially during periods of peak demand.

In the Djerba Declaration (2003), the United Nations World Tourism Organization (UNWTO) acknowledged the need to align the tourism sector's activities with the concerns, objectives and activities of the United Nations (UN) system in relation to climate change, and more generally with respect to sustainable development. The Kyoto Protocol is a first step in the control of greenhouse gas (GHG) emissions. Tourism has an important role to play in achieving and moving beyond the Kyoto targets.

As a key part of the follow-up activities to the Djerba Conference, the UNWTO has initiated a series of pilot projects in Small Island Developing

States, in order to develop and demonstrate measures that reduce climate-related risks in tourism-dependent countries highly vulnerable to climate change impacts, including their beach destinations, tourism-dependent communities and coastal ecosystems. For this purpose, the UNWTO is undertaking pilot studies with support from the Global Environmental Facility, under its Focal Area on Climate Change. The first pilot countries are Fiji and Maldives, with the authors of this book involved as lead experts.

The World Meteorological Organization (WMO), in partnership with national meteorological services (NMSs) and the international meteorological community, is making an important contribution by providing relevant information to the tourism sector in order to reduce the adverse consequences of weather and climate extremes for tourism operators. At the same time the WMO is joining with UNWTO and the tourism sector to maximise the benefits of favourable weather conditions and changes in climate. In this way both organisations are raising awareness levels about the sensitivity of tourism to weather and climate variability and change, including extremes. They are also providing guidance on how key actors in the tourism system might best respond in order to reduce risks and maximise benefits.

There has been increasing cooperation between the UNWTO and WMO, manifested in the First International Conference on Tourism and Climate Change in Djerba in 2003, a special issue of *World Climate News* on tourism (Vol. 27) was published in 2005, and an Expert Team on Climate Change and Tourism has been established. The WMO will continue to spearhead international efforts to monitor, collect and analyse climate data and, in collaboration with the UNWTO, provide timely, relevant and reliable climate information services and products for use by operators, policy- and decision-makers in the tourism sector, and by travellers themselves.

This book, *Tourism and Climate Change – Risks and Opportunities*, is very timely. It supports and adds value to the ongoing work of the WTO and WMO with relation to climate change and tourism. It is the first in-depth and comprehensive analysis and integrated assessment of the interactions between tourism and climate change. The book will be of use to other international organisations, tourism and land use planners, policy- and decision-makers in the tourism sector, and senior students and academics in the areas of tourism policy and management, resource management and climate science and policy. The book achieves a pragmatic balance between the theoretical underpinnings of both tourism and climate change science, and the application of this knowledge in practical, real-world contexts. The text boxes are particularly useful as they present brief and appropriately targeted summaries of case examples and current and emerging issues that are, or should be, debated by policy makers and practitioners from the tourism and climate communities.

The identification of 'climate–tourism hotspots' is novel. They highlight where international organisations such as the WTO and WMO should focus their efforts when addressing the consequences of climate change for tourism. The book also identifies where more effort is needed if the interactions between climate and tourism are to be better understood, and how the UNWTO, WMO and other organisations might engage more effectively with the tourism sector in terms of both mitigation of and adaptation to climate change. The volume also makes an important contribution by demonstrating why a wider perspective is required to capitalise on the many synergies with other development and environmental imperatives, including increasing the sustainability of tourism-related activities that currently impact adversely on the quality of human life and well-being of the planet.

Francesco Frangialli
Secretary-General UNWTO

Michel Jarraud
Secretary-General WMO

Chapter 1

Introduction

Risk is potential loss and opportunity is possible gain. Over the last few years there has been increasing recognition of the risks and opportunities that climate change brings to tourism.

Tourism takes place in a wide range of places that are often closely linked to the natural environment and, as a consequence, to local climatic conditions. Changes in the climate, including climate-related hazards, have the potential to affect tourism businesses and tourist experiences alike. The tourism industry has to face headlines such as:

- 'Cool Season Dampens Fun, Pinches Profits: Summertime Blues' *The Detroit News* 27 August 2004
- 'Unseasonably Warm Weather Brings US Ski Season to Disappointing End' *The Press-Enterprise* 17 April 2004
- 'Wilma Slams Mexico Resorts' *CNN* 24 October 2005
- 'Dutch Seek to Defend Coastal Resorts from Sea Rise' *Planet Ark* 23 January 2006
- 'Tourism Experts Say Hot Summer Means Higher Turnover' *Deutsche Welle* 31 July 2006

Tourism is increasingly recognised as a significant activity, with a range of economic, social and environmental consequences. Tourism is one contributor to the build-up of greenhouse gases (GHGs), which are now recognised as causing unprecedented changes in the global climate. Such links between tourism and climate change are now acknowledged by key players in the tourism industry. Mark Ellingham, the founder of Rough Guides, and Tony Wheeler, who created Lonely Planet, want fellow travellers to 'fly less and stay longer' and donate money to carbon offsetting schemes. They urge their readers to: 'join to discourage "casual flying"'. A "Rough Guide to Climate Change" has appeared in late 2006.

As with the headlines that highlight climate-related risks and opportunities to tourism, the sector also has to deal with a growing number of headlines that sensitise tourists about their impacts on the global climate.

- 'Night flights much worse for global warming' *The Independent* 3 August 2006
- 'Aviation "huge threat to CO_2 aim" ' *BBC News* 21 September 2005
- 'It's a sin to fly, says church' *The Sunday Times* 23 July 2006

The last headline relates to an announcement by the Bishop of London. He said it was sinful for people to contribute to climate change by flying on holiday, driving a 'gas-guzzling' car or failing to use energy-saving measures in their homes. Similar calls have been made elsewhere, for example by a group of Anglican bishops in New Zealand and Fiji. They committed themselves to carbon management and offsetting.

In the last year or two there have been some dramatic changes in the perception of climate change and recognition of the risks if these are not addressed. Research into the risks, as well as opportunities, helps decision makers to implement effective and efficient measures. In the area of tourism, little research has been undertaken with respect to climate change. Early publications include those by Giles and Perry (1998), Viner and Agnew (1999), Wall and Badke (1994) and Wall (1998). An Organisation for Economic Cooperation and Development (OECD, 2003) report on the global impacts of climate change, and the associated benefits of an effective climate policy noted that 'In some key sectors, such as recreation, tourism, and energy, there has been little research conducted that characterises the relationship between climate change and impacts at a global scale.' Development of appropriate policies has been limited as a result.

This book aims to bring together current understanding regarding the interactions between climate change and tourism and to highlight both the policy implications as well as the repercussions for tourism businesses, policy and decision makers, and tourism practitioners. The audience for this book is diverse, as the issue of tourism and climate change is cross-cutting and influences many spheres of life, including both planning and decision making. The topics covered are also of interest to university students of both the environmental sciences and tourism. However, the book mainly targets those decision makers who must take into account the impacts of climate change on tourism or consider the GHG emissions caused by tourism. These decision makers will be working in national, regional or international governmental organisations, and in non-governmental organisations including tourism industry organisations. Importantly, the structure of the book is such that it will appeal to any reader who seeks specific information on the tourism–climate interactions and their practical implications.

The book is structured into two major parts. Chapters 1–3 (Introduction, The Tourism–Climate System, and Tourism–Climate Case Studies) introduce and elaborate the interactions between tourism and climate, including identification of climate–tourism 'hotspots'. Practical considerations are illustrated in four case studies. The first two focus on Alpine Europe and small island states. These have been identified as among the most vulnerable types of destination with respect to climate change. However, both mitigation and adaptation responses are discussed. The insurance industry case study highlights again the need for risk management in relation to climate change. Key concerns for tourism include natural disasters and the increasing difficulty to obtain insurance cover for many tourism businesses and infrastructure. International aviation is discussed in the fourth case study. It highlights the political difficulties of implementing fair and equitable climate change mitigation policies.

Chapters 1–3 thus provide the reader with a comprehensive overview of the key issues related to climate change and tourism, along with a more in-depth and practical understanding of tourism–climate issues for four more specific situations. The remainder of the book provides more in-depth information on the key components of the tourism–climate system. Forward referencing in Chapters 1–3 enables the reader to identify which of the subsequent chapters are of particular interest. Summary bullet points at the beginning of each chapter also assist the reader to gain a quick insight into chapter content and assess its relevance.

Chapter 4 describes tourism and details the recent trends that are relevant to any discussion of climate change issues. An argument will be made as to the special nature of tourism and why it is important to pay attention to these specific characteristics when dealing with climate change issues. This chapter also discusses major tourist flows worldwide and the growing economic importance of tourism. Attention will be paid to the emergence of 'responsible tourism', as this trend is consistent with the need for the tourism industry to reduce GHG emissions.

Chapter 5, 'Global and Regional Climate Change', provides more detailed explanations of the greenhouse effect and global climate change. The key changes in climate, including increased surface and water temperatures and tropic storm intensities, and sea-level rise, are discussed in terms of the relevance to tourism. Scenario analysis is an important tool to explore possible paths for the future. The scenarios developed by the Intergovernmental Panel on Climate Change (IPCC) are used as a starting point to developing plausible futures for tourism. It is also possible that climate change will include 'surprises' – abrupt and pervasive changes that would have catastrophic consequences for tourism and other economic, social and environmental systems.

Given that tourism is a major contributor to GHG emissions, Chapter 6 on 'Methodologies for Greenhouse Gas Accounting' discusses approaches for energy and GHG accounting in tourism. The chapter provides an overview of major emission sources associated with the combustion of fossil fuels, different forms of energy and emission coefficients to convert energy use into carbon dioxide (CO_2) and other gaseous emissions. Two different approaches are discussed: bottom-up and top-down analyses. The former details methodologies for transport and other tourism businesses, as well as for different tourist types. The latter uses the tool of input–output (IO) analysis to derive tourism-related energy use and CO_2 emissions at a country level. Chapter 6 is critical to understanding the energy and emission analyses and response measures described in Chapter 7.

Chapter 7, 'Climate Change Mitigation Measures', presents the current knowledge of energy use and GHG emissions associated with various tourist activities. The main emission categories are air travel, surface transport, tourist accommodation and recreational activities. GHG reduction options to mitigate tourism's contribution to climate change are detailed for each of these categories. The application of renewable energy sources is discussed, for both mobile and *in situ* tourism activities. Finally, the chapter provides an overview of carbon compensation schemes: projects that provide alternative and more cost effective ways to reduce or offset emissions. There are already a number of schemes that offer tourists ways to offset their GHG emissions. These include investing in energy efficiency, renewable energy sources or carbon sink projects.

The risks of climate change to tourism are at the core of Chapter 8 ('Climate Change-related Risks and Adaptation'). Potential impacts, vulnerabilities and adaptation measures are discussed. There are many options for reducing climate-related risks through adaptation in the tourism sector. These can be implemented for all three of the major components of the tourism system – the source region for tourists, for tourist travel and at the destination – and at all levels, ranging from the individual tourist, operator and tourism-dependent community through to global initiatives.

A recent edition of *The Economist* featured a cover with the headline: 'Why aviation will be the next green political battlefield'. Clearly there is an increasing need for policies and actions that address the various roles tourism plays with respect to climate change. Increasingly we see the integration of mitigation strategies, adaptation and disaster management into a common approach of sustainable tourism development. Such 'Climate Change Policies and Practices' are discussed in Chapter 9. The chapter also provides an overview of key international institutions relevant to climate change and tourism as well as the major international agreements and initiatives. An iterative policy framework and process for mitigation and adaptation is proposed.

The 'Conclusion' (Chapter 9) highlights the importance of considering tourism as part of a much bigger system in which the interactions between tourism and climate have repercussions for a wide range of other sectors and social activities, and vice versa. The increasing incidence of weather extremes and anomalous climatic conditions, and the mounting consequences for tourism and tourism-dependent countries and communities, are timely warning of the need to give serious attention to the risks climate change poses to the tourism sector. Climate change will also provide some opportunities, though overall these will be far outweighed by the adverse impacts. This book not only describes both the risks and opportunities but also provides the reader with practical guidance and examples for managing the risks and taking advantage of the opportunities. As a significant emitter of GHGs, in the longer term the tourism industry can also help slow the rate of climate change. Again, practical guidance and examples are provided.

The book is intended to build the knowledge and understanding needed to generate the commitment by key players in tourism to initiate and sustain the actions that will reduce both tourism's contributions to climate change and the climate-related risks to tourism. While individual responsibility and action is required, it must be supported and coordinated by policies and plans at national, community, business enterprise and other relevant levels. For this reason the book targets tourism policy and decision makers in both the public and private sectors, as well as those who have leadership roles in tourism-dependent communities.

Chapter 2
The Tourism–Climate System

Key Points for Policy and Decision Makers, and Tourism Operators

- Those working in tourism are mainly concerned about climate-related risks. At the same time tourism is a very energy-intensive activity that contributes to GHG emissions and the build-up of these gases in the atmosphere.
- Interventions or responses designed to reduce climate-related risks fall into two categories: mitigation and adaptation. Mitigation includes initiatives for reducing GHG emissions, whereas adaptation refers to interventions that reduce the vulnerability to climate change impacts.
- Tourism has been described as a system; earlier system approaches assumed a linear nature of tourism, which failed to capture the true complexity of tourism. The open and complex nature of both the climate and tourism systems makes it is extremely difficult to predict, manage and control future changes.
- There are several key agents involved in tourism, namely the private sector, public sector, 'destinations' (with numerous agents) and tourists themselves. Interest groups are also important agents in tourism.
- It is useful to analyse the tourism agents and try to understand certain behaviours, relationships and developments. Especially for short-term planning, the pragmatic approach is to assume some sort of linearity and predictability. However, the discussion on complexity and chaos should raise awareness among professionals that they cannot control or predict tourism's evolution.
- The climate system has been modified by humans as a result of GHG emissions. These 'trap' solar energy and raise the temperature of the Earth's surface. This raised temperature is termed the *natural greenhouse effect*. Climate projections are associated with considerable uncertainties.
- One manifestation of uncertainty is the ability of the climate system to undergo abrupt and pervasive changes; such changes are termed 'surprises', and present a significant challenge to tourism and other policy makers and planners.
- Climate – and in particular temperature – is an important factor in destination choice, although recent research indicates that destination choice is more complex than assumed in current models.

- Several 'climate–tourism hotspots' – parts of the world where high tourist arrivals are forecast, or tourism will be a major contributor to the national economy, and significant changes in climate are projected for the near term – can be identified; these are Alpine Europe, Western Europe, Central/Eastern Europe, the Mediterranean, North- and South-eastern USA, Mexico and the Caribbean, China, and the small islands in the South Pacific and Indian Ocean.
- In a chaos model of tourism, the climate (system) can be interpreted as an externality. This suggests that key agents with a vested interest in maintaining the stability of the tourism system should be working proactively to avoid, or at least delay, major changes in the climate system. By doing so, these agents are in fact internalising the externality.

Introduction

There are multiple interactions between tourism and the climate. In the first instance climate is a resource for tourism and it is an essential ingredient in the tourism product and experience. At the same time, climate poses a risk to tourism. For example, as a result of climate variability, weather conditions at a given location and time may prevent tourists from engaging in their planned activities. This is the case for skiers when there are snow-poor winters in alpine tourist destinations. A similar situation exists when conditions are unseasonably cool and wet at beach destinations. Climate can also pose a severe risk in relation to extreme events such as hurricanes and floods. These put both tourists and tourism-oriented businesses at risk, including damage to tourism infrastructure and increased financial costs combined with lower incomes.

Those working in the tourism industry are mainly concerned about such climate-related risks as those described above. However, there is another important link between tourism and climate that is causing increasing concern. Tourism is a very energy-intensive activity that contributes to GHG emissions and the build-up of these gases in the atmosphere. One result is an exacerbation of risks due to a changing climate, with detrimental impacts on tourism.

There are many opportunities for the tourism sector to reduce emissions of GHG gases. But if these opportunities are not taken up proactively, the opportunities can soon turn into additional risks. With the increasing recognition of climate change as a major environmental issue that must be addressed in a concerted manner, there is a developing consensus on policies such as taxing airlines for their emissions of GHGs.

These diverse relationships between tourism and climate are visualised in a simplified manner in Figure 2.1. The consequences of climate change for tourism manifest as risks. The decision maker and planner

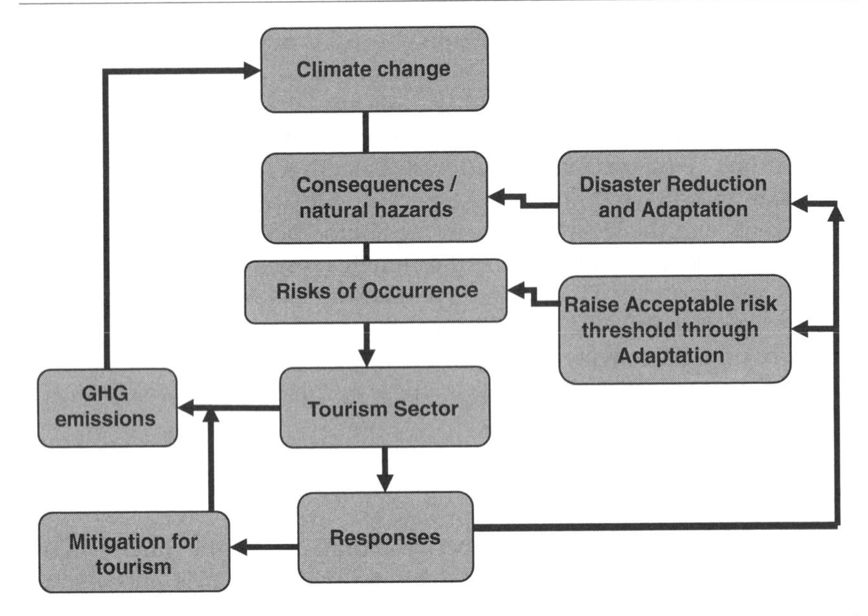

Figure 2.1 Framework for assessing the relationship between tourism and climate change, including possible responses by tourism to mitigate impacts or adapt to changes in the climate

will find it useful to differentiate between acceptable and unacceptable risks. For example a certain level of variability in snow conditions will be acceptable to the operator of a ski facility, but the risk of a permanently higher snowline (say above the operating elevation of the facility) will not be. Such considerations facilitate the prioritisation of the unacceptable risks as well as of the interventions, which will reduce them to acceptable levels.

Interventions or responses designed to reduce climate-related risks fall into two categories (Figure 2.1). First, *mitigation* initiatives reduce GHG emissions (see Chapter 7 for detailed examples) and, as explained in Chapter 5, the likelihood of a specific climate hazard or condition. But present efforts related to mitigation, such as those under the Kyoto Protocol, are proving to be woefully inadequate. Emissions reductions of at least 60% are required to stop further increases in the greenhouse effect. In comparison, the Kyoto Protocol (which includes commitments only to 2012) will achieve emission reductions of only 5.5% reduction in emissions, and only if there is full compliance! Thus further climate change is inevitable. Moreover, even if the necessary reductions in emissions were to be achieved tomorrow, changes in the climate will still occur. This is due to inertia in the climate system. Thus mitigation brings climate benefits only in the longer term. However, mitigation of GHG emissions does generate short-term opportunities and benefits for tourism, for example through energy conservation and increased use of renewable energy.

Given that mitigation measures will reduce climate-related risks only in the longer term, actions that fall into the second category will be required, at least in the shorter term. Unlike mitigation, these interventions deal with the consequence component of risk. As explained in detail in Chapter 8, these interventions reduce risk through *disaster reduction* and wider *adaptation* initiatives. Illustrative examples of relevance to tourism include: (a) ensuring that both operational staff and guests are well informed of the actions to be taken should, say, a tropical cyclone be predicted to pass close to or over the tourist facility and (b) installing and maintaining a desalination plant so that water requirements can be met during a severe drought.

Figure 2.1 assumes that the relationships are linear and that the interactions between tourism and climate are deterministic. In this logic it is possible to control the effects, i.e. reduce the risks and maximise the opportunities, if we put the appropriate policy instruments and management procedures in place. The implicit assumption is that when tourism operators and others working in the sector respond to climate change, their actions will result in a direct reduction in the level of climate-related risk.

However, as has been pointed out in the tourism literature (albeit infrequently, for example by Faulkner & Russell, 1997), tourism is often wrongly characterised as a linear system, much in the way the climate–tourism system is described in Figure 2.1. This fails to capture the true complexity of tourism. In fact, both tourism and climate operate individually and jointly as open systems that are also non-linear, non-probabilistic and non-deterministic as a result of the complex, dynamic relations between and among them and their constituent elements. Thus it is inappropriate to pursue a reductionist approach to understanding the interactions between the tourism and climate systems, as would be the case if each component and linkage shown in Figure 2.1 was to be analysed individually. Importantly, characterising and combining the individual interactions will in itself not provide a fully integrated understanding of the relationships between tourism and climate. The open and complex nature of these systems also means that it is extremely difficult to predict, manage and control future changes with any level of practical significance and relevance. What this means for tourism will be discussed in the next section.

This chapter will outline the interpretation of tourism as a system, first from a deterministic perspective and subsequently by applying the more recent understanding of tourism as a complex system. The characteristics of tourism and climate (encompassing the atmosphere, hydrosphere, cryosphere, land and biosphere) as complex systems will be discussed, including the implications for managing the interactions with the climate system, which have either detrimental or beneficial consequences for tourism. While the interactions between and within two systems of such complexity often go beyond what can be explained with current methods and understanding, some examples will be used to illustrate the nature of these interactions and the implications for policy making, planning and management.

The Tourism System

Evolution of the understanding of tourism

Over the last 20 years a number of models have been developed to try to explain what tourism is, its makeup, how it works and what kinds of relationships exist (see also Chapter 4). These models have been useful for research on climate change and tourism. For example, defining who a tourist is, and what forms part of tourism, is important to the climate change discussion. Any assessment of tourism's relationship with the global climate, including its changes, requires an understanding and agreement of the variety and kinds of activities and stakeholders involved. When one attempts to analyse tourism's vulnerability to extreme weather events or climate change (see Chapter 8) it is useful to understand where the boundaries between tourism and the rest of the economy are. This makes it possible to look at the tourism system in isolation or in relation to interactions with other systems. Similarly, a clear distinction of tourism – as opposed to other human activities – is required when quantifying GHG emissions by tourism activities (see Chapter 6).

The UNWTO defines tourism as the activities of persons travelling to and staying in places outside their usual environment for not more than one consecutive year for leisure, business and other purposes not related to the exercise of an activity remunerated from within the place visited. UNWTO uses the term 'visitors' to describe those persons travelling and also specifies that visitors are the sum of same-day visitors and overnight visitors (also called tourists).

Furthermore, UNWTO argues that tourism includes total consumption expenditure made by a visitor, or on behalf of a visitor, for and during his/her trip and stay at a destination. Hence, the basic assumption is that tourism is not defined by a specifically delivered product or service, but rather by those who consume such goods and services. For example, when they visit a tourist attraction tourists demand services from the commercial sector. Tourists are also key users of the transport sector. Tourism also involves intangible elements, such as experiences, enjoyment, excitement and relaxation. The entirety of businesses involved in providing tourism products or services can be called the *tourism industry*. A wider perspective considers that there is considerable input by governments (both national and local), the environment and local communities into tourism. The combination of all those inputs into tourism has been referred to as *the tourism sector*.

It was argued, early on in tourism research, that by analysing disaggregated components of tourism it is possible to build-up an understanding of tourism as a whole (Pearce, 1989). This and similar approaches are in the first instance very reductionist. As a result they fail to explain more complex relationships, interactions, interdependencies and impacts. Carlsen (1999: 322) argued that 'a systems approach would best be applied to problem solving in tourism research because it can accommodate social and environmental processes' and that the tourism system would be 'an open system in that it responds to changes in the social, natural and economic environment and is evolving toward an increasing state of complexity'.

The first to apply a systems approach to tourism was Leiper, in 1979. The core idea of Leiper's model was that the tourist-generating region and the host destination are linked via the transit route. This applies to each individual tourist. The approach has been taken up and developed further in a multitude of models of tourism systems. The three main elements of a tourism system are of a geographical nature (i.e. tourists travelling at least from their home to their destination and back), involve the human dimension (i.e. the visitors and the hosts) and include all those businesses that are involved directly or indirectly in tourism activities. In addition, external factors interact with the tourism system. An example in the context of this book is the role that climate plays for tourism, both as a desirable resource, or a risk (Figure 2.2).

Traditional tourism models such as that shown in Figure 2.2 assumed that tourism players function in a coordinated manner, and as a result tourism can be controlled in a top-down approach (McKercher, 1999). In response to those simplistic models, there has been a call to extend the current conceptualisation of tourism systems to include understanding gained from ecosystems research, in particular in relation to complex adaptive systems (CASs). The core idea is that tourist systems are analogous to natural ecosystems in that they are very dynamic and changeable and non-linear in their behaviour. This means that simple linear cause-and-effect relationships are unlikely to explain (temporary) equilibriums and changes (Farrell & Twining-Ward, 2005). The traditional approaches to tourism fail to acknowledge the element of uncertainty and chaos. Farrell and Twining-Ward argued that new science (post-normal or non-linear science) is needed to better understand tourism as a CAS. The term *post-normal science* (Funtowicz & Ravetz, 1993) is increasingly being used to describe science that deals

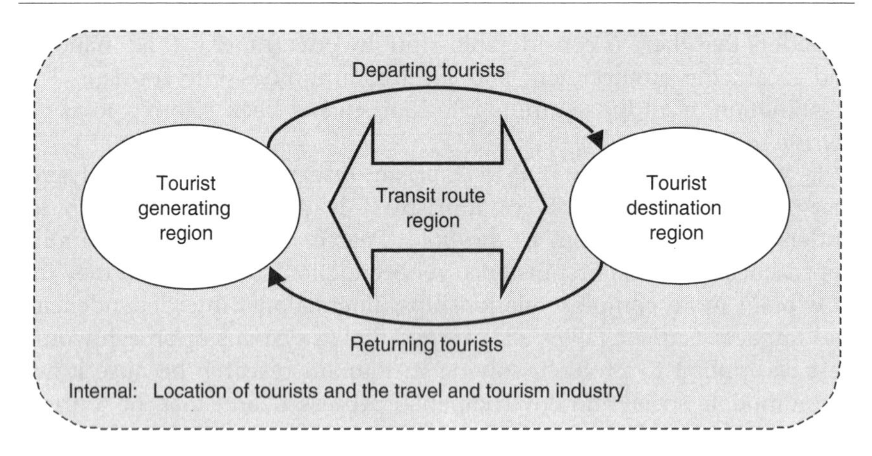

External: Environments: human, sociocultural, economical, technological, physical, political, legal, etc.

Figure 2.2 A model of the tourism system
Source: after Leiper (1995)

with facts that are uncertain, that incorporates values and where the stakes are high in relation to outcomes.

There is an increasing body of literature on complex (adaptive) systems and chaos theory. Tourism may be described as a CAS. CASs are made up of agents, for example tourism businesses, tourists and other agents external to the tourism sector. These agents are semi-autonomous, but they pursue some common measure of good. Each agent interprets their situation and develops rules for their actions and behaviour. These interpretations are often based on incomplete information and depend on the agent's context (see also Elliot & Kiel, 2004).

Agents in tourism

There are several key agents involved in tourism, namely the private sector, public sector, 'destinations' (with numerous agents) and tourists themselves. Interest groups are also important agents in tourism.

The private sector – or the tourism *industry* – usually consists of a few large companies and a large number of small- to medium-sized enterprises. In many countries an industry association represents the tourism industry. Often, however, the smallest businesses are not members of those organisations and are therefore ill represented. There is a risk when businesses work independently and are as a result suboptimal. For example, the flow of information is likely to be inadequate for those disconnected from the 'greater goal' of the industry.

The public sector typically consists of national or central government, and regional and local governments, depending on how the tourist destination is structured. The different levels of government have different responsibilities and interact in different spheres with tourism. The national government is typically responsible for transport networks, border control, security, health services and general infrastructure. Tourism businesses and tourists themselves make extensive use of these public sector services. In turn, the national government taxes tourism, for example, value added tax, excise tax (e.g. for fuel or alcohol), user charges and business taxes through the tourism industry. National government is also responsible for policies relating to energy supply and GHG emissions, including, for most developed countries, national commitments under the Kyoto Protocol (Chapter 9).

Regional or local governments are more concerned with the management of tourism at the meso- or micro-levels. This involves, for example, water supply and wastewater treatment, waste management and the provision of other services, facilities and amenities. The development of a place as a tourist destination, and its promotion (often by businesses), does not always go hand in hand with planning and management; similarly, coordination between different agents is not always optimal to achieve sustainable outcomes. Russell and Faulkner (1999) analysed such underlying dynamics of tourism development by examining entrepreneurial activity on the Gold Coast in Australia. They found that entrepreneurs play a critical role as 'chaos-makers' (see further below), but also as initiators of adaptive responses to chaos caused by external events.

Without tourists there is no tourism. In this sense tourists are the main agents in the tourism system. Tourists exert substantial influence through their purchasing behaviour, for example by supporting businesses that seek to achieve best practice in terms of energy efficiency and GHG emissions. Ecolabels are one means to guide consumers in making their decisions, although there is little evidence to date that tourists recognise tourism ecolabels and base their decisions on such labels (Reiser & Simmons, 2005). The role of tourists and their awareness of climate change issues are further discussed in Chapter 4.

Interest groups also play an important role in the evolution of tourism. Interest groups are usually non-governmental, and in many cases have the specific objective of influencing and monitoring the governmental processes of policy and decision making, as well as the regulatory and service functions of governments. Some interest groups also endeavour to assess and influence the efficacy of activities undertaken by the private sector. It is possible to identify several categories or groups with an interest in ensuring that the tourism sector, and its constituent parts, takes appropriate action to address the influence of climate change on tourism and/or the influence of tourism on climate change. The basic categories are:

- Private sector – includes organisations within the tourism sector, such as the World Travel and Tourism Council (WTTC) and Pacific Asia Tourism Association (PATA), and those having an intimate relationship with that sector, such as the WMO and the International Civil Aviation Organisation (ICAO).
- Environmental organisations, which wish to ensure that adaptation and mitigation initiatives have minimal adverse impact on the environment, both natural and managed.
- Sustainability organisations – these act as advocates for the adoption of adaptation and mitigation policies and measures that enhance the sustainability of tourism, from the individual operation to the global tourism system.
- Civil society, where the focus is on representing the interests of individuals, communities and larger groupings of people in order to ensure that tourism-dependent communities and other such stakeholders are able to share in the benefits that come from adaptation and mitigation while minimising any potential adverse consequences.
- Academia and other education and research organisations, which seek to ensure that the tourism sector has the capacity to, and does, adopt good practices in mitigation and adaptation.
- Labour organisations – these will seek to ensure that worker rights are protected when adaptation and mitigation policies and plans are developed and implemented and will endeavour to ensure that workers have the knowledge and skills necessary to achieve successful outcomes.

Tourism as a complex system

Tourism displays all the characteristics of complexity: self-organisation, non-linearity, chaos and emergent properties. The use of the Internet in tourism is a good example of self-organisation in order to increase efficiency and effectiveness. The emergence of the Internet as a means for gathering information, as well as displaying (marketing) information, meant that an increasing number of businesses started to use this communication channel to reach their customers. In response to this (a positive feedback loop), tourists increasingly made use of this option. Now, the majority of tourism providers maintain their own websites and the Internet allows booking of almost any tourism product and service that tourists want. This leads to substantial changes in the behaviour of all agents involved in tourism, and in the system as a whole.

It has long been known that certain aspects of tourism undergo 'threshold' effects; that means that an increase in a certain parameter does not necessarily relate to a linear increase or decrease in another parameter. A good example is that of tourism impacts on both the natural environment and host communities. Research shows that these systems are resilient to a pressure up to a certain level, but once a threshold has been passed the consequences may catastrophic. The more non-linear relationships in a system there are, the less predictable it is.

In fact, the tourism system has been described as chaotic (Faulkner & Russell, 1997). Chaos does not mean a total lack of order. Rather it refers to a situation where components of a system seem to act independently, but the system as a whole functions with some sort of order (McKercher, 1999). In a chaotic system it is difficult to predict the future in a probabilistic way; however it is possible to identify certain orders that guide the system within broad parameters. While there are periods of relative stability (and even linearity), chaotic systems can undergo abrupt change and move from one state to another within a very short period of time. The terminology from chaos theory is that a system moves toward 'the edge of chaos' until a trigger throws the system into disarray and requires it to reorganise in a possibly very different way. The ability to do this is called adaptive capability. There are many examples of such chaotic changes within tourism, evident from individual businesses to whole destinations. The impact of a natural disaster could be a trigger to lead to a complete reorganisation of a tourism destination. Running out of fossil fuels could be another factor that pushes tourism – or certain subsystems within tourism – over the edge of chaos, requiring restructuring as a very different system.

As a result of chaotic upheaval it is possible that new properties of the tourism system will emerge. These might not have been foreseen under the previous conditions. For example, the introduction of jet engines triggered unprecedented movements of tourists around the globe to distant destinations, with irreversible effects on local communities and environments in the destination (often developing) countries. At the same time those global tourist flows made tourism one of the largest redistributors of wealth between industrialised and developing countries.

It is useful to analyse elements (the agents) of the tourism system and try to understand certain behaviours, relationships and developments. Especially for short-term planning, the pragmatic approach is to assume some sort of linearity and predictability. However, the discussion on complexity and chaos should also raise awareness among professionals in the tourism sector that they cannot control tourism's evolution, nor can they predict with absolute accuracy what might happen as a result of a specific intervention or decision not to act. As will be shown in the next section, uncertainty also results in the inability to specify, with a high degree of confidence, the climatic conditions under which the future development of tourism will occur.

The Climate System

To understand the climate of the Earth, including its variations and changes over time and its interactions with tourism, we must have some understanding of the *global climate system*. The system consists of the atmosphere, oceans, ice and snow masses, land surfaces, rivers, lakes and the biosphere (including humans), as well as the mutual interactions and hence changes that are a consequence of the large variety of physical, chemical and biological processes taking place in and between these components (Figure 2.3, for more detail refer to Chapter 5). Another key feature of the system is the existence of various external forcing mechanisms, the most important being the Sun.

Tourism is one of many contributors to changes in the climate system. As with other human activities, there are many ways and spatial scales at which tourism contributes to climate change (Figure 2.3). For example, changes in land cover and use, such as replacing forest with resort buildings and other structures, or maintaining a well irrigated golf course in an arid environment, can modify the local climate. Local changes may also be caused when particulates and other air contaminants are emitted by incinerators, stationary and mobile engines, and during land-clearing activities. Cumulatively, over space and time, even these locally focused human activities are known to change the climate, regionally and globally. They work in tandem with more global scale forces such as those related to emissions from aircraft carrying tourists to and from their destinations.

The atmosphere is the most rapidly changing part of the wider climate system, its composition having changed as the Earth has evolved. Despite their low concentrations relative to gases such as nitrogen (N_2) and oxygen (O_2), a collection of atmospheric gases, notably CO_2, methane (CH_4), nitrous oxide (N_2O), ozone (O_3) and water vapour (H_2O), play a critical role in determining the temperature of the Earth. These so-called *greenhouse gases* absorb little or none of the short-wavelength radiant heat energy reaching the Earth from the Sun, but are highly effective absorbers of the longer-wavelength radiant heat energy emitted by the Earth. These gases, in turn, emit longer-wavelength radiant heat energy both upward and downward (Figure 2.4). The resulting energy-'trapping' process raises the temperature of the Earth's surface to an average of 14°C, some 33°C above what it would be if the Earth's atmosphere contained no GHGs. This raised temperature is termed the *natural greenhouse effect.*

The natural climate system is in *dynamic equilibrium* – that is, variations occur, but large systematic changes are suppressed through negative feedbacks. For example, if the heat content of the tropical ocean increases in a persistent manner, the resulting increases in evaporation and convection (manifest as tropical cyclones and thunderstorms) will soon move the

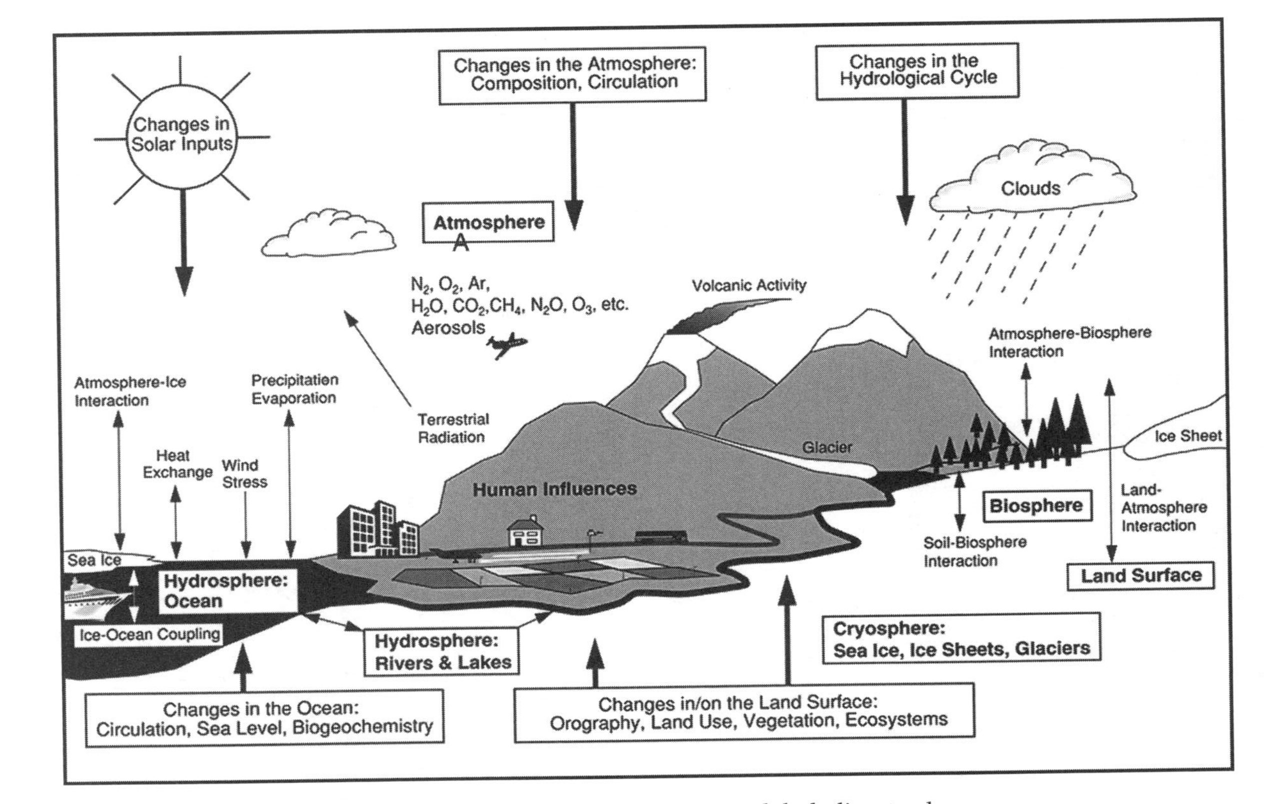

Figure 2.3 Contributions of tourism and other human activities to global climate change
Source: after IPCC (2001)

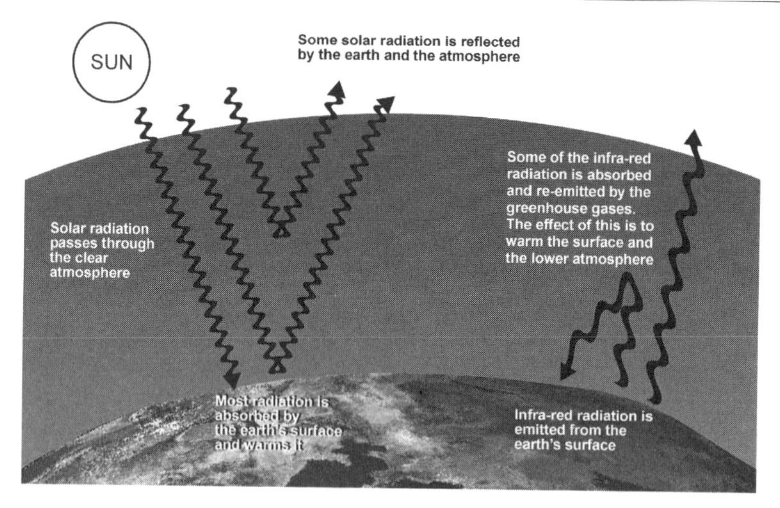

Figure 2.4 The greenhouse effect

excess heat energy into the atmosphere, cooling the ocean, and hence suppressing the initial change.

For centuries prior to the Industrial Revolution the concentrations of GHGs in the atmosphere remained relatively constant. But since then the concentrations have increased, largely as a result of the combustion of fossil fuels (coal, oil, natural gas) for industrial and domestic purposes, and due to biomass burning and deforestation. For example, the CO_2 concentration has increased by more than 30%, and is still increasing at the unprecedented annual average rate of 0.4%. The concentrations of other GHGs, such as CH_4 and N_2O, are also increasing, due mainly to agricultural and industrial activities, including those that are tourism related. The concentrations of the nitrogen oxides (NO and NO_2) and of carbon monoxide (CO) are also increasing. While these are not GHGs, they are involved in chemical reactions that result in higher ozone concentrations in the lower atmosphere. Ozone is a GHG, and its concentration has increased by 40% since pre-industrial times. This increasing concentration of GHGs is exacerbated by chlorofluorocarbons (CFCs) and other chlorine and bromine compounds. These do not occur naturally, but are manufactured for industrial and domestic use. In addition to being strong GHGs, they destroy stratospheric ozone.

The increased concentration of GHGs in the atmosphere enhances the absorption and emission of long-wavelength radiant heat energy. The overall effect is a reduction in the long-wavelength radiant heat energy lost to space. This must in turn be compensated for by an increase in the temperature of the Earth's surface and lower atmosphere. Computer models that simulate many of the complex interactions of the global climate system project a temperature increase of 1.4–5.8°C during this century. The range is large, reflecting in the main uncertainties in future emissions of GHGs as well as the poorly understood influence of clouds. To place it in context, this increase should be compared to the global mean temperature difference of around 5–6°C between the middle of the last Ice Age and the present.

The uncertainties in characterising future climate are such that climate models are said to produce projections, not forecasts. Projections are reasonable examples of how the climate might change. Another important manifestation of the complexities and non-linearities in the climate system is only now being understood, albeit in a preliminary manner. This is the capability of the climate system to undergo abrupt and pervasive changes. Such climate changes have occurred in the past, when the Earth system was forced across thresholds. When this happens, changes in the climate system are no longer controlled by the conditions forcing the change, but by the internal processes of the system. The changes resulting in very different climatic conditions can either be much faster than the forcing ('abrupt climate change'), or significantly slower, as well as temporary or permanent. Such unanticipated transitions to a new state, at a rate determined by the climate system itself and faster than the cause, are termed 'surprises' (Figure 2.5).

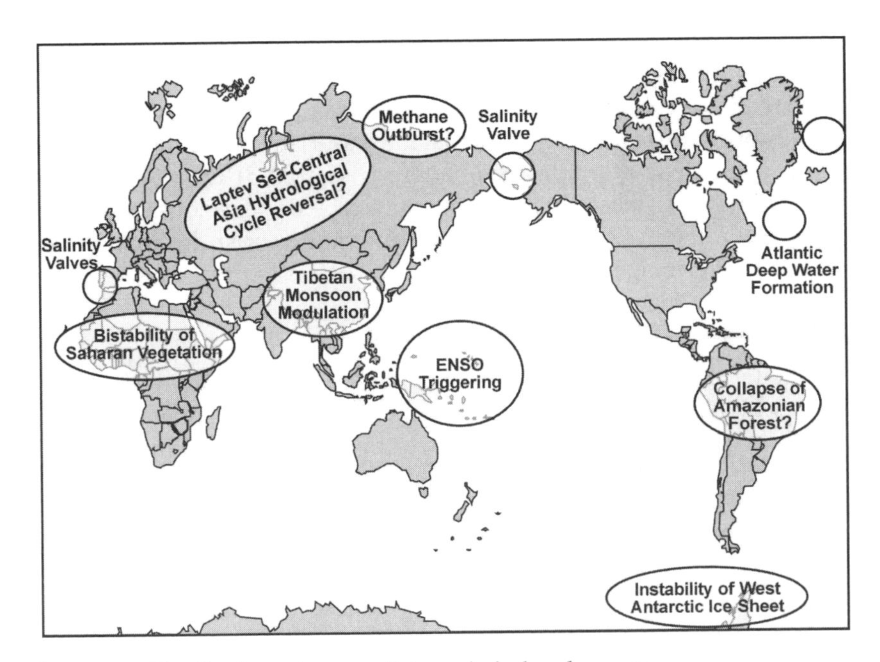

Figure 2.5 The Earth system: switch and choke elements
Source: after Schellnhuber (2001)

The large climatic changes that have occurred in the past had little net forcing, and global climate models have often underestimated the magnitude, speed or extent of such changes (Alley *et al.*, 2003). Abrupt climate changes of up to 10°C in a decade have been identified in some regions. Although abrupt climate change can occur for many reasons, it is conceivable that human forcing of climate change is increasing the probability of such large and abrupt events. Available evidence suggests that abrupt climate changes are not only possible but likely in the future, potentially with large impacts on ecosystems and societies (US National Research Council, 2002).

The most prominent possibility for a large and relatively abrupt change in the Earth system as a consequence of increases in the greenhouse effect is the 4–6-m rise in sea level that would occur should the West Antarctic ice sheet collapse (Figure 2.5). The large uncertainty that characterises the projection arises from a number of different sources: incomplete scientific understanding of the physical phenomena and their interrelationships; incomplete data and measurements; and overly simplistic and unvalidated models. Climate change introduces further uncertainties to the problem of rapid and large sea-level rise. The relationship between West Antarctic ice sheet collapse and global warming, for example, is highly uncertain (Kasperson *et al.*, no date). However, recent disintegration of Antarctic ice shelves due to enhanced surface melting or thinning, and the resulting acceleration of the flow of ice streams that feed the shelves, has given impetus to efforts to reduce the uncertainties (Rignot *et al.*, 2004) and to consider the consequences of such a rise in sea level. Sea-level rise is of major importance to tourism given that most tourism activity takes place in coastal zones. For example, Poumadère *et al.* (2005) have studied the implications and stakeholder responses to a 5–6-m sea-level rise in the Rhone Delta, France, a popular nature-oriented tourist destination and important pilgrimage site.

Examples of Tourism–Climate Interactions

Climate is relevant to tourists in various ways (for more detail see Chapter 8). Aesthetic aspects of climate, for example, are relevant for the tourists' experience and will influence their enjoyment. Physical attributes of the climate may lead to annoyance or even danger, possibly resulting in injury or other damage. Finally, thermal facets of climate determine the tourist's comfort and, in negative cases, could result in stress of various sorts (Table 2.1). Some tourist activities are more sensitive to weather than others, for example beach activities, skiing or playing golf. Indoor entertainment facilities allow the tourist to be independent of weather conditions. Substitution of some indoor activities for outdoor sports has been discussed by Perry (2004). New forms of

Table 2.1 Facets of climate and impact on tourists

Facet of climate	*Impact on tourists*
Aesthetic	
Sunshine/cloudiness	Enjoyment, attractiveness of site
Visibility	Enjoyment, attractiveness of site
Day length	Hours of daylight available
Physical	
Wind	Blown belongings, sand, dust etc.
Rain	Wetting, reduced visibility
Snow	Participation in activities
Ice	Personal injury, damage to property
Severe weather	All of the above
Air quality	Health, physical well-being, allergies
Ultraviolet radiation	Health, suntan, sunburn
Thermal	
Integrated effects of air temperature, wind, solar radiation, humidity, long-wave radiation, metabolic rate	Environmental stress, heat stress Physiological strain, hypothermia Potential for therapeutic recuperation

Source: after de Freitas (2001)

tourism may emerge under changing climatic conditions, either gradually or abruptly.

Climate as a factor in destination choice and holiday experience

Comparatively little is known, other than in very general terms, about the relative contribution of climate to tourism's resource base. The relationship between temperature and tourist comfort is usually based on assumptions and linear modelling rather than empirical (non-linear) evidence (de Freitas, 2005). To date such relationships have failed to reflect the complexity of destination choice. The question can be asked, for example, whether there is a threshold effect of, say, several very hot summers (like the 2003 heatwave in Europe), that leads to the decline of a tourist destination because it becomes 'too hot', or at least the risk becomes too great for the tourist to be exposed to unbearably hot temperatures. Along those lines, Gössling and Hall (2006) used results from two behaviour-focused case studies in Israel and Tanzania to demonstrate that the role of climate in destination choices is more complex than assumed in current models. Consequentially, they caution against the use of top-down models incorporating a few selected climate-related parameters to predict tourist flows.

Tourists frequently base their decision to travel to a particular destination on the climatic conditions they expect at the destination. It is often assumed that the warmer the climate the more attractive it is for tourists – up to a certain point. As a result, many players, as well as promotional and marketing initiatives, are involved in the formation of a perception of whether a destination is 'warm' or 'cold'. New Zealand, for example, attempts through purposeful marketing of beaches and sunny conditions to counteract the perception of it being a cold and unappealing destination. There are feedbacks between the settings and tourists' actual experiences of a destination, modifying or amplifying earlier perceptions of climate conditions.

The importance of climate among other factors was revealed in a study by Hamilton and Lau (2004). They surveyed 400 tourists at various departure points in Hamburg, Germany. Nearly 60% said they had been tracking the weather at their destination during the week before departure. The majority identified climate as the most important destination attribute, with maximum temperature being the dominant climatic attribute (Table 2.2).

The importance of temperature for both international and domestic tourism was also noted by Agnew and Palutikof (2001). For the four European countries they studied, a given increase in summer temperatures led to a growth of domestic holidays by between 1 and 5%. In general, wetter weather in the home country encouraged holidays abroad in both the current and following years. Similarly, Bigano *et al.* (2004) found that during extremely hot summer months increased numbers of Italian tourists travelled abroad, particularly to more northern countries. Domestic tourism also increases, due to more weekend trips, not only in a year with a hot summer, but also in the following summer. Conversely, high temperatures in January result in less domestic tourism, likely due to the negative effect of such temperatures on the ski season. Higher temperatures in spring and fall also resulted in increased domestic tourism, suggesting a relatively higher elasticity of domestic tourism to climate factors in the intermediate seasons. Precipitation in July reduces domestic tourism in that month. Sometimes the relationship between climatic conditions and attractiveness is not so clear.

According to Maddison (2001), the maximum acceptable daily temperature for tourism is close to 30°C, while Lise and Tol (2002) estimate the optimal mean daily temperature to be around 21°C. Some destinations are

Table 2.2 Importance of destination and climate attributes

Destination attributes	*Rank*	*Climate attributes*	*Rank*
Climate	1	Maximum temperature	1
Access to sea/lakes	2	Water temperature	2
Nature/landscape	3	Duration of sunshine	3
Cultural/historical attractions	4	Number of rainy days	4
Price	5	Average temperature	5
Hospitality	6	Minimum temperature	6
Accommodation	7	Amount of precipitation	7
Sport/leisure activities	8	Humidity	8
Ease of access	9	Cloudiness	9
Cuisine	10	Wind conditions	10
		UV radiation	11
		None of above	12

Source: after Hamilton & Lau (2004)

chosen despite unfavourable weather conditions (e.g. cool temperatures in the seaside resorts of the Baltic Sea) (Lohmann & Kaim, 1999).

Projected climate changes for 'climate–tourism hotspots'

According to the UNWTO (2001), for the coming few decades the major tourism regions in terms of total international arrivals will be Europe (France, Spain, Italy, the UK and the Czech Republic), the Russian Federation, East Asia and the Pacific (China), and the Americas (the USA and Mexico). Table 2.3 presents a synthesis of information available from a large number of sources, including Canada CCME (2003), Gao *et al.* (2002, 2004), IPCC (2001), Kjellström (2004), Lemmen and Warren (2004), Parry (2000), Räisänen (2005), Ruosteenoja *et al.* (2003) and US NAST (2001). While trends in both mean temperature and, to a lesser extent, precipitation conditions will unlikely impact the tourism sector for many decades, it is clear that variability and extremes will likely have more immediate and significant impacts. These consequences will be examined in detail in Chapter 8.

In 'climate–tourism hotspots', the interactions between and within the complex tourism and climate systems combine in ways that could result in large adverse impacts of climate change for tourism. These hotspots are where climate change can have a major adverse effect on tourism

Table 2.3 Climatic information on the three climate–tourism hotspot regions

	Europe/Russia	*East Asia and the Pacific (principally China)*	*Americas (principally USA and Mexico)*
Temperature (mean °C)			
Mean 2010–2039	1.0–2.0	1.5–1.5	1.0–3.5*
2040–2069	2.5–4.0	2.5–3.5	2.0–5.0*
2070–2099	3.5–6.5	3.5–6.0	3.0–8.5*
Daily extremes	Higher maximum and minimum temperatures; lower diurnal temperature range		
Heatwaves	Increase in frequency and severity of heatwaves		
Cloud cover			Increase in Western areas in all seasons other than summer; small decrease in NW and SW USA
Precipitation (mean%)			
Mean 2010–2039	−10 to +5	2 to 3	−5 to +10
2040–2069	−10 to +10	6 to 8	−10 to +10
2070–2099	−15 to +10	9 to 15	−10 to +20

(*Continued*)

Table 2.3 (*Continued*)

	Europe/Russia	***East Asia and the Pacific (principally China)***	***Americas (principally USA and Mexico)***
Extremes	Increase in extreme rainfall, even where average annual precipitation decreases; longer droughts, esp. for countries bordering the Mediterranean	Increased frequency of extreme rainfall events, and of drought, in most areas	More extreme precipitation along the Gulf Coast, in the Pacific NW, SW Canada and East of the Mississippi; more dry spells, separated by heavier rainfalls; summer drying in mid-continental regions due to increased evaporation, sometimes coupled with decreased precipitation
Storms and winds	More and more intense Atlantic storms in winter; increased windiness and extremes, especially in W. and N. Europe		More intense storms in Gulf of Mexico, Atlantic coast and Gulf of St. Lawrence; more intense, and periods of more frequent, hurricanes
Snow and ice	Decreased duration of snow cover; by end of century most alpine glaciers gone; areas below 2500 m ice free		Substantial reductions in snow cover and snow-pack by mid-century
Sea-level	0.1–0.7 m by end of century	0.1–0.4 m by end of century	0.1–0.5 m by end of century

*Temperature range is large due not only to uncertainties in model-based projections but also to the large and diverse geographical area

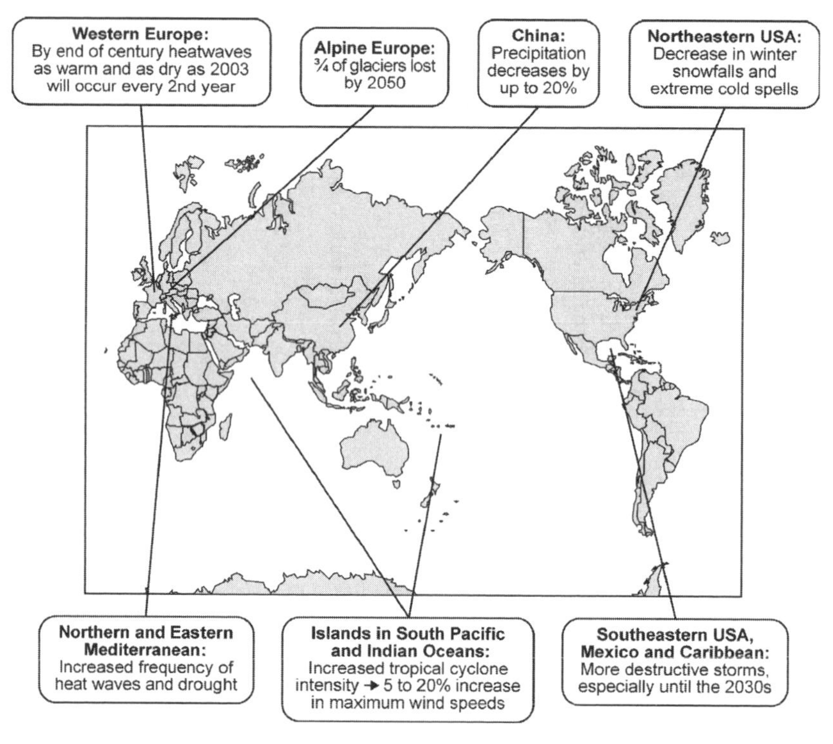

Figure 2.6 Climate–tourism hotspots

'pull' factors and on tourism operations.[1] 'Climate–tourism hotspots' are defined here as those parts of the world where high international tourist arrivals are forecast, or where tourism will be a major contributor to the national economy, and significant changes in climate are projected for the near term. Based on this definition, several such hotspots can be identified (Figure 2.6).

Alpine Europe

Today the Alpine region is the most important tourism region in Europe (see König, 1998 for a description of the situation in Austria). The European Alps account for an estimated 7–10% of the annual global income from tourism. Some 100 million tourists visit the Alps each year. In winter, skiing is the main attraction, while year round the snow- and ice-covered mountains lure people to the region. But such landscapes are already in decline. Since the mid-1980s the length of the snow season and snow amount have substantially decreased (Beniston, 1997). Since 1850 the glaciers of the European Alps have lost about 30–40% of their surface area and about half of their volume (Beniston, 2003). Paul *et al.* (2004) report an accelerating loss of glacial ice, with an 18% reduction in area between 1985 and 1999. On average the ice masses lost around a third of their surface and half of their volume between the mid-19th century and 1975. Since then, a further 20–30% of the ice volume has melted. By 2050 three quarters of today's Alpine glaciers may well be lost. Conservatively, regions below 2000 m above sea-level (a.s.l.) will be ice-free by 2050 and those below 2500 m a.s.l. by 2100. Regional models indicate that summers will become progressively as hot as the 2003 heatwave, when in only three months some mountain glaciers in the Alps lost up to 10% of their mass still present before 2003 (Beniston & Diaz, 2004). Simulations of future temperature scenarios for Northern Switzerland suggest that about every second summer could be as warm or even warmer (and dry or even dryer) than 2003 towards the end of this century (Schär *et al.*, 2004).

Western Europe

Currently Western Europe is the second most visited region in Europe, attracting 117 million visitors in 1995, and 131 million in 1998. While the growth rate for arrivals in Western Europe will be the lowest in Europe over the period 1995–2020, at only 1.9% per year, 185.2 million people are forecast to visit the region in 2020 (UNWTO, 2001).

Greenhouse-induced warming will be greatest in summer, when temperatures could be up to 10°C warmer than present, with increases in the highest yearly maximum temperature being even greater, as well as a very large increase in the lowest yearly minimum temperatures. In the latter part of the 21st century conditions experienced during the 2003 heatwave are likely to become the norm (Beniston & Diaz, 2004). During that heatwave the very high minimum temperatures had an unprecedented compounding heat stress effect. A general decrease in precipitation is projected, despite increased precipitation in winter. The main decrease in precipitation is in summer, with a reduced number of precipitation days rather than reduced precipitation intensity. Thus the extreme daily precipitation increases. Interannual variability of both temperature and precipitation increases substantially in the summer. Cloudiness and snow cover decrease, the latter by at least 80% (Giorgi *et al.*, 2004; Räisänen *et al.*, 2004; Rowell, 2005). For the last decade windstorm events were the second highest cause for insured loss due to natural catastrophes in Europe. Since the beginning of the last century the number of winter cyclones has increased. The increase in the frequency of extreme cyclones is projected to continue, bringing more extreme wind events (Leckebusch & Ulbrich, 2004). Relative sea-level is expected to rise by between 0.1 and 0.6 m over the 21st century (IPCC, 2001). This would particularly affect coastal tourism in France, the Netherlands and the UK.

Central/Eastern Europe

Tourism in Central and Eastern Europe in 2005 is expected to generate US$234.8 billion of economic activity (direct and indirect) in 2005, and account for 9.6% of gross domestic product (GDP) and 12.3 million jobs (8.3% of total employment). Tourism is forecast to increase by 6.8% per annum, in real terms, between 2006 and 2015. Out of 13 regions, the Central and Eastern European travel and tourism economy is ranked fourth regionally in absolute size worldwide, ninth in relative contribution to regional economies and second in long-term (10-year) growth. Visitor numbers to Central/Eastern Europe are expected to grow faster than for other parts of Europe, such that by 2020 the subregion will attract almost 40 million more visitors than Western Europe (223.3 versus 185.2 million) (UNWTO, 2001).

Unlike other parts of Europe, maximum warming over Eastern Europe, by about 7°C, will occur in winter (Rowell, 2005). Large increases in the lowest minimum temperatures are anticipated (Räisänen *et al.*, 2004). Precipitation will increase in winter, but decrease in summer (Rowell, 2005). There will likely be a significant increase in the interannual variability of temperature in spring and summer, and substantial reductions in winter snow cover. Both the precipitation intensity and frequency in summer will increase, while in other seasons there will be an increase in the intensity of daily precipitation events, but a decrease in the frequency. Cloudiness will decrease (Giorgi *et al.*, 2004). Relative sea-level is expected to rise by between 0.1 and 0.6 m over the 21st century (IPCC, 2001).

Northern and Eastern Mediterranean

The Mediterranean is currently the world's most popular and successful tourist destination. Of the 637 million international tourist arrivals worldwide in 1995, over 166 million (almost one-third) visited the Mediterranean region, spending close to US$100 billion. Over the period 1995–2020, it is expected that the average annual growth rate of tourist arrivals to the 21 countries bordering the Mediterranean Sea will be 3%. By 2020, visitor arrivals will grow to just over 200 million, representing a 30% share of European arrivals. Growth will concentrate in some of the Eastern destinations, particularly Turkey, Croatia and Slovenia (UNWTO, 2001). Therefore, despite its continued growth as the world's largest tourist-receiving region, the Mediterranean's share of global tourism is expected to decrease over the next two decades.

The greatest warming is projected to occur in summer, when temperatures could be up to 8°C higher than at present. Maximum temperature is one of the most sensitive variables to greenhouse warming in the Mediterranean basin, with the largest changes (over 6°C) occurring in summer and the lowest in winter. Projected mean maximum temperatures increase more than minimum temperatures, for all seasons. The daily temperature range will be greater than current climate values. Heatwave frequency is likely to increase substantially in the Western and Eastern parts of the region (Sánchez *et al.*, 2004). Increases in the highest yearly maximum temperature may be even greater, while the lowest yearly minimum temperatures will also increase. There will be a decrease in precipitation in all seasons except winter, but an increase in precipitation intensity (Giorgi *et al.*, 2004). Precipitation extremes show an increasing trend in many Mediterranean areas, particularly in the warmest seasons (summer and early autumn). The number of rain days will decrease (Sánchez *et al.*, 2004). Maximum wind speeds are likely to decrease in a warmer world while a marked change in the frequency of cyclonic disturbances is highly unlikely (Leckebusch & Ulbrich, 2004). Relative sea-level is expected to rise by between 0.1 and 0.6 m over the 21st century (IPCC, 2001).

North-eastern USA

Temperatures are projected to increase by 2–3°C by 2100. Winter minimum temperatures show the greatest change, with projected increases being as much as 5°C by 2100. The largest increases will be in coastal regions. Maximum temperatures are likely to increase much less than minimums, again with the largest changes in winter. Winter snowfalls and periods of extreme cold are very likely to decrease. Some models suggest that precipitation will increase by as much as 25% by 2100, while others indicate little change or small regional decreases. The variability in precipitation in the coastal areas is projected to increase. Over the coming century, winter snowfalls and periods of extreme cold will likely decrease. In contrast, past increases in heavy precipitation events are likely to continue into the future (US NAST, 2001).

South-eastern USA, Mexico and the Caribbean

From a total of 13 regions, the Caribbean tourism economy is ranked number one in relative contribution to regional economies (WTTC, 2005). In 2005, tourism in the Caribbean is expected to generate US$45.5 billion of economic activity. The direct and indirect impact on the economy is expected to account for 15.4% of GDP and 2.4 million jobs (15.1% of total employment). Tourism is expected to grow by 3.3% in 2005 and by 3.4% per annum, in real terms, between 2006 and 2015. The Caribbean is forecast to increase from 20 million tourists in 2000 to 40 million by 2020 (UNWTO, 2001). Similarly, tourism in Mexico is forecast to grow from current levels of around 30 million visitors to nearly 50 million visitors by 2020, though this represents a slight decline in market share.

In the South-east USA, by 2030 maximum summer temperatures will increase by 1–3°C, while maximum winter temperatures will increase by at least 0.6°C. The projected increase in mean annual temperature is

1–2°C by 2030 and 2–5.5°C by 2100. The heat index (a measure of thermal comfort, based on temperature and humidity) is likely to increase by as much as 8°C in summer, by 2100. Water quality and availability are likely to decline significantly during the early summer months over the next 30 years (US NAST, 2001). Heatwaves are predicted to increase in frequency and severity. Similar changes will occur in Mexico and the Caribbean. For all three areas, projections of precipitation increases are as common as those of decreases (Ruosteenoja *et al.*, 2003). Global warming will likely result in more frequent El Niño-like conditions, favouring cooler winter temperatures, wetter than normal winters (except in the north) and springs, more frequent severe storms and flooding, and less frequent, but possibly more intense, Atlantic hurricanes (Timmerman *et al.*, 1999). There is a likelihood of more destructive storms in the Gulf of Mexico and along the Atlantic seaboard (Emanuel, 2005). But any such trend in the frequency of Atlantic hurricanes is likely to be overshadowed by decade-to-decade changes in ocean currents, and hence in sea surface temperatures. As a result, periods of higher hurricane activity, such as 1941–1965, may persist for decades (Bengtsson, 2001; Goldenberg *et al.*, 2001). Thus the period of high hurricane frequency that began in 1995 is thought likely to last until around 2030. Higher sea-levels, especially when accompanied by storm surges, will have adverse impacts on infrastructure and accelerate coastal erosion (Titus, 2002).

China

By 2020, China will be the number one tourist destination in the world, with 130 million international arrivals. Its growth rate of 7.8% will be the fifth highest average annual growth rate in the East Asia and Pacific region (UNWTO, 2001). Inbound tourists to the Chinese mainland reached 109 million in 2004, breaking the 100 million mark for the first time. This earned foreign exchange revenue up to US$25.7 billion. Tourism in China in 2005 is expected to generate US$265.1 billion of economic activity, while direct and indirect economic impact in 2005 is expected to account for 11.7% of GDP and 64.6 million jobs (8.6% of total employment). Tourism is expected to grow 10.1% in 2005 and by 9.2% per annum, in real terms, between 2006 and 2015. Out of 174 countries, the China travel and tourism economy is ranked number seven in absolute size worldwide, 74th in relative contribution to national economies, and second in long-term (10-year) growth (WTTC, 2005). Further information is provided in Text Box 3 (Chapter 4).

By the end of the century temperatures in China will likely have increased by 3–8°C, with greater increases in winter. However, heatwaves will be more common in summer, while cold spells in winter will be less common. The extreme temperature range will increase. Precipitation will likely increase by up to 20%, with the greater increases being in the winter.

The change of the Pacific Ocean and climate to a more El Niño-like state will increase the likelihood of drought, forest fires and flooding in summer. The number of rain and heavy rain days will increase, while the number of frost days will decrease. The frequency of tropical storms will increase, with the dominant path somewhat northward of its present-day position. In the 20th century the area of mountain glaciers in China has declined by 21%. This trend will continue, likely at an accelerated rate. Relative sea-level rise could range from 31 to 65 cm by 2100, aggravating coastal erosion (Government of China, 2004; Li & Xian, 2003; Ruosteenoja, *et al.*, 2003; Uchiyama *et al.*, 2005; Yihui *et al.*, 2004; Xuejie *et al.*, 2004).

Small island states in the South Pacific and Indian Oceans

Tourism dominates the economies of the Cook Islands and Palau and generates a substantial part of GDP in Fiji, French Polynesia, New Caledonia, Samoa and Vanuatu. Tourism contributes as much as 47% to the national GDP, with visitor arrivals to the region now being about a million people annually. But in global terms tourism in the South Pacific is small, with the region receiving less than approximately 0.2% of world tourist arrivals. The tourism industry is now the number one foreign exchange earner for about half the South Pacific. Total revenue for South Pacific Tourism Organisation members from tourism exceeds US$500 million, although a substantial part of that bleeds out of the region's economies to foreign-owned hotels, tour companies and airlines (South Pacific Tourism Organization, 2005). Regionally, tourism is expected to grow at the rate of 6.5% per annum, and reach over 14 million visitor arrivals by 2020 (UNWTO, 2001).

Tourism in Maldives in 2005 is expected to generate US$573.7 million of economic activity. Direct and indirect economic impact in 2005 is expected to account for 62.6% of GDP and 60,696 jobs (54.2% of total employment). Due to the Indian Ocean tsunami and other factors, Maldives' tourism industry is expected to decline by 14.2% in 2005, but it will recover and grow by 6.2% per annum, in real terms, between 2006 and 2015. Out of 174 countries, Maldives' tourism industry is ranked 136th in absolute size worldwide, fifth in relative contribution to national economies, and 29th in long-term (10-year) growth.

Tourism in Mauritius in 2005 is expected to generate US$506.3 million of economic activity, contributing directly and indirectly 31.6% of GDP and 170,857 jobs (33.9% of total employment). Mauritius's travel and tourism is expected to grow 12.7% in 2005 and by 4.8% per annum, in real terms, between 2006 and 2015. Tourism in Reunion in 2005 is expected to generate US$3961.5 million of economic activity (5.8% of GDP and 6.5% of total employment). Reunion's travel tourism is expected to grow 1.8% in 2005 and by 8.3% per annum, in real terms, between 2006 and 2015. In Seychelles, tourism in 2005 is expected to generate US$51.0 million of economic activity, with a GDP contribution of 60.2%. Growth rates are 14.0% in 2005 and 3.8% per annum, in real terms, between 2006 and 2015 (WTTC, 2005).

In both small island regions mean temperatures are projected to increase by about 2°C by the end of the century, with little difference between the four seasons. On the other hand, annual or seasonal precipitation amounts may increase slightly or not at all (Ruosteenoja *et al.*, 2003). Changes in rainfall variability and tropical cyclone characteristics in the South Pacific are strongly dependent on El Niño Southern Oscillation (ENSO) (see Chapter 5). As this is not well understood, there is little confidence in projections of either change, except that tropical cyclone intensity is likely to increase, resulting in a 5–20% increase in maximum wind speeds by the end of the century. Also for the Pacific Islands region, sea level is projected to rise by between 0.3 and 0.4 m by the end of the century (Hay *et al.*, 2003).

Climate change and tourism – internalising an externality

To date the contributions of tourism to climate change, as well as the consequences of climate change for tourism, have been largely ignored by the tourism industry. Thus climate change can also be described as a non-tourism-related externality. In his chaos model of tourism (Figure 2.7), McKercher (1999) argues that some non-tourism-related externalities have the potential to plunge tourism into chaos, precipitating rapid change. Climate change certainly has the potential to impact tourism in this way, as has already been demonstrated by the way climate variability and extreme weather events affect tourism today, and have done so in the past.

McKercher identifies several other non-tourism-related externalities, namely natural disasters, oil shocks, global economic crises, global terrorism and the outbreak of war. Despite the uncertainties in characterising future climates, such changes are inherently more predictable than climate change externalities. Thus there is greater opportunity to influence the nature and extent of the consequences for the tourism system.

In principal, all the components shown in Figure 2.7 are linked and changes in one part induce changes in all the others. Climate change impacts on tourists, transit routes, destinations, agencies and tourism-related externalities, for example competing destinations. Feedback loops exist. For example, an increasing number of tourists visiting an island destination consume increasing amounts of freshwater. When the water resources are depleted, no further growth of tourism is possible, or adaptation measures such as a desalination plant are required to maintain growth. Climate change, including changed precipitation patterns and higher temperatures, will influence the relationship between tourist numbers and water availability in a given location.

The different components of the chaos model of tourism can themselves be seen as systems, nested within the larger complex system of tourism as a

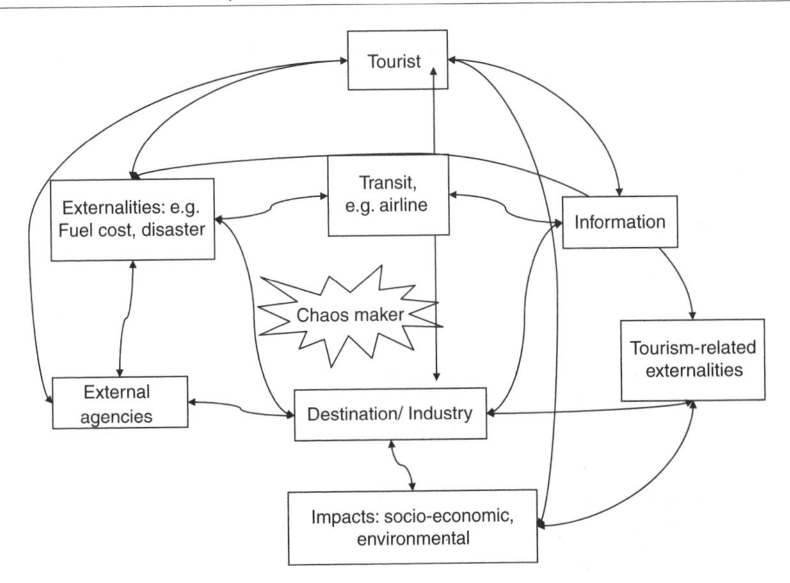

Figure 2.7 Chaos model of tourism
Source: after McKercher (1999)

whole. As will be shown in Chapter 3, the destination is a system that consists of the tourism industry and policy makers, among others. These agents are involved in developing and implementing integrated responses to climate change, namely through adaptation and mitigation policies and initiatives. Similarly, and as already noted, the climate system – which is here considered to be an externality to tourism – is a system in itself, with complex processes and a substantial amount of uncertainty.

McKercher (1999) also introduced 'chaos makers' in his model. These are typically internal to the tourism system. Low-cost air carriers are an example of chaos makers, as their actions have the potential to push a system over the edge of chaos, and force restructuring.

The ability of climate change to disrupt the tourism system, albeit temporarily until a new order emerges, suggests that key agents with a vested interest in maintaining the stability of the tourism system should be working proactively to avoid, or at least delay, such a situation. By recognising that tourism is a contributor to climate change, and endeavouring to reduce or even negate that contribution, the agents of tourism are in fact internalising the externality – that is, taking ownership of the problem, rather than leaving the problem to others to resolve. Importantly, this results in the financial and other repercussions of limiting or preventing tourism-induced climate change being borne by tourists and other appropriate agents of tourism, rather than more widely by those who may or may not be engaged in tourism-related activities.

Internalising such externalities is consistent with the four key principles that generally underpin international and national climate policies: (a) the precautionary principle, (b) polluter pays, (c) sustainable development and (d) equity. The principles will be discussed in more detail in Chapter 9.

Conclusions

The multiple interactions between tourism and climate are complex, with the consequences of change being both difficult to predict and manage. This is because the component systems are themselves now recognised to be non-deterministic and non-linear. It is thus difficult to predict the nature of future climate–tourism interactions in a deterministic or even probabilistic way, including how components of the tourism system will change over time, as the system responds to climate change. As a result, implementing apparently appropriate policies and management procedures, be they tourism or climate related, will not necessarily reduce climate-related risks and maximise opportunities created by climate change. However, the tourism system as a whole will likely respond in a more orderly and hence predictable manner, governed by a number of underlying principles and relationships that determine how the overall system will respond, within broad parameters.

Managing the resulting changes takes place on two fronts. On the one hand, mitigation reduces GHG emissions and hence the rate of climate change and the likelihood of climatic conditions that are hazardous to the tourism industry. On the other hand, adaptation reduces the consequences of the climate-related hazards (see Chapter 9). To date efforts to reduce GHG emissions globally have been largely unsuccessful. Inertia in the climate system also means that even if meaningful reductions in GHG emissions are achieved, the rate of climate change will continue unabated for many decades at least, placing even more importance on adaptation.

The complexities and non-linearities in the tourism–climate system also result in a propensity for abrupt and pervasive changes, including 'surprises'. These have been observed in the past, for the two subsystems as well as collectively. Such abrupt changes are particularly challenging for those who seek an orderly evolution of tourism, at national or business enterprise levels. This is especially so in the case of the climate–tourism 'hotspots' such as the European Alps, the islands of the South Pacific and Indian Oceans, and China.

Note

1. The possibility of 'push' factors, such as changes in the climate in the tourists' country of origin, has not been considered in the identification of climate–tourism hotspots.

Chapter 3

Case Studies of the Tourism–Climate System

Key Points for Policy and Decision Makers, and Tourism Operators

- Two case studies – small island states and Alpine Europe – take a geographic approach, whereas the two other case studies – insurance sector and international aviation – provide a sectoral perspective.
- The four case studies illustrate the complex relationships between the tourism and climate systems and highlight the need for policy interventions and implementation of adaptation and mitigation strategies in an integrated way at various levels.
- Climate change is highly relevant to those destinations that have been identified as 'climate–tourism hotspots'; impacts are already affecting tourism activity and economic benefits.
- The two geographic case studies highlight the importance of integrating climate change adaptation and mitigation into wider sustainable development strategies. Increasing air conditioning as a response to hotter temperatures is an example of an unsustainable adaptation measure, whereas the conservation of water (when facing a drier climate) is likely to reduce the carbon footprint as an ancillary benefit.
- Insurance cover against climate change impacts is increasingly important for vulnerable destinations; insurances need to be readjusted to take climate risks into account. This can make them potentially unaffordable or unavailable for certain businesses in high-risk locations.
- The insurance case study illustrates how climate change is increasingly becoming a business consideration; economic viability depends on minimising risks for the business, destination and the tourist, including those related to climate change.
- Climate change also impacts on vulnerable destinations and sectors through climate change policies that affect the current competitiveness and relative costs of energy. International aviation is a prime example of how policies can be potentially far-reaching and raise questions of equity and sustainable (economic) development.
- GHG emissions from aviation are one example where tourism (in this case the air travel component) requires different treatment compared

with other activities. This results from the atmospheric impact of emissions in the upper troposphere and also from the international nature of aviation, which makes regulation very difficult.

Introduction

This chapter aims to explore and discuss the complex interactions between climate change and tourism through a case study approach. It will use the conceptual model of tourism as a complex system shown in Chapter 2 and apply it to the specific situation of four case studies. The first two case studies – small island states and Alpine Europe – take a geographic approach, where climate change impacts on tourism and responses to climate change can be described for a defined type of destination. The other two case studies are sector-specific. First, we discuss the importance of insurance for tourism and the impacts that climate change might have on the existing relationship between tourism and insurers. Second, we analyse the issue of climate change and international aviation and give particular emphasis to the specific effects of aircraft emissions in the atmosphere and strategies for managing those emissions in order to reduce aviation's contribution to global climate change. Policy options are the focus of this last case study.

Climate Change and Tourism in Alpine Europe

Introduction

In Chapter 2, Alpine Europe has been identified as a 'climate–tourism hotspot', principally because tourism is a major contributor to the national and local economies of the countries that make up Alpine Europe and significant changes in climate are projected for the region, with major consequences for tourism. But this case study was prepared not only because the region is a climate–tourism hotspot, but also because it illustrates very clearly the complex interactions between climate change and tourism, and how these could be or are being managed (Figure 3.1).

Major changes in the climate (shown as an externality in Figure 3.1) of Alpine Europe are already occurring, and even more significant changes are projected to occur within the time horizon of planning tourism in the region. Based on Beniston (1997), Fink *et al.* (2004), Giorgi *et al.* (2004), Räisänen *et al.* (2004), Schär *et al.* (2004), Rowell (2005) and the Swiss Agency for Environment, Forests and Landscape (2001), and from an alpine tourism perspective, these changes can be summarised as follows:

- Over the last 50 years there have been periods with abundant snow, and other periods with a dearth of snow; the interannual variability is linked to temperature and precipitation anomalies, which are in

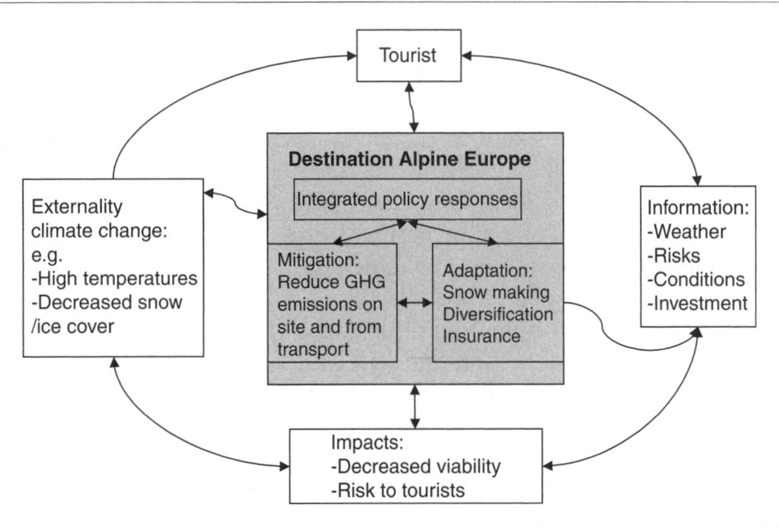

Figure 3.1 The tourism–climate system with a particular focus on Alpine Europe. The main relationships are shown as lines between the system's components

turn linked to the strength of the westerly flow over the Atlantic Ocean, as measured by the North Atlantic Oscillation Index (NOAI).

- Since the mid-1980s the length of the snow season and snow amount have decreased substantially; the sensitivity of the snowpack to climatic fluctuations diminishes above 1750 m.
- There has been widespread and generally consistent retreat of glaciers since 1880, with the rate increasing markedly since the mid-1980s.
- By the 2050s surface temperatures are likely to have increased by 2–4°C in summer and 0.5–1.5°C in winter.
- Winter precipitation is projected to increase by as much as 15–45% before the end of the century, while summer precipitation may well decrease; in both seasons there will be an interannual variability, along with increased frequency of heavy precipitation events.
- A slight increase in average wind speeds is likely to occur in winter, while there will be at least a slight decrease in spring and autumn.

Climate consequences in Alpine Europe

Based on the sources above, the specific future climate consequences for Alpine Europe can be summarised as follows:

- By the end of the century about every second summer could be as warm or warmer, and as dry or dryer, than 2003.

- Duration of snow cover is expected to decrease by several weeks for each degree Celsius of temperature increase (Hantel *et al.*, 2000; Wielke *et al.*, 2004).
- Most glaciers are likely to disappear during the 21st century, and areas below 2500 m will be ice-free by the end of the century (Haeberli & Burn, 2002; Paul *et al.*, 2004).
- The combination of rising temperatures with increased snowfalls will lead to more frequent avalanches.
- As glaciers disappear, flood discharge will significantly increase in the upper and middle course of rivers (Lehner *et al.*, 2006) while spring and summer discharge will decrease (Hagg & Braun, 2005).
- Lowest river flows will decrease by up to 50% after glaciers have melted (Schneeberger *et al.*, 2003).

Consequences for tourism in Alpine Europe

Both the local and national economies of countries in Alpine Europe depend heavily on tourism, and for many alpine areas winter tourism is the most important source of income, with snow reliability being the key to maintaining tourism flows and hence financial viability. Other important factors for the success of tourism have been the length of season and the scenic value of glacial landscapes. Over 80% of Swiss tourists questioned in a survey said snow reliability was amongst the top ten requirements (Bundesamt für Statistik, 1999). Earnings of cableway companies decreased by 20% in the late 1980s due to significant snow shortages. The companies operating at lower and medium altitudes were particularly affected, while those above 1700 m were much less affected or even had an increase in business. Hotels and holiday apartments were less affected as non-skiers make up part of their clientele. Moreover, accommodation tends to be booked long in advance. But demand will fall off eventually, if tourists repeatedly experience poor snow conditions (Elsasser & Buerki, 2002).

A ski resort in Switzerland is considered 'snow-reliable' if in seven out of ten winters there is a snow covering of at least 30 cm on at least 100 days between 1 December and 15 April. Currently 85% of Switzerland's ski resorts are considered to be snow-reliable. With the line of snow-reliability rising to 1500 m, as is projected to occur by 2030–2050, the number of snow-reliable ski resorts drops to 63%. A rise to 1800 m results in only 2% of small ski areas and 44% of larger ski resorts qualifying as snow-reliable. For every 1°C rise in temperature there will be about 14 fewer skiing days (Schwarb & Kundzewicz, 2004).

Climate change is anticipated to lead to a new pattern of favoured and disadvantaged ski tourism regions as well as to concentration of winter tourism in the midwinter months of January and February. The favoured regions will have transport facilities that provide access to altitudes higher than 2000 m. A survey of skiers revealed that during a snow-poor season 49% would change to a more snow-reliable ski resort, while 32% would ski less often. Only 4% would not ski in such a season (Buerki *et al.*, 2003).

However, due to increasing variability in climate, including more extreme events, planning must also consider the possibility of winters with an overabundance of snow, such as occurred in Switzerland in 1998/99. In that winter the direct losses incurred by mountain cableways as a result of avalanches and large quantities of snow amounted to around US$10 million. Total losses were closer to US$130 million, the major portion of which was indirect losses. In total, 36 facilities were damaged. Mountain railways spent an additional 77% on snow clearing and around 25% more than usual was spent securing ski slopes (Buerki *et al.*, 2003).

Increasingly tourist resorts have been developed in areas where there is a high risk of damage from avalanches, rock falls and flash flooding. However, damage by avalanches has remained at similar levels since the 1950s. This might be due to technical and planning measures, such as design of structures and infrastructure, induced release of snow, closing of roads and hazard zoning. Such measures have generally kept pace with development and significantly reduced the probability of catastrophic avalanches. This highlights the fact that many climate-related risks can be managed effectively.

Adaptation

The complex nature of the tourism–climate system in Alpine Europe means that climate change-related risks and opportunities can be managed but certainly not controlled. Some tourism representatives in Central Switzerland recognise climate change as a problem for winter tourism, due to the high dependency on snow and the risk of snow-deficient winters; but overall the importance of climate change is considered to be relatively low. Some even consider climate change and its consequences to be exaggerated by the media, and also in science and politics. Notwithstanding this, there is a consensus that winter sports can only survive in the European Alps if snow reliability is guaranteed. Climate change is already affecting the strategies and plans of winter sport resorts. For example, climate change is being used as a reason to construct artificial snow-making facilities and as an argument to justify opening up high-alpine regions to tourism, even though this increases the pressure on these ecologically sensitive areas (Buerki *et al.*, 2003).

Thus climate change represents a new challenge for winter tourism in the European Alps. It is viewed as a catalyst that will reinforce and accelerate the pace of structural change in the industry and clearly highlights the risks and opportunities inherent in tourism development today. The range of adaptation strategies available to winter tourism in Alpine Europe is shown in Figure 3.2. In addition to trying to maintain

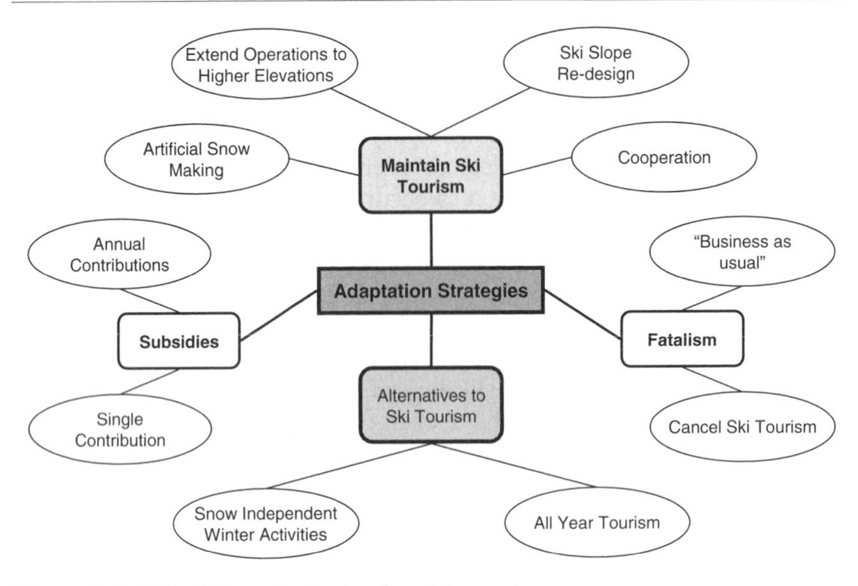

Figure 3.2 Adaptation strategies for ski resorts
Source: after Buerki *et al.* (2003)

snow-based tourism through technological interventions or subsidies, there are several other proactive or reactive options (see also Chapter 9). These include diversification of the type and seasonal focus of the activities available to tourists in winter and taking advantage of new opportunities in the summer season. In general there is now a trend to creating the infrastructure required for such alternative activities, but to date there is no long-term strategy for adapting to climate change. Moreover, increased summer tourism could also come at a cost to the Alpine landscape resulting from, for example, the retreat of glaciers. Another option is to remain inactive or fatalistic with respect to climate change. Fatalism means that neither suppliers not consumers are willing to modify their behaviour. They endeavour to continue with business as usual, despite the changing conditions, or withdraw from ski tourism without adequate planning. This might involve closing down and dismantling facilities without attempting to promote and in other ways encourage other types of tourism.

As noted above, higher air temperatures and diminished snow cover will result in the closure of lower altitude ski resorts and the concentration of skiing on higher ski fields, many of which would be forced to use expensive snow-making equipment, at least for as long as climatic conditions remain within appropriate limits. Currently, smaller ski fields at lower altitudes in Alpine Europe are having difficulty financing the investments needed to mitigate the impacts of reduced snow cover, including snow-making machines, levelling out ski slopes and opening higher-elevation runs. For example, investment cost for snow-making equipment is approximately US$0.6 million per kilometre of ski slope, while annual operating costs are between US$18,000 and 30,000 (Ammann, 1999).

The conventional snow-making system uses compressed air and water. Snow guns (Figure 3.3) must be closely spaced – about 15–30 m apart – to give adequate coverage, meaning that both capital and operating costs are high – for larger resorts power costs alone can amount to several hundred thousand dollars a month. The preferred technology is now airless 'fan guns' comprising small nozzles arrayed around a large fan blade. The water is atomised, resulting in higher trajectory and thus more time to freeze and improved coverage. Nucleation chemicals are added to the water to encourage quicker freezing, and at temperatures some 6°C higher than would be the case without nucleation. About 15 times less air is used than in a conventional snow gun, saving on power costs. The guns, which are much quieter, can also be programmed remotely, allowing real time adjustments for changes in temperature, humidity and wind conditions. The result is about 25% more snow, relative to a conventional gun, which had to be adjusted manually and was thus unable to respond to changing weather conditions. Currently some 15% of American ski resorts use such automated snow-making equipment, compared to almost 100% in Europe (*The Economist*, 26 November 2005). However, artificial snow-making also raises environmental concerns because of the quantities of energy and water required, the disturbances generated during the operation of the

Figure 3.3 Artificial snow-making using snow guns

equipment and the damage to vegetation observed following the melting of the artificial snow cover.

Financial institutions are already responding to the growing risks faced by low-elevation ski resorts in Switzerland. For example, they are unwilling to provide funding under the usual terms. But the loss of low-elevation facilities could affect the entire industry in Alpine Europe. Such fields play a key role in promoting the importance of skiing. Thus there are divergent views on how best to manage the risk. Some people regard healthy shrinkage of the sector as necessary. These people favour the closure and dismantling of non-profitable cable-ways and ski-lift facilities. Others believe there is an obligation to retain these ski fields, for regional economy reasons. Thus there is increasing pressure for cable-way companies to be subsidised (Buerki *et al.*, 2003; Elsasser & Buerki, 2002).

The UNWTO (2003c) identified the following additional adaptation measures:

- As a result of less stable (wetter) snow, greater avalanche protection will be required.
- Resorts that are no longer within the reliable winter snow belt may need to reinvent themselves and address alternative markets.
- As mountain summers become drier and warmer, the tourist summer season may be extended into the shoulder months. Changing demographic patterns, particularly an ageing population, may prove beneficial to attracting this market.
- Introduction of built attractions to replace natural attractions if the appeal of the latter diminishes, e.g. the installation of an ice rink, spa facilities, etc., if skiing becomes less reliable at lower altitudes.
- The development of alternative marketing strategies to cope with an expanding or a diminishing market (including stronger promotion of domestic tourism).
- Recognition of the increasing vulnerability of some ecosystems and the adoption of measures to protect them as far as possible.
- Introduction of fiscal incentives (e.g. accelerated depreciation) or financial assistance for changes to the built tourism infrastructure, to deal with the consequences of climate change.
- Changes to the school year in order to change peak holiday times (e.g. if traditional mid-summer periods become too hot or ski resorts have shorter snow seasons).
- Provision of direct training to the tourism sector in dealing with the consequences of climate change, including assistance with practical issues e.g. the 'hazard' mapping of sites and zones.

Mountaineering and hiking may provide compensation for reduced skiing, and thus certain mountain regions could remain attractive destinations. However, global climate change has wider implications for traditional holiday breaks, with destinations other than mountains in winter becoming increasingly competitive. Higher temperatures may imply longer summer seasons in mid-latitude countries. As a consequence, a new range of outdoor activities may emerge (Beniston, 2003).

Mitigation

As an activity that is severely affected by climate change, snow-based tourism has an interest in keeping climate change to a minimum. This can only be achieved through mitigation strategies, the benefits of which may be felt only in the longer term. In this case mitigation would include reducing emissions in relation to both activities at the ski fields (see Text Box 1) and to transportation systems that connect to ski fields and other winter attractions.

As is also noted in Chapter 7, 'car-free' tourism communities are becoming increasingly common in the European Alps. The example of Bad Hofgastein and Werfenweng (Text Box 12) illustrates the implementation of integrated policy initiatives with respect to the environment, transport, tourism and technology, as formulated, for example, in the National Environment Plan, the European Union (EU) Environment Action Programme and the Alps Convention. Local community projects promote an improvement of living quality for inhabitants and guests by reducing transport-related environmental impacts (e.g. air quality, congestion, noise, safety) as well as reducing GHG emissions. A wide range of local measures and initiatives have been implemented, including improved environmentally sound transport connections (for both journeys to and from the destination and local/regional transport), baggage/shuttle services, availability of alternative vehicles in the communities (e.g. rental service with electric cars, scooters and bikes, car-sharing) and electronic travel information systems that cover all modes and regional/interregional services of transport. High acceptance of the concept is illustrated by the rapid growth in overnight accommodation (+11% in the participating communities compared with an average of 2.5% in non-participating communities).

Mitigation of GHG emissions from alpine tourism also involves the accommodation sector. The most important areas for improving energy efficiency relate to heating and hot water supply. Measures for reducing energy demand and emissions will be discussed further in Chapter 7.

Integrated responses

For alpine communities and ski resorts it is important to adapt to climate change without increasing GHG emissions. Snow-making systems may reduce the risk of snow-scarce winters in the short term, but in the long term they constitute unsustainable adaptation measures

Text Box 1 Aspen Skiing Company
Source: Aspen Snowmass Sustainability Report (2004), Schendler (2003)

The Aspen Skiing Company (ASC) attracts 1.3 million visitors each winter to almost 5000 acres of skiable terrain on four mountains – Snowmass, Aspen Mountain, Buttermilk and Aspen Highlands – and year-round visitors to 15 restaurants and two hotels. ASC has become the first American ski resort to achieve ISO 14001 certification of their environmental management programme. ASC was also the first ski/snowboard resort to announce a policy to protect the climate. Annual CO_2 emissions have decreased markedly between 2000 and 2004. Specific actions taken to reduce CO_2 emissions include:

- Operation of a wind turbine; overall, 12% of electricity comes from renewable energy sources.
- Use of a 80/20 blend of diesel/biodiesel in all company-owned snowcats.
- The new compressors for snow making are 30–40% more efficient than the old ones. This translates into annual savings of US$17,000–22,610 at Buttermilk and US$23,375–35,078 at Aspen Mountain.
- Installation of a computerised energy management system in the 5-star Little Nell hotel. This enables accurate monitoring of the mechanical system. Early data analysis suggests that the system has saved $48,000 in gas costs annually and $2500 monthly in electricity.
- Certification of the Sundeck restaurant through the US Green Building Council's 'Leadership in Energy and Environmental Design' programme.
- Improved energy efficiency of many appliances (using EPA's Energy Star criteria), including printers, copiers and beverage machines. The latter has a special electronic sensor that shuts it down when it's not in use. The company installed energy-efficient washers in many employee housing units. These horizontal axis washers cut gas or electricity use by 63% over the standard top-load washers.
- Replacement of 1500 60-W lamps at the Club Commons (Snowmass), Heatherbed (Aspen) and Snow Eagle (Aspen), resulting in savings of US$8952 annually in energy alone.
- Party to several innovative national emissions reduction programmes, including a municipal tax on energy use in new buildings.

as a result of high energy consumption and other impacts on the natural environment such as changing run-off patterns. Adaptation measures that do not generate additional GHG emissions are preferable. These include the development of a wider spectrum of activities as well as reflecting potential shifts in seasons (peak snow) in marketing strategies.

The car-free glacier resort of Saas-Fee, for example, is well known for winter tourism. Saas-Fee also markets itself successfully as a summer and autumn destination, emphasising the health benefits of a dry climate with few allergens. Marketing campaigns target people who suffer from allergies and asthma. Summer activities include hiking (on the 350-km-long network of hiking paths), summer skiing and snowboarding, mountain biking and other activities. Autumn activities promoted by Saas Fee are hiking in autumn colours and 'culinary delights'.

Similarly, mitigation strategies, such as the development of new transport systems, benefit from integration with adaptation strategies. An example is increased emphasis on summer tourism. High tourist numbers all year round mean that new public transport connections are more viable, and therefore can be offered at more competitive prices. These strategies require coordination and collaboration between local decision-makers and national agencies or companies (e.g. railway companies and tour operators).

Climate Change and Tourism in Small Island Countries

Introduction

Chapter 2 identified small island countries in the South Pacific and Indian Oceans as 'climate–tourism hotspots', for two principal reasons: (a) the importance of tourism to many of their national economies; and (b) the major consequences of climate change, and especially changes in variability and extreme events. Such changes are already being experienced by the tourism sector in those countries. It is projected that these consequences will increase dramatically in coming decades. Island destinations are also included in other tourism–climate hotspots, including the Northern and Eastern Mediterranean and the South-eastern USA, Mexico and the Caribbean. These islands face similar issues. But as tourism and climate change are also important to most small island countries, this case study encompasses all small island countries.

An important feature of small island countries is the existence of very strong two-way interactions between climate change and tourism. The case study is intended to show how these could be and, in some cases, are being managed. Thus the case study addresses all the components of the climate change-tourism system (Figure 3.4).

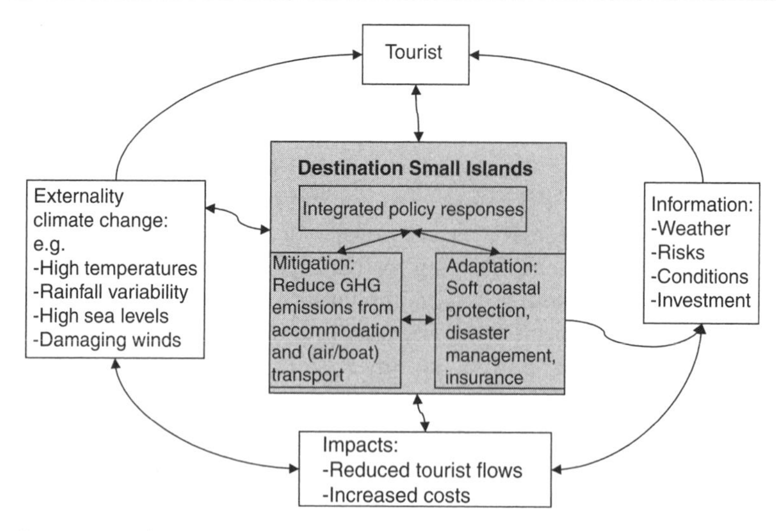

Figure 3.4 The tourism–climate system with a particular focus on small island destinations. The main relationships are shown as lines between the system's components

Based on Hay *et al.* (2003), IPCC (2001), Shea *et al.* (2001), and from a small island tourism perspective, the observed and projected changes are summarised in Table 3.1 and as follows:

- The intensity and frequency of many extreme events, such as heavy rainfall, drought, high air and ocean temperatures, strong winds and storm surges have increased, especially in the last two decades.
- These same extreme events, and also interannual variability, are likely to be more common in the coming decades.
- Most small island countries have experienced a rise in sea level, though for a few countries sea level has fallen due to tectonic uplift.
- Sea level is projected to rise, at an escalating rate.

Climate consequences in small island countries

Based on the same sources, it is possible to identify numerous indirect consequences of climate change for small island countries (Figure 3.5). This includes discomfort, ill health and coral bleaching as a result of excessively high temperatures and coastal erosion and land loss due to high wave incidents and elevated sea-levels (see also Chapter 8). As will be described in the next section, many of these consequences flow through to, and affect, tourism. One example is the increasing concentration of CO_2 in the oceans, associated with an increase in atmospheric concentrations. This will in turn increase the acidity of the water, impacting on marine ecosystems such as coral reefs, and representing

Table 3.1 Projected increases in (a) air temperature and (b) changes in precipitation (%) for small island regions

Region	*Forecast period*		
	2010–2039	*2040–2069*	*2070–2099*
(a)			
Mediterranean	0.60 to 2.19	0.81 to 3.85	1.20 to 7.07
Caribbean	0.48 to 1.06	0.79 to 2.45	0.94 to 4.18
Indian Ocean	0.51 to 0.98	0.84 to 2.10	1.05 to 3.77
Northern Pacific	0.49 to 1.13	0.81 to 2.48	1.00 to 4.17
Southern Pacific	0.45 to 0.82	0.80 to 1.79	0.99 to 3.11
(b)			
Mediterranean	– 35.6 to 55.1	– 52.6 to 38.3	– 61.0 to 6.2
Caribbean	– 14.2 to 13.7	– 36.3 to 34.2	– 49.3 to 28.9
Indian Ocean	– 5.4 to 6.0	– 6.9 to12.4	– 9.8 to 14.7
Northern Pacific	– 6.3 to 9.1	– 19.2 to 21.3	– 2.7 to 25.8
Southern Pacific	– 3.9 to 3.4	– 8.23 to 6.7	– 14.0 to 14.6

Source: after Ruosteenoja *et al.* (2003)

another threat to their long-term survival as well as to tourism businesses dependent on healthy reef systems.

Consequences for tourism in small island countries

Climate change will likely affect many components of the tourism system of small island countries. Uyarra *et al.* (2005) demonstrate this in terms of the environmental assets of two Caribbean islands, Bonaire and Barbados, and the importance of the quality of such assets in determining the choice and holiday enjoyment of tourists. Warm temperatures, clear waters and low health risks were found to be the most important environmental features influencing the choice of holiday destination. Thereafter, tourists in Bonaire prioritised marine wildlife attributes (such as coral and fish diversity and abundance) over other environmental features, while tourists in Barbados showed stronger preferences for beach features and other components of the terrestrial environment. The willingness of tourists to revisit these islands was strongly linked to the state of the preferred environmental attributes – more than 80% of tourists in Bonaire and Barbados would be unwilling to pay the same price for a

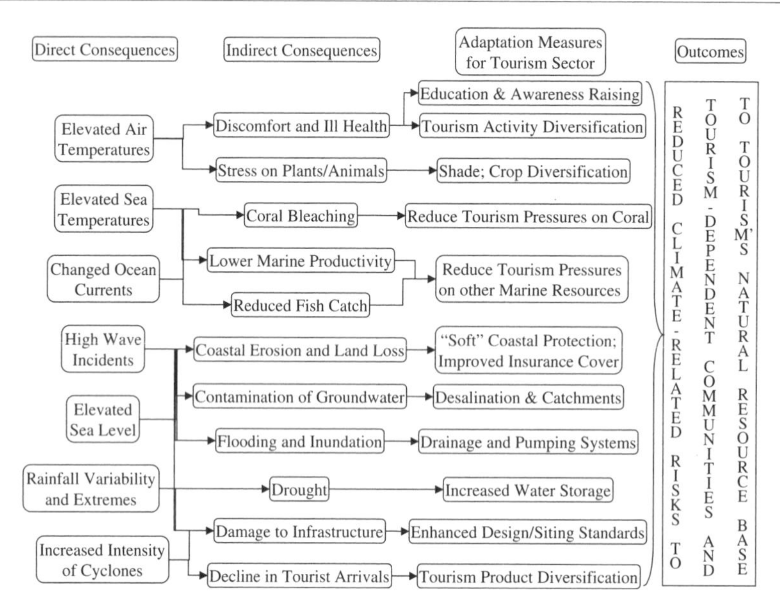

Figure 3.5 Direct and indirect consequences of climate change for tourism and adaptation measures

holiday in the event of a coral bleaching event resulting from elevated ocean temperatures or reductions in beach area as a result of sea-level rise.

Generalising from these findings, climate change may well have a significant influence on the tourism economies of small island countries by way of degradation of environmental features important to destination selection. For example, during the El Niño event of 1998–1999, near-surface temperatures in the coastal waters of Palau exceeded 30°C from June through November, 1998. This resulted in a massive coral-bleaching event that killed one-third of Palau's reefs. While no species became locally extinct, some populations declined to as much as 99% below pre-bleaching levels. The associated economic loss was estimated at US$91 million, partly because of a 9% drop in annual tourism revenues (Hay *et al.*, 2003).

Chapter 8 discusses the significant consequences of coastal erosion and land loss for tourism facilities located close to or on the coast. For example, of the 77,000 hotel rooms in the Commonwealth Caribbean, well over 65% are located in coastal areas. In Barbados over 90% of the 6000 hotel rooms are located within 1 km from the high-water mark and less than 20 m above sea-level, with over 50% of the rooms at risk from a category 3 hurricane. Replacement costs would be up to US$550 million. In Jamaica, around 85% of the 23,000 hotel rooms are in coastal areas. Around 85% of the 6000 hotel rooms on the Montego Bay coast would likely be affected by a 1-in-25-year hurricane-related storm surge. The

Caribbean is also one of the world's outstanding regions for yacht cruising, but a major obstacle to expansion of the industry is the increased threat of damaging hurricanes and the attendant risks of damaging waves and storm surges (Jackson, 2002).

In Maldives, an estimated 50% of all inhabited islands and 45% of tourist resorts currently suffer from beach erosion. This is a serious situation given that more than 70% of the inhabited islands have buildings less than 30 m from the shoreline, while buildings are within 15 m of the shoreline on over 55% of the islands. Some 2% of the islands have buildings at the shoreline. Like many other small island countries in the tropics, in Maldives the effects of sea-surface warming on coral reefs are reflected in the increased incidence of coral bleaching and mortality events. Coral bleaching events occurred in 1977, 1983, 1987, 1991, 1995, 1997 and 1998, with the latter being the most severe. Almost all the shallow reefs in the country were impacted. Average live coral cover before and after the bleaching was approximately 45% and 5%, respectively. Even as late as 2002 live coral coverage remained around 10%. Bleaching events, as well as slow recovery, have significant consequences for the tourism sector, as well as for global biodiversity (Government of Maldives, 2003).

Hurricane Ivan was a 'category 4' hurricane system when it reached Grenada in September 2004. It was accompanied by sustained winds of approximately 140 mph, with gusts exceeding 160 mph. Prior to the passage of the hurricane, an economic growth rate of 5.7% was forecast. Negative growth of at least -1.4% is now projected. An official damage assessment (Organisation of Eastern Caribbean States, 2005) reported the following:

- 28 persons killed;
- 90% of hotel rooms damaged or destroyed, totalling US$108 million or 29% of GDP;
- heavy damage to ecotourism and cultural heritage sites, accounting for 60% job losses in this subsector;
- 90% of housing stock damaged, totalling US$517 million or 38% of GDP;
- telecommunication losses equivalent to 13% of GDP;
- damage to schools and education facilities amounting to 20% of GDP;
- losses in the agricultural sector equivalent to 10% of GDP – the two main commercial crops, nutmeg and cocoa, are expected to make no contribution to GDP or earn foreign exchange for at least 6–8 years;
- damage to electricity installations amounting to 9% of GDP; and
- overall damages estimated at US$824 million, or two times current GDP.

The rapid growth of tourism in many small island countries is placing added stress on already overloaded water supplies, reducing the availability of water for general consumption as well as for water-hungry tourism businesses. For example, during 1982–1983 rainfall in many parts of the Western Pacific was just 10–30% of the long-term average. At the start of 1983 the reservoir in Majuro, Marshall Islands, held 23 million litres, but by May 1983 the total storage had declined to 3 million litres, most of which was reserved for hospital use. At the height of the drought associated with the 1997–1998 ENSO event, municipal water on Majuro was available for only 7 h every 14 days. Some small islands, despite receiving more than adequate rainfall under normal circumstances, may also suffer from periodic water shortages. These may be exacerbated by inadequate rainwater catchment facilities, or the sandy nature of the soil, allowing the limited rain that falls to infiltrate rapidly and become difficult to access (Hay *et al.*, 2003). The problem on these islands is largely one of water management. Dominica in the Caribbean, and Seychelles in the Indian Ocean, are almost entirely dependent on surface water from ephemeral and perennial streams. Both countries experienced serious water shortages during the 1997/1998 El Niño event.

Adaptation

Figure 3.5 also identifies adaptation options that may be undertaken by the tourism sector in order to reduce unacceptable consequences of climate change. Many of these represent 'no-regrets' measures and hence, at one level, appear rather generic as opposed to being specific to the tourism sector. However, special circumstances pertaining to tourism need to be taken into account when these and other interventions, such as mainstreaming adaptation in national and tourism planning, are made by the tourism sector. A more complete list of possible adaptation measures is provided in Table 3.2, along with examples of their specific relevance to tourism, barriers to their implementation and examples of ways in which these barriers can be removed. For small islands the challenge is to identify and implement 'softer' adaptation options that retain the aesthetics, integrity and functionality of coastlines. For example, it is desirable for management of accelerated erosion at resort and community levels to be based around the use of 'soft' options such as planting of soil-binding vegetation rather than the use of 'hard' options such as sea walls. Best practice is to plant and maintain intact native vegetation communities. These are ideal for stabilising shorelines, as native plants have evolved to survive in tropical environments, tolerating tropical heat, humidity, salt water, extreme sunlight and storms. Native vegetation communities function as soil binders, maintaining coastal

Table 3.2 Possible adaptation measures for tourism in small island countries and barriers to implementation

Adaptation measures	*Relevance to tourism*	*Barriers to implementation*	*Measures to remove barriers*
Mainstreaming adaptation in planning	Currently adaptation is not mainstreamed in tourism planning	Lack of information on which to base policy initiatives	Improve targeted information, e.g. climate-risk profile for tourism
Include climate risk in tourism regulations, codes	Currently such risks are not reflected in tourism-related regulations	Lack of information on which to base regulatory strengthening	Improve information, such as climate-risk profile for tourism
Institutional strengthening	Shortfall in institutional capacity to coordinate climate responses across tourism-related sectors	Lack of clarity as to the institutional strengthening required to improve sustainability of tourism	Assess options and implement the most appropriate strategies
Education/awareness raising	Need to motivate and mobilise tourism staff and also tourists	Lack of education and resources that support behavioural change	Undertake education/awareness programmes
Shade provision and crop diversification	Additional shade increases tourist comfort	Lack of awareness of growing heat stress for people and crops	Identify, evaluate and implement measures to reduce heat stress
Reduce tourism pressures on coral	Reefs are a major tourist attraction	Reducing pressures without degrading tourist experience	Improve off-island tourism waste management
Reduce tourism pressures on other marine resources	Increased productivity of marine resources increases well-being of tourism-dependent communities	Unsustainable harvesting practices and lack of enforcement of regulations and laws	Strengthen community-based management of marine resources, incl. land-based issues

(Continued)

Table 3.2 (*Continued*)

Adaptation measures	*Relevance to tourism*	*Barriers to implementation*	*Measures to remove barriers*
'Soft' coastal protection	Many valuable tourism assets at growing risk from coastal erosion	Lack of credible options that have been demonstrated and accepted	Demonstration of protection for tourism assets and communities
Improved insurance cover	Growing likelihood that tourists and operators will make insurance claims	Lack of access to affordable insurance	Ensure insurance sector is aware of actual risk levels and adjusts premiums
Desalination, rainwater catchments and storage	Tourist resorts are major consumers of fresh water	Lack of information on future security of freshwater supplies	Provide and ensure utilisation of targeted information, based on climate risk profile
Drainage and pumping systems	Important services for tourist resorts and for tourism-dependent communities	Wasteful practices; lack of information to design adequate systems	Provide and ensure utilisation of targeted information, based on climate-risk profile
Enhanced design and siting standards	Many valuable tourism assets at growing risk from climate extremes	Lack of information needed to strengthen design and siting standards	Provide and ensure utilisation of targeted information
Tourism activity/ product diversification	Need to reduce dependency of tourism on 'sun, sea and sand'	Lack of credible alternatives that have been demonstrated and accepted	Identify and evaluate alternative activities and demonstrate their feasibility

berms and forests. These plant communities are part of the dynamic coastal process, and are well adapted to shifting shorelines.

A recent study (Belle & Bramwell, 2005) explored the views of policy makers and tourism industry managers in Barbados with respect to their views on the added risk of climate change for the island's tourism sector, and their preferred policies to manage those impacts. Many individuals felt that damage to coastal tourism facilities was very likely, with a significant number considering it likely that there will be beach change, sea-level rise and damage to marine ecosystems. Both categories of respondents shared similar views on these impacts. However, more managers were also of the opinion that there will be impacts on the volume of tourists to the island and hence on air traffic to the islands. Both groups identified increasing public awareness as the most appropriate policy response to the impacts of climate change on tourism in Barbados. But the industry managers generally were less inclined to regard the policy responses as very appropriate, perhaps because they were more cautious about applying such policy interventions at an early stage.

The important role of information in facilitating adaptation to climate by the tourism sectors of small island countries has been emphasised by Boodhoo (2003) and others. For example, many meteorological services issue seasonal climate forecasts towards the beginning of each season. The tourism industry uses such information to devise appropriate marketing strategies well in advance and hotels begin planning for cyclones and other extreme events. In the case of Mauritius, the Comoros and Seychelles, hotels respond to drought warning by assessing water needs and securing alternative sources of supply, and to cyclone warnings by taking such precautionary measures as:

- removal of surplus branches from trees, especially those near to buildings and electric and telephone installations;
- harvesting of fruits such as coconuts, so as to avoid accidents and to provide emergency food;
- inspection of the internal communication system, entertainment facilities, water pipe and underground cables; and
- servicing of the sewerage system.

In the Cayman Islands preparing for tropical storms has largely followed the disaster risk management process, i.e. plan, prepare, respond and then recover, but it has been adapted to suit local conditions. Factors that led to the development and adoption of the Cayman Islands disaster risk management strategy were:

- mental link made between hazard events, vulnerability and impacts;
- important respected individual takes responsibility for the issue;

- respected individual persuades others to engage with the issue;
- small group of concerned individuals formed;
- small group pushed for preparedness action in their areas; and
- National Hurricane Committee formed.

When the Cayman Islands' Government began to prepare for climate change, it recognised that plans are not perfect. They require constant upgrading and improvement. Acceptance that no plan is perfect allowed each department to revise its plan on an annual basis, without any fear of criticism or sense of failure (Tompkins *et al.*, 2005).

Many accommodation businesses already adapt to climate change; for example, a survey conducted in Fiji showed that 19 out of the 25 respondents indicated that their buildings are adequate to protect against climate change impacts, typically related to cyclones (Table 3.3). Those managers who were interviewed also indicated that the risk of a cyclone and resulting damage to the tourism facilities constitutes a natural part of running a tourism business in Fiji. To prevent damage from storm surges and sea-level rise, buildings in Fiji are now built at least 2.6 m above mean sea-level. But one interviewee commented that rather than generalising a minimum height it is more useful to consider the specific geographic situation of a development site, such as slope and topography. Some

Table 3.3 Adaptation measures taken by accommodation providers in Fiji

Measure	*Frequency*
Construction-based	
Adequate building structures	20
Own water storage	19
Replanting trees/mangroves	18
Pollution control	17
Self-sufficient for energy supply	13
Setting back structures	12
Behaviour-based	
Guest education	18
Reef protection	17
Evacuation plan	13
Indoor activities	10

Source: Becken (2004a)

coastal resorts have set back their structures as a result of cyclone damage. The construction of new resorts, however, still focuses on coastal areas, mainly in the already developed areas. Some of the new developments include tourist accommodation that is built right on the reef (over-water bungalows). The practice of building new resorts on elevated areas seems less popular. Resorts commonly adapt to increased levels of erosion and the risk of storm surge by constructing seawalls, as well as by planting trees, mainly coconut palms or mangroves. Seawalls – as well as other 'hard' coastal protection – have the disadvantage that they accelerate erosion elsewhere, often leading to the need for further erosion protection measures. This is especially the case when the construction does not take into account latest knowledge or technologies. The loss of sand as a result of cyclones and sea-level rise is a major problem, especially on low sandy islands. Despite knowledge about the importance of mangroves for shore protection, large areas are still cut down to provide space for further development. Tourism development on Denarau Island (Nadi area), for example, involves the deforestation of large areas of mangrove forest, which then results in sedimentation and considerable pressure on coral reefs in the adjacent islands (Becken, 2004a).

Adaptation of buildings to reduce climate-related risks is best undertaken at the design stage. It is usually more difficult to retrofit existing structures. For example, old concrete accommodation blocks are less suited to hot temperatures. Thus there is often a need for air conditioning in such buildings. When designing or upgrading a resort it is often possible to reduce climate-related risks by using more appropriate materials and structures, for example wooden bungalows that can be constructed in a way to maximise airflow, negating the need for air conditioning. Similar considerations apply for retrofitting existing buildings with rainwater collectors or solar hot water systems.

Mitigation

While small islands contribute very little to GHG emissions, currently accounting for less than 1% of total emissions, some islands have very high per capita emissions. For example, the US Virgin Islands is the largest per capita producer of CO_2, with each citizen on average producing 99 metric tons of CO_2 each year (Tompkins *et al.*, 2005). The Pacific Islands region accounts for approximately 0.03% of global emissions of CO_2 from fuel combustion, despite having 0.12% of the world's population. The per capita emissions of Pacific islanders are approximately one-quarter of that of the average person worldwide (Hay & Sem, 1999). International tourist arrivals in small island countries represent a minimal percentage of the world total; the total number of international arrivals by air and sea to small island countries was estimated at 27.2 million in 2000, representing 3.9% of the world total of 697.8 million arrivals. Thus tourists travelling to small island countries make a relatively minor contribution to climate change. However, it is also important to remember that most island tourism involves international air travel, which is currently not accounted for as part of the country's GHG emissions. In some cases, the emissions from international air travel could be substantial.

In small island states, energy is a major cost factor in the operation of a tourism accommodation business, especially when energy is derived from fossil fuels either for transport or electricity generation. Becken (2004a) reports that the operation of diesel generators is resource-intensive, because of inherent inefficiencies, high maintenance requirements and transportation costs. Thus, managers have an economic interest in keeping electricity consumption low. A critical issue with diesel generators is that the optimum range of electricity generation is about 80% of the maximum performance. Electricity conservation efforts may result in an existing generator running below this threshold. In such instances the diesel is not combusted completely, reducing the lifetime of the generator. Replacing the existing generator with one of smaller capacity is often unwarranted due to the high capital cost. For these reasons some managers see little incentive in conserving energy if this results in loads falling below the optimum range of the existing generator. One option used in some tourist establishments is to operate a smaller generator at night-time.

As discussed in Chapter 7, there are several energy-efficiency measures that can be applied in tourist accommodation. In small island states the most commonly installed technologies relate to cooling. Bohdanowicz and Martinec (2001) note that about 50% of overall energy consumption in hotels is due to air conditioning. A warming climate may result in increased energy usage and reduced profitability of hotels. Some budget resorts do not offer air conditioning or only use it at night. This keeps energy costs low and reduces the need for a high-capacity generator. Adequate building materials and structures, and planting trees for shade, help minimise the need for air conditioning. Setting the room temperature at between 22 and 24°C, rather than 18°C, is becoming common in many hotels. This saves substantial amounts of energy.

Other energy-use-reduction measures centre around lighting, including energy-efficient light bulbs, sensor lighting in the garden, solar panel lights and room keys used to operate lights and air conditioning in each room. Energy-efficient bulbs may pose a problem to some operators, as they are expensive and do not last because of fluctuating power supply from generators. In the case of smaller islands the energy costs of shipping are substantial, and managers seek to maximise load factors by combining passenger vessels with transporting food, waste or water.

Guest education initiatives can focus on reducing GHG emissions. While one resort manager in Fiji commented that 'tourists are not here to worry about air conditioning', another manager noted that tourists are surprisingly supportive and open (Becken, 2004a). Some managers try to educate their staff, but this proves very difficult and requires considerable reinforcement and supervision, for example in the area of recycling.

A range of renewable energy sources is available to tourism businesses in small island countries. Accommodation businesses often operate solar hot water systems (see Figure 3.6). Photovoltaic (PV) systems are less common. Many resort managers consider that, from a long-term perspective, investment in solar energy and other renewable energy sources is economically sensible, as energy costs are typically a substantial portion of the operating costs of a resort.

Tompkins *et al.* (2005) present some examples of financial tools that can facilitate mitigation in the tourism sectors of small island countries, including:

- encouraging longer stay visitors, through a tax on international travel;
- promoting ecotourism through investment in preservation of buffering ecosystems, through incentive payments such as those provided to tourism businesses that preserve natural ecosystem buffers;
- encouraging visitors to an island destination for a weekend break, through subsidies such as government promotion of the island as 'green'; and
- promotion of low carbon-intensive activities such as sailing, surfing, shore diving, walking, swimming and cultural activities, through a

Figure 3.6 Backpacker accommodation in the Cook Islands with a solar hot water system on the roof

Table 3.4 Adaptation measures for tourism and their positive or negative ancillary effects

Adaptation	*Impact on mitigation*	*Wider opportunities*
Trees protect against storms, improve microclimate, reduce erosion	Trees are carbon sinks and reduce net CO_2 emissions	Benefits biodiversity, water management, soils; could be included in a carbon-trading scheme
Conserving water	Reduces energy costs for supplying water (e.g. pumps, transportation)	Saves costs, especially when water is shipped to islands; reduces conflict in areas where water is a limiting factor
Increasing the share of renewable energy sources	Reduces CO_2 emissions	Less polluting than fossil fuels; reduces dependency on fuel imports
Using natural building materials (e.g. wood)	Smaller carbon footprint for locally produced materials	Depends on sustainability of plantations; stimulates national forestry sector
Water tanks and rainwater collection	Saves transport energy for supplying water	Possibly interrupts the natural water cycle; saves costs
Setting back building structures	Neutral	Positive effects on integrity of dune systems
Diversifying markets	Positive if new markets are more eco-efficient ($ spent/kg CO_2 emitted)	Depends on the environmental impact of new markets
Weather-proofing tourist activities	Depends on the type of activities	Depends on the type of activities
Water desalinisation	High energy costs	Takes pressure off freshwater resources; currently expensive
Increasing air conditioning	Increases CO_2 emissions; counteracts mitigation	Other pollution in the case of diesel generation
Beach nourishment	Energy-intensive to transport sand	Disturbs coastal ecosystems; high costs involved

(*Continued*)

Table 3.4 (*Continued*)

Adaptation	*Impact on mitigation*	*Wider opportunities*
Reducing beach erosion with seawalls	Neutral	Disturbs currents and causes erosion elsewhere; requires ongoing maintenance

Source: after Becken (2005)

tax on high energy intensive activities or on the fuel used in these activities.

Integrated responses

As will be outlined in Chapter 9, it is recommended that win–win measures be implemented, i.e. measures that provide benefit for sustainable development even in the absence of climate change. Adaptation measures that counteract sustainability goals should be avoided, for example as a result of increased energy consumption (e.g. for air conditioning or desalination) or impacts on natural systems (e.g. extraction of sand for use in coastal protection). Reforestation of native forest and coastal ecosystems, the conservation of water and the use of local, natural building materials are good examples of adaptation measures that can also reduce the direct or indirect GHG emissions of a tourist resort. Table 3.4 considers linkages between adaptation and mitigation measures, as well as wider environmental and economic considerations. Priority has to be given to maximising the benefits from both adaptation and mitigation initiatives (Dang *et al.*, 2003).

Insurance, Climate Change and Tourism

Insurance is critical for tourism. In many cases it allows the tourism industry to spread the residual climate-related risks that cannot be avoided by other adaptation measures. Risk management is nowhere more at the core of decision making than in the insurance industry. Insurers understand the concept of risk and its economic implications. Climate change has been recognised as an additional risk that will alter the known risks that insurers seek to manage. In the tourism context, as elsewhere, there are four broad areas in which climate change might affect the insurance industry: changing customer needs, changing patterns of claims, new and tightening regulation, and impacts on reputation (The Association of British Insurers (ABI), 2004).

In this case study the consequences of climate change – in particular disasters – for tourism will be discussed. The insurance industry responds to such new risks by adapting their products and services

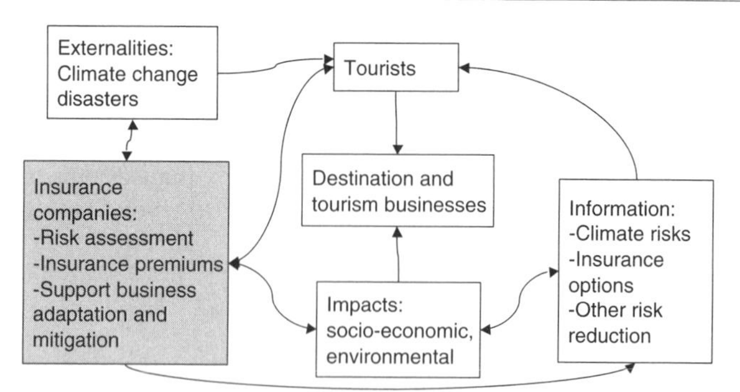

Figure 3.7 Climate change impact on the insurance industry and tourism. The main relationships are shown as lines between the system's components.

(e.g. increasing premiums in response to higher risks). Moreover, insurers are increasingly active in supporting implementation of climate change adaptation and mitigation measures by their customers (Figure 3.7). Information exchange between insurance companies, the tourism sector and those at risk from climate change is essential in order to shape adaptation measures and also to reduce uncertainty associated with future risks. In this case study the risks will be discussed in relation to the economic losses borne by tourism businesses and tourists, and responses by the insurance industry.

Climate consequences, risks and natural hazards

Climate change impacts on tourism in many ways. In particular, climate-related hazards are of increasing significance, leading many tourism businesses to seek protection through insurance. In turn, the insurance companies have to manage their exposure to such additional risks. Munich Re (1998) identified the following hazards, all of which have significance for the tourism sector:

- Geological: earthquakes, seismic sea waves (tsunamis), volcanic eruptions
- Storm events: tropical cyclones, extra-tropical storms (winter storms), tornadoes, regional storms, monsoon storms
- Flood/rain events: storm surges, high waves, heavy rain, hailstorms and lightning
- Pack ice and iceberg drift
- El Niño, climate change

Most of the above natural hazards are related to weather or climate, with the El Niño phenomenon being paid particular attention to in relation to climate change. As will be noted in Chapter 5, El Niño and La Niña events have global consequences. The most severe ones are extreme

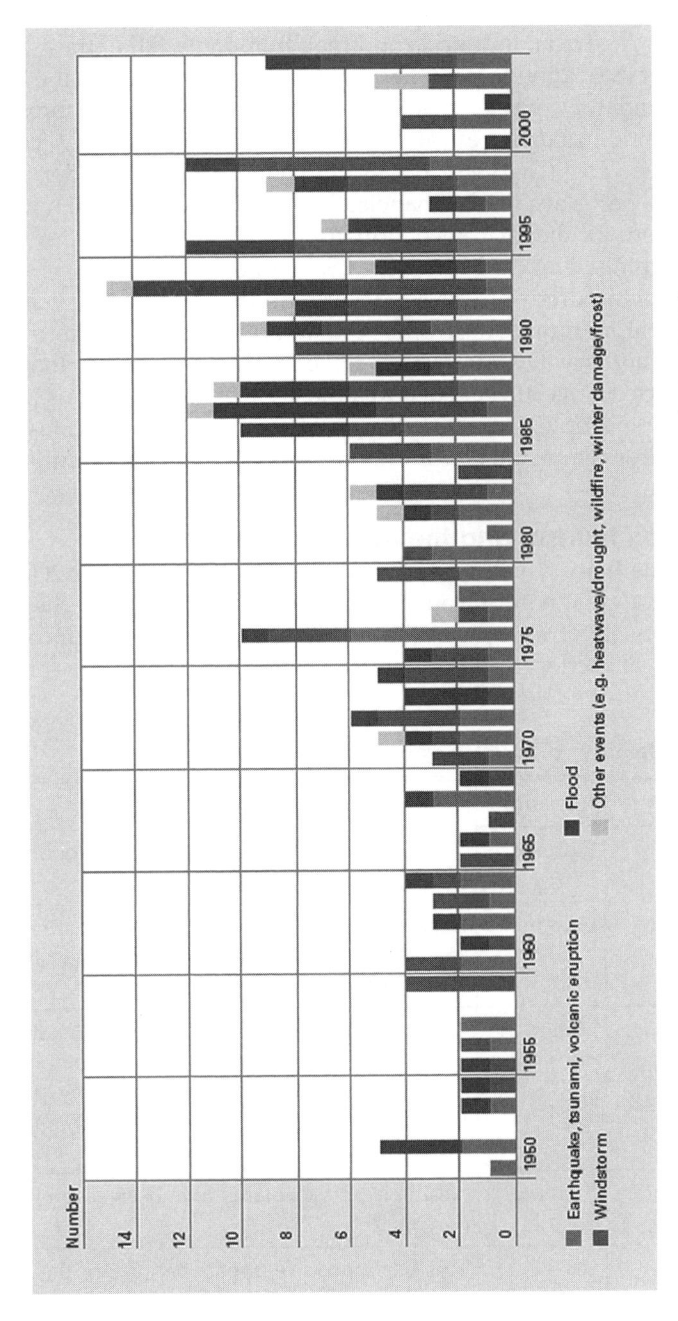

Figure 3.8 Number of great natural catastrophes per year, by type of event (Munich Re, 2005)

rainfall, flooding and storms on the Pacific coast and East Africa, increased cyclone activity in the Eastern Pacific, and high risk of drought and fires in Australia, Indonesia and the Philippines. All of these impacts are relevant in relation to personal and property insurance in the tourism industry, and for tourists themselves. Some disasters are foreseeable, given historic occurrences and trends, while others may come as surprises (see also Chapter 2 on abrupt changes). Iwan Stalder, head of catastrophe perils at Zurich Financial Services in Switzerland, noted 'It is a real concern for the insurance industry that no models exist for some of the worst things that could happen' (Schiermeier, 2005).

There are hundreds of loss events around the world every year, but it is the 'great natural catastrophes' that are of most concern to the insurance industry. Disaster claims have increased from less than 1% of all insurance claims 30 years ago to almost 5% of today's claims (Kovacs, 2006). For example, in 2004 there were nine natural catastrophes worldwide; seven resulted from floods and wind events (Figure 3.8).

Impacts on tourism and the insurance industry

Hurricane Ivan, which passed over Grenada in September 2004, is an example of a great natural catastrophe impacting directly on the natural and built assets of a popular tourist destination, and therefore on the livelihoods of the local population (see Table 3.5). The native forest and

Table 3.5 Damage to livelihoods by Hurricane Ivan

Livelihoods	*Intensity*	*Comments*
Hunting	Extreme	The wildlife habitats and sources of food have been destroyed
Tour-guiding	Extreme	Access routes to the ecotourism sites are impassable; it is envisaged that the sites have also sustained some damage
Craft-making	Major	Although the bamboo and screw pine have sustained damage, material can still be obtained from the damaged stock; unfortunately, preservation and storage of the material may prove problematic
Fruit gatherers	Extreme	All the trees and plants have been severely damaged
Charcoal burners		This has a positive impact because of the abundance of wood from the fallen trees

Source: Organisation of Eastern Caribbean States (2005)

other vegetation were almost completely destroyed, with some of the damage being irreversible. As a consequence, native wildlife, most notably endemic bird species, were severely affected. Intensive environmental management will now be required for the island to recover from the event. Mangroves and beaches were also affected. However, their recovery time is expected to be shorter compared with that of forest ecosystems. Coral reefs and fisheries sustained relatively minor damage.

The insurance industry keeps detailed records on economic impacts associated with natural hazards (Figure 3.9), although no information is specifically available on costs related to tourism. Extreme climatic events caused a trillion dollars in economic losses in the 1990s (Kovacs, 2006). Even before the tsunami in the Indian Ocean 2004 was a record loss year for insurance companies. Munich Re estimated that as a result of natural catastrophes the insurance industry lost US$40 billion in 2004, with tropical storms in the Caribbean and the West Pacific generating the biggest losses. Economic losses in 2004, including insured and uninsured damage, amounted to US$130 billion. Even without the tsunami, natural disasters caused over 15,000 fatalities (Munich Re, 2005). The tsunami was responsible for at least 275,000 deaths, mostly in Indonesia. Up to 9000 foreign tourists (mostly Europeans enjoying the peak holiday travel

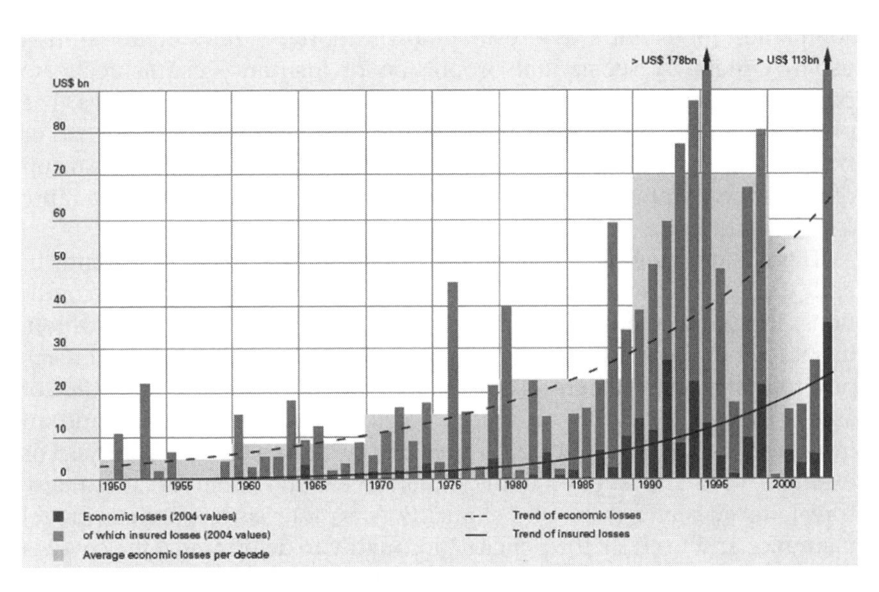

Figure 3.9 Economic losses and insured losses – absolute values and long-term trends – adjusted to present values. The trend curves verify the increase in catastrophe losses since 1950.
Source: Munich Re (2005)

season) were among the dead or missing. Sweden alone reported more than 500 dead or missing.

In terms of economic losses, 2005 was even worse than 2004. Costs as a result of weather-related natural disasters alone amounted to more than 200 billion dollars. Losses in excess of 70 billion dollars were covered by insurance companies. Most of this was related to the three hurricanes, Wilma, Rita and Katrina (Text Box 2), that affected the South-eastern USA and Mexico. The hurricanes hit important tourist destinations such as Cancun, the Bahamas and Miami. The unusual severity of these storms impacted both the local tourism industries and insurers, with many companies experiencing major losses. Hotel operators in Mexico tried to restore rooms and facilities quickly after Wilma hit in October, but most tourists decided to take their winter (December) vacations elsewhere.

Berz (1999) notes that non-catastrophic climate anomalies can also cause substantial damage, much of which is also covered by the insurance industry. For example, a warmer than normal summer in the UK – as experienced in 1995 – could cause damage in excess of £1.5 billion. The mortality rate that summer increased by 5% in August and 1% in July (Palutikof *et al.*, 1997). This had implications for insurance companies in relation to health, and life insurance claims increased in those conditions. Gradual changes in climate may result in reduced risks for insurance companies, for example as a result of lower mortality rates in milder than usual winters. A substantial proportion of insurance claims relate to personal automobile insurance and other types of transport systems. The physical damage policy for cars, for example, covers damage from hail or flood, as well as impacts and flying objects. Insurance claims rise as a result of higher accident rates in adverse weather conditions, for example rainy weather and hot periods (Mills *et al.*, 2005).

There is ongoing debate about the reasons for the increasing trends in both insured and total economic losses. Changing environmental conditions, including climate, is certainly one important factor. In addition, increased numbers and densities of population (especially in coastal areas), new tourist resorts in high-risk areas, higher living standards and value of assets, increased mobility and participation in leisure activities, and an increased use of insurance also play important roles (Berz, 1999). For example, since the tsunami in the Indian Ocean in December 2004, many travel websites and embassies explicitly warn tourists to purchase travel insurance and to check the policies thoroughly to determine if the cover is adequate. Many insurances exclude injury or death through acts of terrorism or nature; for other risks compensation is usually restricted to medical expenses and lost baggage.

The costs of an adverse weather event are spread among several parties. Insurances cover damage that has been insured and that meets

Text Box 2 Hurricane Katrina, 29 August 2005.

While Katrina did not pass directly over New Orleans, the associated storm surge resulted in the breaching of flood walls in several places, and within 24 h about 80% of the city was under flood waters up to 6 m deep.

Tourism is the largest industry locally, generating about 80,000 jobs. More than 10 million tourists came to New Orleans in 2004, to visit the historic French Quarter, the casinos or attend one of the many conventions and festivals, for example the Jazz Festival. Many tourist facilities, including the waterfront hotels and 13 floating casinos, were damaged or completely destroyed; even by December 2005 many hotels remained closed and airport services continued to be disrupted and operations limited.Estimates of economic damage vary widely, but could reach well over $120 billion; insurance claims may be in the order of $60 billion. Insured losses relate to property damage, cars, ports, refineries, public property and the disruption of businesses. The insurance industry expects over two million claims; it is unknown what proportion of claims relates to tourism businesses.

Standard insurance policies in the USA cover damage caused by storms, but not by floods. Flooding, however, was the major cause of damage as a result of Katrina.

Only one-fifth of the homes and businesses in Mississippi in flood-prone areas were covered by flood insurance before Katrina. Businesses and homeowners who did not sign up to the National Flood Insurance Program were unable to claim any compensation for flood damage. There is now a debate as to whether storm surge should be treated the same way as flood, or whether it is a result of the hurricane and therefore to be covered under the standard policies. If a lawsuit claiming this is the case is successful, as much as $15 billion could be added to the costs for insurers from Katrina.

It will take a long time to resolve all the legal issues and to rebuild New Orleans, including tourist facilities; in the meantime tourists are likely to substitute other destinations for New Orleans, for example Las Vegas. Katrina's impact extends to the wider tourism industry serving New Orleans. For example, many airlines and tour operators waived cancellation fees for tourists who had booked a trip to New Orleans before the hurricane. The economic damage from lost income from tourism activities has yet to be determined.

the criteria for the claim. Uninsured damage is borne by the affected businesses and tourists, unless governments, international organisations (e.g. UN) or NGOs (e.g. aid agencies) provide resources, for example in the case of disaster recovery (Figure 3.10).

Figure 3.10 Costs of damage-causing weather events may be spread among various parties
Source: after Epstein and Mills (2005)

Responses by the insurance industry

The insurance industry faces the prospect of a growth in the number and size of claims from the tourism industry and tourists, as a consequence of increases in extreme events and climate variability. There are many insurance claims that relate to climate and tourism, including mortality and sickness while travelling, increased accidents between home and destination or at the tourist destination, structural damage to tourist facilities or hotels, service interruptions and large-scale bush or forest fires. In short, climate has the potential to affect all insurance categories (Brauner, 2002). In turn this has repercussions for tourism.

Insurance companies do not reduce risk, but spread the consequences more evenly across the insured community. In this sense they provide an important mechanism of risk transfer. However, if losses resulting from climatic events become regular and all members suffer damage, there are no non-victims to share the burden. Insurance then becomes irrelevant. For this reason, the aim of insurance companies is to ensure that damage does not become the norm. The important steps for considering climate change in insurance policy are (Brauner, 2002):

- *Identifying risk*. Risks need to be recognised, even when tourists, tourism operators or other decision makers do not yet perceive a risk; the insurance industry can contribute to communicating new risks to those who might be affected by them, for example tourism operators in coastal environments.

- *Analysing risk*. Improve understanding of probability analyses of risk events and factors influencing the ramifications and frequency of an event.
- *Mitigating risk*. The insurance industry can support practices that reduce the risk in the short term (e.g. reduce tourism or tourist vulnerability to climate-related impacts, such as by requiring that buildings are set back a minimum distance from the shoreline) and in the longer term (e.g. emission reductions by hotels); insurance companies can demand compliance with specific adaptation measures to increase resilience to certain risks.
- *Transferring risk*. Adapt to changing risks and ensure that insurance cover is available.

The last step of transferring risk constitutes a particular challenge in the context of climate change. Both the changing needs of customers, and new patterns of claims, require new products and underwriting conditions. Underwriting is the process of deciding the conditions (e.g. premiums) under which a customer can receive insurance protection. Traditionally insurers have based their policies for underwriting on past records of events, including their likelihood and their severity. Climate change poses a challenge because the past history becomes largely irrelevant. Tol (1998: 258) goes as far as saying that it is not possible to insure a risk that cannot be quantified. A lack of historical experience also makes it difficult to insure new items or new technologies. For example, new renewable energy technologies (e.g. offshore wind farms providing electricity to tourist resorts and other consumers) face unknown or unquantifiable risks from climate change (including extreme events). Climate change also poses a challenge in that times between events may decrease and the geographic distribution of events might also change. Many popular tourist destinations, often in developing countries, are particularly at risk (e.g. small island developing states). The identification of hotspots (see Chapter 2) is an important way to assess comparative risks on a global scale.

While it is difficult, if not impossible, to insure an unknown risk, it is also not practical to insure risks that are 'certain'; it is difficult to spread the risk from those who are likely to be affected to those who will probably remain unaffected. An example is insurance against sea-level rise in places where scientific certainty is such that it is possible to predict the rate at which the sea level will rise. For high-risk situations, such as tourist resorts in cyclone zones (Figure 3.11), premiums have already become unaffordable. In some cases insurers have already withdrawn cover. In other instances insurance companies are providing incentives to implement risk-reducing adapta-

Figure 3.11 Complete devastation of Niue after cyclone Heta (January 2004) (photo by David Poihega)

tion measures. Premiums for cyclone insurance in Fiji, for example, can be reduced substantially if the insured party introduces measures such as improving building structures, having emergency plans in place, conducting staff training (e.g. evacuation), and making sufficient provisions for water, food and fuel storage (Becken, 2004a)

A recent study by the Ceres investor coalition of the USA recommends a range of actions by key players to reduce anticipated substantial losses to the insurance industry due to climate change (Mills *et al.*, 2005). The key players are insurers, insurance regulators, tourist destinations and tourists (Table 3.6).

Insurance companies are also taking increasing interest in energy efficiency and renewable energy sources. One reason is to contribute to climate change mitigation and therefore reduce the overall risk of adverse climatic changes in the long-term. There are, however, more short-term and direct benefits. Mills (2003) identified a wide range of energy management measures that provide risk reduction benefits. An example is efficient refrigeration in a tourist resort. This can maintain cool temperatures longer in the event of power cuts and thus reduces costs resulting from business interruptions and potential food poisoning. The use of renewable energy sources for power and hot water reduces the risk of disruptions in tourist accommodation in the event of extreme events (e.g. hurricanes) and power outages. In the transport sector, energy use and transport risk (e.g. accidents) are closely related and some companies offer incentives to reduce the use of personal automobiles (Mills, 2003). Some insurance companies charge based on

Table 3.6 Actions to reduce climate change-related loss to the insurance industry

Insurers	*Insurance regulators*	*Destinations*	*Tourists*
• Improve loss data collection and enhance the actuarial analysis. • Analyse the implications of climate change; share the results with shareholders. • Promote advanced building codes and tools to mitigate potential losses. • Engage in weather/climate research and promote the use of scientific methods for climate modelling. • Lead by example: reduce corporate climate footprint. • Encourage policy action and technical measures to reduce GHG emissions, especially where there are direct co-benefits for the insurance core business.	• Review the standards of insurability; identify new challenges. • Incorporate climate risks in solvency and consumer-impact analysis. • Encourage insurers to collect more comprehensive data on weather-related losses. • Elevate the practice of catastrophe modelling. • Assess exposures for insurer investments and adequacy for capital and surplus to weather extremes. • Explore the feasibility of developing a weather exposure questionnaire.	• Foster and participate in public–private partnerships of risk spreading. • Reduce disaster losses through improved planning and post-event response. • Assess the government's overall financial exposure to changing patterns of weather disasters. • Expand research on climate change and loss modelling, and issue climate hazard maps. • Take policy action to reduce GHG emissions.	• Minimise disaster losses through preparedness and planning. • Reduce GHG emissions. • Support destinations and businesses that engage in GHG reduction.

Source: after Mills *et al.* (2005)

Table 3.7 Potential climate change challenges and opportunities

Class of business	*New challenges*	*New opportunities*
Travel	Less demand, as more holidays taken in warmer UK, as temperature in Mediterranean destinations become increasingly uncomfortable	More active lifestyles, warmer summers encouraging more active holidays. Increased demand to protect customers from weather-related cancellations.
Motor	Increased frequency of claims due to weather-related accidents, both in heavy rain and hot weather. Increased mileage due to longer, warmer summers	Decreased weather-related claims due to less frequent fog and frost. Improved driving technology may reduce the overall frequency and severity of accidents. Reduced mileage due to higher fuel prices, environmental concern, use of public transport.
Construction	More claims from wind and rain. Unanticipated stress on materials and failures of novel systems.	Reduced claims due to cold weather and new technologies adopted in response to climate change risks
Health	Greater heat stress, summer mortality. Greater incidence of food poisoning due to longer, hotter summers. Increased occurrence of endemic malaria due to warmer climate.	Greatly reduced winter illness and mortality. More active lifestyles with warmer summers. Greater availability of fresh food.

Source: ABI (2004)

distance travelled; this could be an interesting option for the pricing policies of rental car companies.

There are a number of challenges and opportunities for the insurance industry in relation to tourism and climate change. ABI (2004) identified these for travel, transport, construction and health in the UK (Table 3.7). In the future fewer British will travel overseas as a result of warmer temperatures at home; as a result, fewer tourists will require travel insurance. If overseas holidays are replaced by activities at home there could be increased demand for insurance cover related to a more active lifestyle.

Insurers worldwide are taking action to manage risks associated with climate change. As Kovacs (2006: 52) pointed out, Hurricane Katrina taught the insurance industry (and possibly society) to 'hope for the best and always, always prepare for the worst'. Close cooperation between the tourism industry and insurance industries is required to ensure effective risk minimisation for tourism businesses and tourists, and at the same time ensure the continuing viability of insurers despite new climate-related risks.

International Aviation

The increasing demand for air travel is a major concern in the context of climate change. It is also one of the major impediments to sustainable tourism. Aviation already contributes substantially to the build-up of GHGs in the atmosphere. Effects could be larger than currently estimated given the uncertainty associated with the climate impacts of aviation-induced cirrus clouds. Policy responses discussed in this case study relate to the mitigation of GHG emissions from air travel (Figure 3.12). Also, very little is known about how climate change might impact on aviation and how the tourism industry could adapt to those changes, for example in relation to extreme weather events (e.g. storms). An example of how the aviation sector is responding to extreme conditions is provided in Chapter 8.

Society and air travel

Over recent decades global air travel has grown at annual rates of about 5%, and future growth rates are expected to vary between 4 and 6% (see also Chapter 4). International air travel, and long-haul in particular, are expected to drive this increase (Royal Commission on Environmental Pollution, 2002; Schafer & Victor, 1999). This is especially the case for regional traffic flows and flights within Asia, between Asia and Oceania or Europe, and flights between North America and Asia/Oceania (Penner *et al.*, 1999). The total amount of passenger-kilometres travelled has grown even more than total number of trips, and will

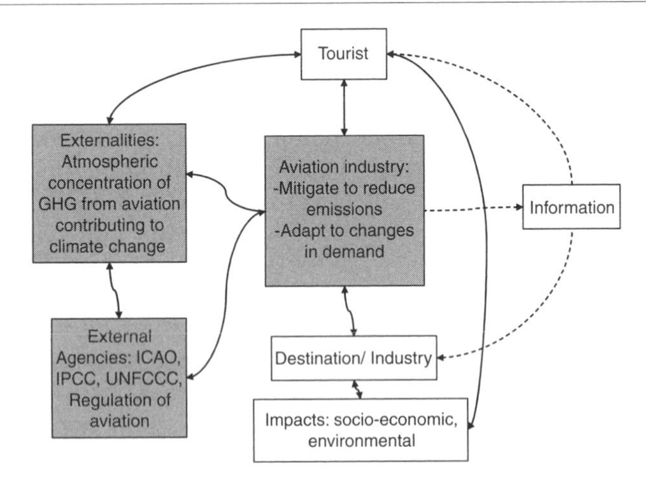

Figure 3.12 GHG emissions from aviation, impacts on the climate, and policy responses for reducing the emissions. The main relationships are shown as lines between the system's components

continue to do so given the trends described above. Air travel's share of all tourist transport has increased considerably in recent decades, especially for international travel. Of all international arrivals, 40% were by air in 2000 (Kester, 2002).

Further developments in the transport sector will continue to influence global travel. In 1998, UNWTO predicted that long-haul travel (between world regions) would grow fastest (Figure 3.13). Environmental impacts will grow accordingly, unless major improvements in aircraft and other transport technologies are achieved.

Greenhouse gas emissions from aviation

International aviation is an important source of global GHG emissions. As noted in Chapter 2, aviation contributes about 3.5% to the total anthropogenic radiative forcing, excluding the potential effects of cirrus clouds. Gössling (2002a) estimated that at present tourists travel about 1179 billion passenger-kilometres (pkm) by air. This is equivalent to about 467 million tonnes of CO_2-equivalents in 2001. It is difficult to estimate the total fuel consumption and resulting GHG emissions attributable to air transport. There are three reasons for this: the absence of relevant data; the difficulty determining actual emissions of an individual flight (in particular for non-CO_2 GHGs); and the scientific uncertainty around some of the effects aviation has on the atmosphere, such as the generation of contrails, as discussed below.

There are several sources of data on aircraft emissions, with a number of organisations (e.g. Deutsche Luft und Raumfahrt, Dutch Civil Aviation

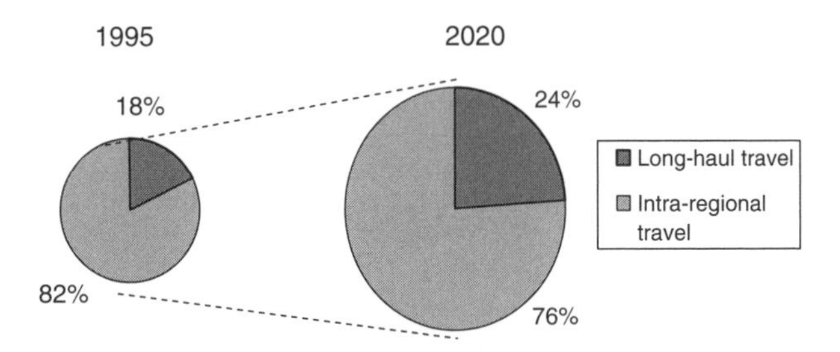

Figure 3.13 International tourist arrivals worldwide: long-haul versus intra-regional trips 1995–2020 (million tourist arrivals)
Source: UNWTO (2001)

Association and Inrets) operating different models for estimating emissions. Aircraft manufacturers supply a performance manual that provides certified information on fuel consumption under different operational circumstances, including climb angle and cruise speed. However, these data are the property of the airlines. They use the tables to predict their fuel consumption and to calculate fuel requirements for a given flight. Airlines document each flight, including the amount of fuel consumed. So far, there is no commonly agreed method for calculating emissions across fleets, let alone the global aviation industry.

Climate impacts of aviation emissions

A large proportion of aircraft emissions occur at an altitude of 9–13 km in the upper troposphere or lower stratosphere (Figure 3.14). The tropopause, located at about 10 km altitude, separates these distinct regions of the atmosphere. The effectiveness of emissions from aircraft depends on altitude and latitude, because of changes in chemical composition (principally the abundance of trace gases) and differences in transport times of gases. The troposphere is turbulent and as a result gases are mixed within several weeks. It is also rich in water vapour and ozone concentrations are low, depending on NO_x and CH_4 levels. The stratosphere is drier and rich in ozone, and more stable due to the temperature increasing with altitude. The resulting lower rates of mixing means that gases emitted in the stratosphere, including water vapour, stay there for a longer time.

The emissions of GHGs differ for the different flight phases; for example NO_x emissions are largely related to running the engine at full power, i.e. in the take-off and climb phases. CO_2 and water vapour emissions are proportional to fuel use (see emission factors in Chapter 7). Thus any reductions in fuel consumption also reduce these types of GHG emissions. Once emitted to the atmosphere, CO_2 is not involved in any further chemical reactions. Therefore the altitude of emissions is not

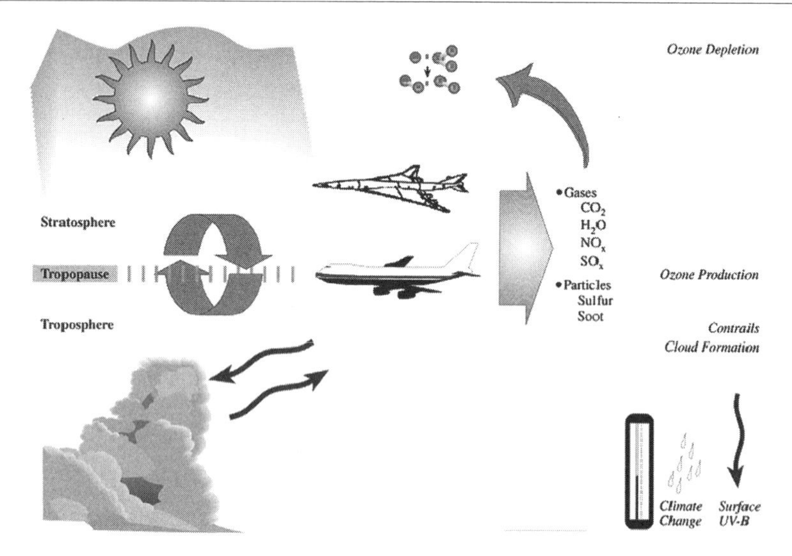

Figure 3.14 Impacts of aviation on the atmosphere
Source: Penner *et al.* (1999)

relevant. But other GHGs (mainly NO_x) influence the atmosphere indirectly by a complex interaction with other compounds. The chemical reactions depend on altitude and therefore their contribution to global warming is difficult to quantify. In the troposphere, NO_x leads to the production of ozone, an effective GHG. In contrast, NO_x emissions in the stratosphere lead to a destruction of ozone. As most flights are in the tropopause, the overall radiative forcing by ozone has been estimated to be in the order of $+0.6$ to $+1.8\,W/m^2$, thus contributing to global warming. At the same time, NO_x emissions lead to a reduction in CH_4, thereby decreasing the radiative forcing from aviation (Figure 3.15).

The emission of water vapour can lead to contrails when the exhaust emissions that are saturated with water vapour mix with low-temperature ambient air, especially when the latter is supersaturated with ice. Under these conditions contrails will form, and can spread as cirrus cloud. Generally, the higher the altitude the greater the likelihood of contrail formation. Contrails and cirrus clouds have two effects: first they reflect some of the shortwave radiant heat energy back into space, resulting in a cooling of the earth; but second they absorb longwave radiant heat energy, leading to a warming of the lower atmosphere. The warming effect of contrails and cirrus clouds is believed to dominate (Royal Commission on Environmental Pollution, 2002). Aerosols emitted by aircraft indirectly affect cirrus cloud cover throughout the atmosphere and cause a positive radiative forcing, thus also adding to the greenhouse

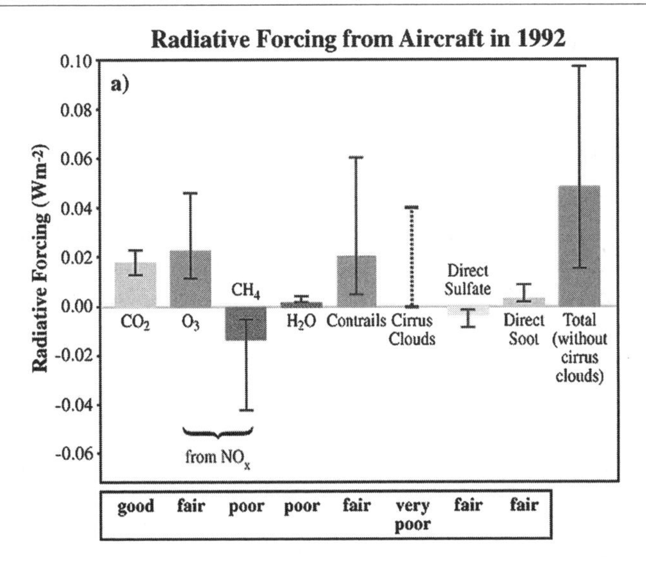

Figure 3.15 Estimates of the globally and annually averaged radiative forcing (W_m^{-2}) from subsonic aircraft emissions in 1992. Degrees of certainty are expressed qualitatively (very poor to good).
Source: Penner *et al.* (1999)

effect (Penner *et al.*, 1999). After the terrorist attack on 11 September 2001, aircraft in the skies over the USA were grounded for three days. This resulted in reduced contrails and hence cirrus cloud cover, affecting both night and daytime temperature regimes (Travis *et al.*, 2002).

Overall, the IPCC (Penner *et al.*, 1999) estimated the total radiative forcing of air travel to be 2.7 times the effect of CO_2 emissions (Figure 3.15). This compares to a factor of between 1 and 1.5 for other activities at the surface. A more recent study of radiative forcing (Sausen *et al.*, 2005) reported much lower radiative forcing from contrails compared with the earlier estimates by the IPCC; however, the study also pointed out that the effect of induced cirrus clouds could be much larger than anticipated. The authors conceded that knowledge on aircraft-induced clouds is still very poor.

Aviation's impact on the global climate would increase if supersonic flights at altitudes of 17–20 km are reintroduced. At those altitudes NO_x emissions decrease ozone levels and therefore decrease radiative forcing. But as a result of water vapour emissions at higher cruising altitudes, the overall forcing will increase. Penner *et al.* (1999: 210) show the total climate impact of supersonic flight to be 5 times greater than for the subsonic flights. In contrast, it is possible that lower cruise altitudes for subsonic aircraft might reduce the effect, due to such effects as fewer contrails and less ozone production; however, these effects might be outweighed by increased fuel consumption at lower altitudes as a result of larger drag (Royal Commission on Environmental Pollution, 2002).

Policy responses

International agreements

Responses to climate change by the aviation industry relate largely to mitigation, i.e. reducing GHG emissions. GHG emissions from fuel consumed for international air transport are recognised in the Kyoto Protocol (Article 2, Paragraph 2, bunker fuels). Annex 1 (developed) countries are encouraged to account for those emissions and reduce them. However, they are reported separately from fuel consumed within national borders (IPCC, 1996) and do not form part of national emission reduction targets under the Kyoto Protocol. This means that to date countries have largely focused on reducing GHG emissions from activities within their national borders. The exclusion of international travel from the Kyoto Protocol's binding targets also means that aviation is isolated from activities such as emissions trading.

The reasons for the exclusion of international transport emissions from the Kyoto Protocol lies in the multilateral nature of aviation and shipping and the long history of special treatment for fuels used in international transport. The Chicago Convention of 1944 is often referred to as a milestone in a number of events that led to tax exemption status for fuels used in international air travel. Specifically, the Chicago Convention outlined that fuel *on board* aircraft is not to be taxed by third states. In 1950 the ICAO resolved that both fuel remaining *on board* and fuel *taken on board* should be exempt from taxation. These considerations are now enshrined in about 3000 different bilateral air service agreements between states (Meijers, 2005). This means, effectively, that international air and ship travel enjoy a competitive advantage over other transport modes, all of which are subject to some form of fuel taxation. It is difficult to renegotiate bilateral agreements. ICAO urges states 'to refrain from unilateral environmental measures that would adversely affect the orderly development of international civil aviation'. Unilateral action would likely lead to undesired effects such as 'tankering', where aircraft fuel up in a country where fuel is subject to less or no taxes. Such practices would likely increase emissions.

In addition to the challenges associated with taxing aviation fuels, little progress has been made in finding ways of allocating emissions to countries involved in international air travel. This is a prerequisite to controlling such emissions. The Subsidiary Body for Scientific and Technological Advice (SBSTA, 1996) of the United Nations Framework Convention of Climate Change (UNFCCC) suggested the following options for allocating emissions:

- no allocation at all;
- allocation of global bunker fuel sales and emissions in proportion to a country's national emissions;
- allocation according to the country where fuel is sold;
- allocation according to the nationality (or country of registration) of the airline;
- allocation according to the country of departure of the aircraft or, alternatively a half share of emissions between country of departure and arrival;
- allocation according to the country of departure or destination of passengers or owner of the cargo;
- allocation according to the country of origin or destination of passengers or cargo;
- allocation to a country of all emissions in its national air space.

So far no agreement has been reached regarding allocation of emissions. In the 35th session of the Assembly of ICAO in October 2004 it was decided that an emission charge will not be introduced internationally for at least three years.

Market-based instruments

Most recently, on 6 July 2006, the European Parliament gave its approval (by a vote of 439 to 74 with 102 abstentions) to recommendations relating to new fuel taxes, the ending of airlines' value added tax exemption and a closed emissions trading scheme. These measures will help level the playing field and bring fiscal as well as environmental benefits. The Parliament's recommendation must now be turned into a bill, which will need the approval of EU member states and the European Parliament.

Around the same time, the environmental body of the Committee on Aviation Environmental Protection (CAEP) of ICAO discussed three market-based mechanisms for aviation, namely voluntary agreements, emission trading, and emission taxes or charges. For all of those it is probably pragmatic to initially address CO_2 emissions and to extend policies to other GHGs once scientific uncertainty associated with the effect of non-CO_2 emissions is reduced. It is the intention of ICAO that market-based instruments are not imposed on developing countries so as not to compromise their economic development. The initiatives by the EU, described above, are therefore not welcomed by the ICAO.

Emissions trading could follow two approaches. One option is to develop a trading scheme in which interested states can participate voluntarily. This would involve determining a maximum amount of GHG emissions for participating states, allocating this total through emission certificates that can be traded among states or airlines. The advantage of this system is that there is some assurance that the determined level of emissions would not be exceeded. Also, emission reductions are made where it is most economical to do so. Another option is to include international aviation in the trading schemes set up by states under the UNFCCC. The lack of binding commitments for international aviation emissions under the Kyoto Protocol makes this option politically more difficult.

In a recent study commissioned by the European Commission (Wit *et al.*, 2005) a number of practical issues associated with emissions trading for aviation were identified, for example deriving a fair way for allocating emission certificates to existing airlines and procedures for allocation to new entrants. Depending on the specific mechanisms, there is some concern that inclusion of aviation in emissions trading would allow the sector to continue its growth, with GHG savings made elsewhere. Lee *et al.* (2005) argue that stabilisation of CO_2 emissions will be difficult if the aviation sector is able to continue growing at the expense of other sectors. For countries with a large international air travel sector the inclusion of international air travel in the emissions trading scheme would mean that a large proportion of national emissions and emission certificates would be used up by aviation. In the case of the UK, if the currently unaccounted emissions from international air travel were added to the national GHG account they would make up 22%, 39% or 67% of the national CO_2 budget in 2050, for assumed growth rates of air travel of 3%, 4% or 5%, respectively.

Another option for regulating international air travel is to impose some form of levy. Such a policy could be primarily targeted at reducing demand, but could also serve as an incentive for the airline industry to reduce fuel costs and minimise emissions. Levies on air travel could also be imposed to generate revenue for the government in general or for climate change-related measures in particular. Such policies, however, have to be formulated with care. Price elasticities of different markets would need to be considered. For example, several studies found that leisure travellers are more price-sensitive than business travellers (i.e. the former have a higher price elasticity). Thus in order to minimise decreases in demand, airlines facing a climate-related levy would increase costs to business travellers relative to economy-class travellers. The possibility of splitting costs unequally between different fare classes means that low-cost airlines face a competitive disadvantage as they typically offer one class where the main selling point is low airfares.

A recent meta-analysis of different price-elasticity studies also showed that long-distance travel is less price-elastic (i.e. less sensitive) than short-distance travel, possibly because the choices and possibilities for substitution are less (Brons *et al.*, 2002). However, an earlier study (Crouch, 1994) noted that long-haul air travel is more price sensitive to increased transportation costs compared with short-haul travel. Brons *et al.* found that air travellers have become less price-sensitive over the last two decades, maybe reflecting the increasing importance of air travel in all parts of society. When analysing price elasticities it is important to distinguish short-term and long-term responses to price changes. Long-term price elasticities have been found to be higher than short-term elasticities. This means that following an increase in airfares travellers are less likely to change their behaviour immediately (i.e. decrease demand), but will react more sensitively in the long run when they have had time to adjust to the new prices and can make alternative arrangements. These findings highlight that price elasticities for air travel need to be understood in the context in which a study was undertaken, especially when findings are used for policy development.

As noted above, it is difficult to introduce levies on air travel because of the clauses in the Chicago Convention and numerous bilateral agreements. However, there are several options to circumvent these constraints (Table 3.8). ICAO pointed out that charges are generally preferred over taxes, as the revenue generated by a charge is specifically allocated to address the impacts of aircraft on the atmosphere. Several global funds could administer revenue raised from an airline emissions tax. Also, some of the raised revenue could be allocated to the Global Environment Facility (GEF), and used for the financing of climate change adaptation measures.

A charge on tickets is the most straightforward and simplest option to internalise the climate-change costs of aviation. Ticket charges could be staggered by distance, as is now done in many instances for security surcharges. For example, the British Government proposed an emission charge of £3 per passenger on short-haul flights and £20 on long-haul flights. While it has been recognised that such charges would not dramatically reduce the demand for travel, it is believed that a charge of about £35 per ticket would have a more marked effect on travel demand (Royal Commission on Environmental Pollution, 2002). Norway has introduced a 'green charge' for national air travel when an alternative by rail exists, and also for all international flights departing from Norway. The most prominent benefit of introducing a charge on the ticket is the reduction of unequal tax burdens of air travel compared with other transport modes; actual reductions of emission are probably small. Overall, demand is likely to continue to grow. Moreover, a ticket charge is unlikely to provide an incentive for airlines to reduce their emissions.

A tax on aviation fuels would provide incentives for reducing emissions and it would also address the currently unequal taxation on different forms of transport. A fuel tax is directly related to the amount of CO_2 emissions as these are proportional to fuel consumption. In terms of non-CO_2 GHG gas emissions, however, fuel consumption is not directly

Table 3.8 Possible schemes for a levy on air travel emissions within the EU

Type of levy	*Operational issues*	*Financing (in the EU)*	*Effect on emissions*	*Legal aspects*
Charge on ticket	Simple and possible to introduce in the short term; airlines could be responsible for collecting the charge.	Assuming a charge of 5% on the airfare, this charge could raise about € 10–16 billion annually.	Probably little effect on demand given estimated price elasticities; no incentive for airlines to reduce emissions. Little overall effect.	Legally feasible.
Fuel tax	A tax could be added as a fixed amount per litre of fuel sold or as a percentage of current fuel price; petroleum companies could collect the tax.	Assuming a tax of €0.32/l of kerosene, a total of about €14 billion could be expected.	Incentive for emission reductions; research into fuel-efficient technologies and operations; reduction in demand less likely.	Problematic, especially concerning the many bilateral agreements stating tax exemption for fuel.
Emission tax	Complicated given the many factors that determine overall radiative forcing. Estimates of emissions possible when considering aircraft type, engines, LTOs and routings	Assuming emission charges per litre of kerosene of €0.12 for CO_2, €0.12 for water vapour and €0.6 for NO_x, the total amount would be around €14 billion.	An emission tax would have the greatest impact on emission reductions and provides an incentive for technological and operational improvements for airlines.	Legally feasible unless the tax is closely correlated to fuel usage, because this could be seen as a hidden tax on kerosene.

Source: German Advisory Council on Global Change (2002)

London to New York: 5,700 km
Aircraft: Boeing 747-400, American Airlines, 310 passengers
Actual fuel consumption: 57,000 kg

CO_2 charge: 57,000 * € 0.3 = € 17,100
(the emission cost for CO_2 is 0.3 €/kg kerosene)
NO_x charge: 57,000 * 14.3 g/kg (NO_x emission index) * € 14.31 = € 11,664
(according to the AREONOX report the NO_x emission index for this aircraft with its specific engines on a distance of this magnitude is 14.3 g/kg; the emission cost for NO_x is 14.31 €/kg kerosene)

Figure 3.16 Example of an emission charge on a flight from London to New York
Source: Whitelegg & Cambridge (2004)

correlated to emissions and, therefore, reducing fuel costs does not necessarily lead to a reduction in those emissions. It is possible to develop a tiered fuel tax that would also take into account the number of landings and take-offs to derive an estimate of NO_x emissions. There is great opposition to a tax on aviation fuels, based on the wording in the Chicago Convention and even more explicitly the exemption of such fuels from any taxes or charges in the many bilateral air service agreements.

Charging emissions rather than fuel circumvents such problems. Also, emissions charges target the source of the damage. In that sense, they would have great potential to reduce GHG emissions from air travel, and technological developments would not only focus on reducing fuel consumption but also consider the effects of other emissions, such as NO_x. Theoretically it is possible to impose a charge that is directly related to the impact the flight has on the climate. Practically, however, it is not possible to measure the different GHG emissions for each flight. Rather, the radiative forcing caused by a given flight has to be estimated, based on fuel consumption, aircraft type, engine type, load factor and flight distance. CO_2 and water vapour can be estimated based on fuel consumption (3.2 kg of CO_2 is produced per kilogram of kerosene), whereas NO_x can only be estimated on the basis of altitude, distance, aircraft and engine type. The example in Figure 3.16 shows the emission charge that could be imposed on a flight from London to New York.

Conclusions

This chapter has used four case studies to highlight the relationships between climate change and mitigation and adaptation policies. The case studies related to tourism in alpine areas and in small island destinations, to the insurance sector and to international aviation. While the two geographic case studies focused on climate-related risks, adaptation measures, GHG reduction options and the integration of adaptation and mitigation, the insurance example highlighted the business dimensions of climate change, and the aviation case study examined policy issues in more detail.

The case studies also showed that to assess and manage tourism–climate change interactions for a given destination or sector it is important to understand the nature of tourism (in general and at the specific destination in particular) as discussed in Chapter 4 and the interaction with the climate system as explained in Chapter 5. GHG accounting (Chapter 6) is critical for planning appropriate mitigation measures (Chapter 5). Risk management and climate change adaptation – as described in Chapter 8 – are important for the long-term sustainability of destinations and should become part of best business practice. Policies are designed to encourage climate change mitigation and assist adaptation (Chapter 9), and are therefore equally relevant to destinations and sectors.

Chapter 4

An Overview of Tourism

Key Points for Policy and Decision Makers, and Tourism Operators

- Tourism encompasses activities by those who voluntarily travel away from their home, for less than one year.
- Tourism differs from other economic activities and social phenomena in many ways, including its fragmentation and the large number of small businesses; as a result it requires specific research and policy development to address climate change.
- Annually, over 700 million people engage in international travel; this number is expected to reach about 1.6 billion by the year 2020. At present, the major international tourist destinations are France (77 million arrivals per annum), Spain (52 million) and the USA (42 million). China is predicted to generate 100 million international tourists by 2020.
- Tourism's contribution to national economies can be substantial and includes much more than the generation of foreign exchange earnings, revenue generated from tourist taxes and employment. Key measures, such as contribution to GDP, are provided through Tourism Satellite Accounts (TSAs).
- While traditional forms of beach-oriented mass tourism are still highly popular, in latter decades the tourism industry has become much more complex and diversified.
- The most recent changes in the tourism transport industry relate to the introduction of low-cost airlines, which facilitated affordable access for most of society, and stimulated growth, especially on new routes to secondary airports.
- New information technologies (e.g. the Internet) present both a challenge and an opportunity for the tourism industry, including provision of in-depth information to increasingly experienced tourists, and in a competitive way.
- Different agents involved in tourism (e.g. tourists, resort operators, transport providers) require different types of weather or climate information at different times; a plethora of information can be accessed, to provide both qualitative and quantitative guidance and decision support.
- Wider shifts or trends in societies – such as emancipation or 'ageing' – could have major impacts on tourism, particularly as

these influence product preferences and the geographic dimensions of tourist flows.
- There is an increasing interest in responsible or sustainable tourism from the tourism industry, tourists and tourism researchers. However, when climate risks to tourism are considered, the sustainability of some forms of tourism is called into question.

Introduction

What is tourism? Business for some, a social phenomenon for others – tourism can be treated as a form of land use, an aspect of mobility and also the subject of psychological studies. There are many aspects to tourism, and the explicit study of tourism is a comparatively recent activity. Tourism attracts the attention of geographers and planners as a result of its spatial nature; sociologists are studying tourism trends and their social implications, and psychologists focus on internal processes of those engaging in various forms of tourism. Economists research the manifold business dimensions of tourism, including marketing and the economic impacts of tourism on destinations. A number of natural scientists are interested in the environmental impacts related to tourism, including that of energy use and GHG emissions. Issues relating to global change such as climate change have been addressed only to a small extent. In fact, tourism has been largely absent from academic and political debates relating to global changes or even more specifically the human dimensions of global change.

There are several reasons why tourism is sufficiently different from other sectors or social phenomena to warrant more specific research in the area of climate change. At present, tourism is often treated as a component of other sectors, for example transport or households. As a result there are few national policies relating to climate change that specifically address tourism. This carries the risk that policies are suboptimal and ineffective. Climate change research and policy making benefits from taking into account the following particularities of tourism:

- The role of the public sector. Tourism is delivered and advanced through the combined operations and investments of the private and public sectors. Management strategies that do not recognise the interaction of these two sectors may not deliver a sustainable outcome. These national-scale strategies include, among others, revenue collection from tourists and appropriate reallocation of resources, energy policies, disaster management and infrastructure provision.
- Business structure. The tourism sector is largely made up of small and medium-sized enterprises (SMEs), typically with fewer than 10

employees. Often these SMEs are run as family-owned businesses. No other industry is determined to that degree by a myriad of small operators, who deliver a wide range of products and services. The diverse and often fragmented nature of tourism makes it difficult to communicate issues such as climate change and to implement policies effectively.

- Lifestyle component of the tourism business. The business motivation for tourism SMEs is often diverse and not necessarily purely economic in nature. This could be both beneficial and adverse to introducing new business models that consider climate change.
- The tourism product is the integration of independent goods and services sold by description. Prior to arrival, visitors rely substantially on the description of the tourism product, which will be a concatenation of numerous, independent elements that are purchased in a variety of ways during their stay. It is sometimes difficult for tourists to assess the sustainability (e.g. carbon footprint) of different products, and the complexity of supply chains makes the implementation of climate policies and monitoring difficult.
- Tourism is an export product. Almost all tourism entities service the export market, i.e. they cater for international visitors. Tourism entities seldom have any choice but to compete internationally, and meet international quality standards. This differs from other non-tourism businesses who can develop their business domestically first, before entering wider export markets.
- Location of tourism SMEs. Because visitors want to experience more remote areas as well as central hubs, services (e.g. accommodation) must be provided outside the main centres. Unlike most other industries, the customer travels to where the service is provided. This offers potential to counteract increasing urbanisation and provides development opportunities for local communities and support in maintaining infrastructure. If tourist mobility continues to be based on fossil fuels, the emissions of GHG from transport can be seen as a trade-off for regional economic development.
- The role of the natural environment. Tourism more than any other industry relies heavily on the natural environment, especially when the services or products are closely linked to natural attractions (e.g. scenic tours). This dependency poses a risk when climate changes are likely to affect the attractiveness of a destination or change the balance between competing destinations.
- The role of the social environment. Tourism takes place largely in communities, and the 'hospitality atmosphere' that communities create is paramount for a quality visitor experience and the sustainability of tourism itself. This means that visitor satisfaction

is heavily influenced by factors that are beyond those attributable to any individual supplier. Communities play a crucial role in climate change adaptation and therefore the viability of many tourist destinations.

- Risks inherent in tourism. Tourism is heavily influenced by factors that are beyond the individual firm. As indicated above, tourism is subjected to external risks that may not apply to other sectors. For example, global tourist flows can be volatile (e.g. SARS), tourism destinations are subject to fashion, and degradation of perceived and actual quality (of product or the environment) may lead to decreased demand. Climate change poses a key risk to tourism globally.
- Diplomatic/political aspects. There is a fundamentally political dimension to tourism, in that visitors to foreign countries can pose threats to those countries (by transmitting diseases, dealing in prohibited substances, etc.); could potentially undermine a country's customs, traditions and ideologies; and are a liability to their country of citizenship in cases of disaster, rescue mission or other incidents that require diplomatic interference. Climate change impacts potentially increase the number of incidents of killed, injured or stranded passengers that require support from their host and home countries.

With these tourism-specific characteristics in mind, this chapter starts by defining tourism and providing an overview of major global tourism flows. These tourist flows represent tourist movements between places of residence and destination, which is important for assessing GHG emissions, but also for understanding economic dependencies on tourism and likely vulnerabilities in different geographical regions, for example as a result of a global tax on air travel or climate change impacts. The economic importance of tourism will be discussed. The development of a TSA was a major step towards the recognition of tourism's role in many economies. Trends in tourism that are relevant to the issue of climate change will be discussed, in particular those of product diversification, technological advances, consumption patterns and responsible tourism.

Who Is a Tourist and What Is Tourism?

There is a recognised need to define the scope of tourism, but there still is much debate about what exactly a tourist is and what tourism includes (e.g. Collier, 1999). Attempts have been made to define tourism on the basis of:

- **Motivation or travel purpose.** A common classification distinguishes holiday and recreation, business, health, study, missions, meetings/conferences, family/friends, religion and sport.
- **Distance travelled or length of trip.** For example, American travel statistics distinguish local and non-local travel, whereby the latter is defined as any travel further than 100 miles; German statistics often refer to tourist trips as those that are longer than 5 days away from home.
- **Nature of trip.** Packaged travel or inclusive tours provide bundled products typically consisting of some transport and accommodation arrangements, whereas independent travel (often referred to as *free and independent tourists*, FIT) means that tourists make their own decisions concerning the elements of their trip; often FITs see themselves as travellers rather than tourists, especially in relation to longer trips. In the USA 'travellers' refers to all people travelling, with tourists being a subset of those who travel for pleasure.

Earlier, in Chapter 2, the definition of a 'visitor' provided by the UNWTO has been put forward. It is probably the most commonly applied definition of visitors or tourists, and therefore tourism. The main criticism of the UNWTO definition is that it does not correspond to the popular understanding of who a tourist is. Tourists are often associated with leisure, which contains the elements of freedom and non-work-related pastimes. The UNWTO definition encompasses leisure tourists as well as business travellers and 'visiting friends and relatives' tourists. The main reason behind this is that from an economic and planning perspective it is useful to include all travellers independent of their purpose of travel, as they all use the same infrastructure and services, for example air travel and accommodation. The UNWTO definition of visitors, tourists and excursionists is applied in this book.

In addition to the UNWTO definition, it might be useful to introduce additional ways of differentiating tourism. One useful distinction relates to tourism relative to a given destination:

- **Domestic tourism:** residents of a given country travelling only within that country, but outside their usual environment
- **International tourism:** non-residents travelling to a given country for longer than one day and less than one year
- **Inbound tourism:** non-residents travelling to a given area that is outside their usual environment; often the area in question refers to a country or state (e.g. California)
- **Outbound tourism:** residents of a given area travelling to and staying in places outside that area (and outside their usual environment)

Tourist Destinations – Main Global Tourist Flows

Together with geographical location, topography, landscape, flora and fauna, weather and climate constitute the natural resource-base of a place for tourism (Martin, 2005). Tourists frequently base their decision to travel to a particular destination on a combination of factual evidence and perceptions that the climatic conditions at the destination will be advantageous relative to any other potential destination, including those in their own locale. It is important to distinguish between weather and climate. Weather is the day-to-day variations of the meteorological conditions experienced at a specific location, whilst climate is the average of that weather. The WMO use a 30-year average (1961–1990) as a definition of climate. As a large proportion of tourist flows is determined or at least influenced by climatic conditions, it is likely that global climate change will have some modifying influence on those flows and the global distribution of tourism.

More and more people can afford to travel. In 2003, over 700 million international tourist arrivals were counted, an average annual growth rate of 6.6% between 1950 and 2003. Growth is predicted to continue at a rate of between 3.8% for interregional and 5.4% for long-haul travel between 1995 and 2020 (UNWTO, 2001). International arrivals are expected to reach about 1.6 billion by the year 2020. The top three receiving regions in 2020 will be Europe (717 million tourists), East Asia and the Pacific (397 million) and the Americas (282 million). At present, the most important tourist destinations for international arrivals are France (77 million arrivals per annum), Spain (52 million), the USA (42 million) and Italy (40 million) (Table 4.1). Developing countries receive only a very small share of international tourism, albeit a growing one. Tourist arrivals to South East Asia amounted to 42 million in 2002, with an average growth rate of 4.9% compared with the previous year, but country-specific growth rates were up to 30% (e.g. Cambodia). Central America received 4.7 million international arrivals, and international tourism in Africa grew at around 3%, reaching a total of 29 million (UNWTO, 2003a).

The UNWTO does not maintain a database on domestic tourism; i.e. a total assessment of worldwide movements is not possible based on UNWTO statistics alone. This is particularly important given the importance of domestic tourism in large countries (see below). To overcome some of these data deficiencies, Bigano *et al.* (2004) compiled arrival data for 181 countries and departure data for 107 countries based on various data sources (e.g. World Resources Institute, Euromonitor, national statistical offices, other governmental institutions or trade associations) and interpolated tourist activities for those countries where no statistics were available.

Table 4.1 Top destinations for international tourism

Destination	*Arrivals 2002 (million)*	*Growth 2002/2001 (%)*	*Market share (%)*	*Population 2002 (million)*
France	77.0	2.4	11.0	60
Spain	51.7	3.3	7.4	40
USA	41.9	– 6.7	6.0	288
Italy	39.8	0.6	5.7	58
China	36.8	11.0	5.2	1279
UK	24.2	5.9	3.4	60
Canada	20.1	1.9	2.9	32
Mexico	19.7	– 0.7	2.8	103
Austria	18.6	2.4	2.6	8
Germany	18.0	0.6	2.6	82

Source: UNWTO (2003a)

When international and domestic tourism are taken together, the USA is by far the most visited tourist destination, with 1042 million trips, 999 million of which are domestic. Table 4.2 shows the top 10 tourism-generating countries when domestic and international tourism demands are aggregated. These top 10 countries cover 73.4% of world tourism demand. Domestic tourism dominates global tourism in terms of trips undertaken. In China, for example, 644 million Chinese travel domestically. Domestic tourism is also important in India (320 million), Brazil (176 million) and the UK (134 million). The ratio of domestic trips (see also Figure 4.1) to international trips (see also Figure 4.2) undertaken by the population of a given country varies markedly (Table 4.2).

Developing countries are characterised by a large domestic tourism market (well over 90% of all trips), among others a result of less disposable income for international travel. The ratio also seems to be influenced by the size of a country; for example the USA shows a high propensity for people to travel within their own country, whereas German residents are more likely to undertake international trips. Germany's location within Europe partly explains the large number of border crossings. The ratio of domestic to international travel has implications for GHG emissions, when an increasing proportion of international trips results in increased distances travelled and a shift to

Table 4.2 Total tourist activity (country of origin of trip): top 10 countries in the world

Country	*Departures (domestic and international trips) (million)*	*Domestic departures as a percentage of total (%)*
USA	1059	94
China	649	99
India	324	99
UK	183	73
Brazil	179	98
Germany	170	48
Indonesia	109	98
Canada	102	79
France	96	77
Japan	96	81

air travel. Only small changes in the travel behaviour of Chinese (Text Box 3) or Indians, for example, could have major effects on global tourist flows and resulting emissions from tourist transport to and from their holiday destinations.

Economic Importance of Tourism

Tourism has become the hope for many developing countries, and it is already an integral part in most developed economies. International tourism represents approximately 7% of the worldwide export of goods and services. This makes tourism the fourth most important export sector after chemicals, automotive products and fuels (UNWTO, 2004). Tourism is the only export sector where the customer travels to the product and not the other way round. Table 4.3 shows those countries that benefit most from international tourism.

Tourism contributes to foreign exchange earnings, revenue generation through taxes and levies, and employment, both directly and indirectly. Tourism generally stimulates infrastructure development (e.g. airports) and contributes to regional development. Tourism activity that occurs in peripheral regions has the potential to counteract trends of agglomeration of industries, population and investment in urban and core development areas. Tourism potentially provides an opportunity for

Text Box 3 China – major emerging player in global tourism.

Many mature and developing tourist destinations closely follow the developments of China's outbound travel – a so far untapped resource in terms of global tourism. There are about 1.3 billion people in China and the Chinese economy has experienced 25 years of 10% annual growth. However, only about 2% of the population (25 million) currently has the same spending power as Europeans. The Chinese middle class is likely to grow to 43 million in 2008, 50 million in 2010 and 100 million by 2020. Accordingly, China's outbound tourism is growing at a fast rate. In 2004, about 20 million Chinese travelled overseas, mainly to Hong Kong, Macau, Thailand, Japan, Russia, the USA, Korea and Singapore. Growth in departures was over 60% between 2003 and 2004. The UNWTO (2003b) estimated that in 2020 China will be the fourth largest (after Germany, Japan and the USA) source for outbound tourism worldwide, with 100 million international tourists.

Travel to and from China is regulated by the China National Tourism Administration. In the past the majority of travellers were government officials, employees of state-owned enterprises and other people sent abroad by the government and issued with a public passport. Currently (2005), private passports can be issued for the purpose of visiting friends and relatives, studying overseas and for travel. Tourists travelling on a private passport have to be part of a tour group and the tour has to be organised by a qualified and authorised travel agency. Travel must go to an approved destination; the Approved Destination Status (ADS) is a bilateral agreement that allows Chinese to visit a specific country.

The regulations concerning China's outbound travel are dynamic. In 2003, only 19 countries had an operational ADS agreement with China, but in early 2004 China opened up a wholesale agreement with the European Union (EU) and European countries outside the EU. Also in 2004, an agreement with a large group of African countries was reached to allow Chinese citizens to travel to Africa. Following this, China and the USA signed a 'Memorandum of Understanding of Tourism Co-operation', which enables Chinese to travel to North America. In late 2004, China initiated an 'Individual Travel Scheme' for its citizens to visit Germany as independent travellers.

Currently, most outbound holiday tourism from China is still in the form of an all-inclusive coach tour (Wen Pan & Laws, 2001). Chinese overseas travellers are mainly from Beijing, Shanghai and Guangzhou.

They are typically from high-to-middle-income groups, but tend to be very price sensitive, in particular relating to accommodation. As the market matures, Chinese tourists could potentially become high spenders for the following reasons:

- When Chinese go on their first overseas trip they have often been waiting to do so for a long time and are willing to spend a lot.
- A Chinese proverb says 'economise at home, but take enough money en route'.
- Incentive tourists are growing in number, and such tourists have a higher ability to spend.
- Chinese are becoming more sophisticated, and are increasingly demanding 4- or 5-star accommodation.
- Shopping is a popular activity with Chinese, especially for souvenirs to bring home to friends and relatives (see also Asia Pacific Foundation of Canada, 2002; Chen, 1998; Wen Pan & Laws, 2001).

people to find employment and business in their region, although it can sometimes be difficult to develop tourism outside main tourist centres or tourism 'honey-pots'.

Determining the economic importance of tourism is not straightforward, because tourism is a 'composite' sector and defined through consumption rather than production. For this reason, information is typically not readily available from national accounts as they are expressed in the System of National Accounts (SNA). Tourism is nonetheless very important in many countries, and governments and other organisations need tourism statistics for many purposes. Consequently, TSAs (UNWTO, 1999) have been constructed by a number of countries in an attempt to capture a variety of production, consumption and other statistics relating to tourism in a structure that is compatible with national accounting and other economic data.

At the most basic level, TSA provides data distinguishing tourist and non-tourist consumption from the conventional industry sectors. Tourism product ratios (TPRs) can be constructed to show the proportion of total supply absorbed by tourists for each commodity. Based on these ratios, the UNWTO (1999) has developed a system for classifying industries as tourism-characteristic, tourism-related or non-tourism industries. These qualitative classifications or the quantitative tourism ratios themselves can be used in the construction of GHG accounts for tourism, as will be shown in more detail Chapter 6.

Tourism is often represented positively because of its contribution to the economy, but tourist activity also has negative economic impacts.

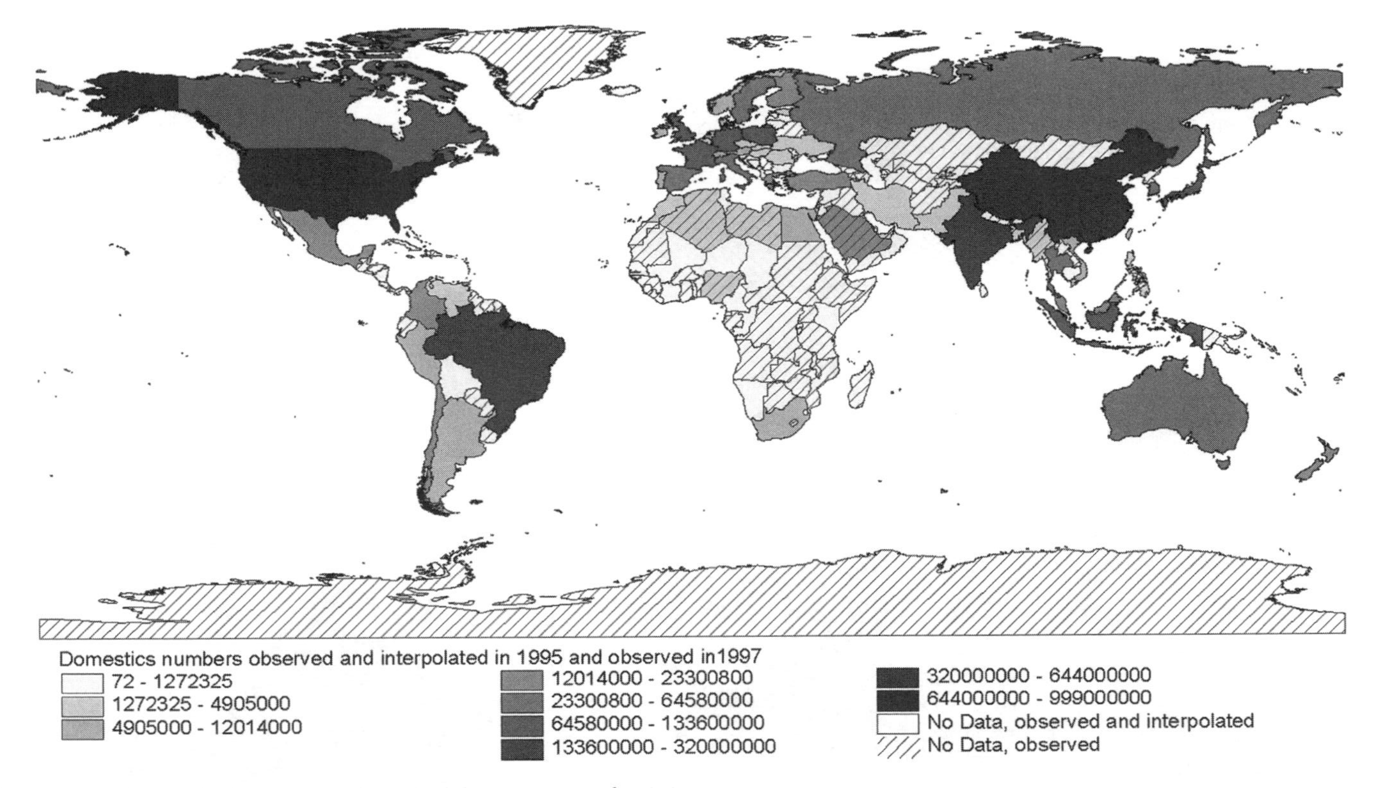

Figure 4.1 Domestic departures (or trips) by country of origin

Source: Bigano *et al.* (2004)

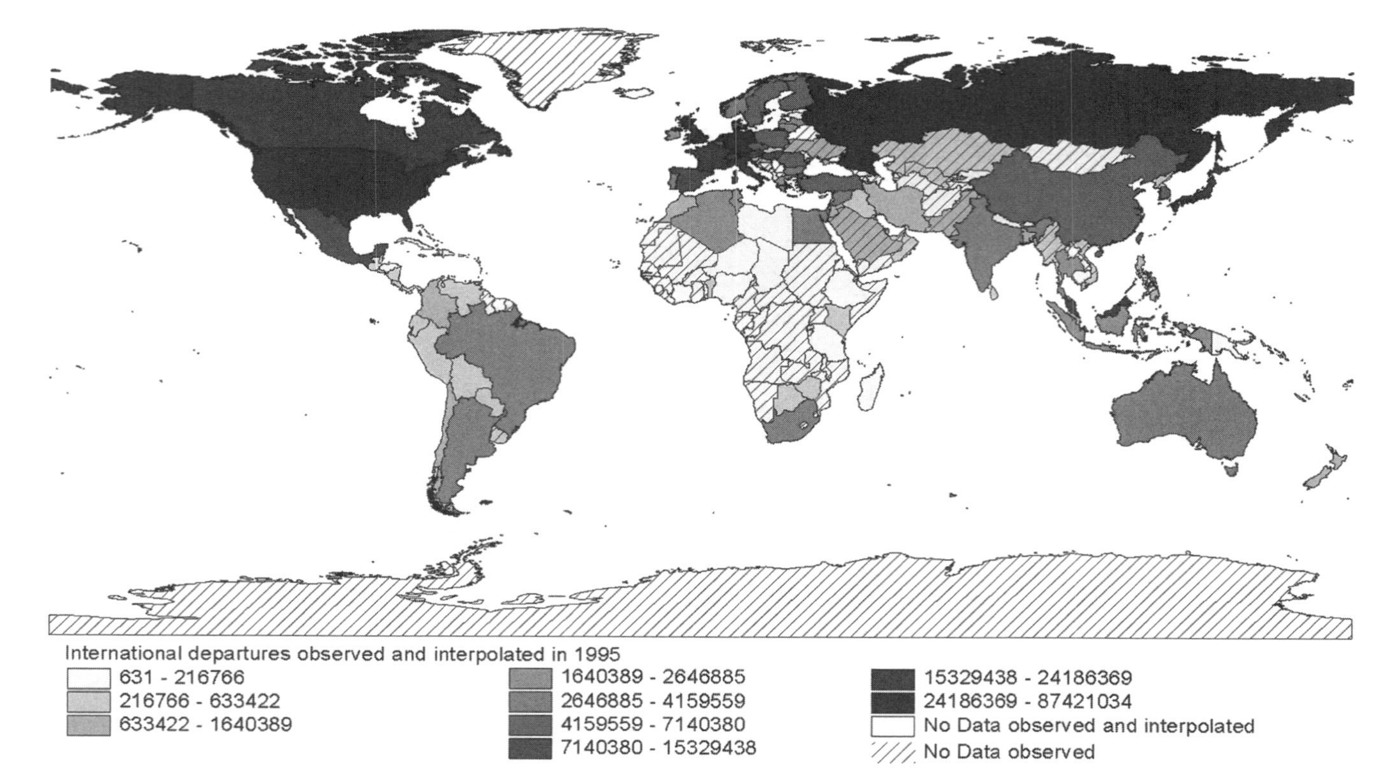

Figure 4.2 International departures by country of origin
Source: Bigano *et al.* (2004)

Table 4.3 World's top tourism earners

	International tourism receipts (US$ billion)%			*Change (%)*		*Population 2002 (million)*	*Receipts per capita (US$)*
	2000	*2001*	*2002*	*2001/ 2000*	*2002/ 2001*		
World	473	459	474	– 2.9	3.2	6228	76
USA	82.4	71.9	66.5	– 12.8	– 7.4	288	231
Spain	31.5	32.9	33.6	4.5	2.2	40	837
France	30.8	30.0	32.3	– 2.5	7.8	60	539
Italy	27.5	25.8	26.9	– 6.2	4.3	58	465
China	16.2	17.8	20.4	9.7	14.6	1279	16
Germany	18.5	18.4	19.2	– 0.3	4.0	82	233
UK	19.5	16.3	17.6	– 16.7	8.0	60	294
Austria	9.9	10.1	11.2	1.9	11.1	8	1375
Hong Kong	7.9	8.3	10.1	5.0	22.2	7	1385
Greece	9.2	9.4	9.7	2.4	3.1	11	915
Canada	10.8	10.8	9.7	– 0.6	– 10.0	32	304
Turkey	9.4	7.4	9.0	– 21.7	22.0	67	134

(*Continued*)

Table 4.3 (*Continued*)

	International tourism receipts (US$ billion)%			*Change (%)*		*Population 2002 (million)*	*Receipts per capita (US$)*
	2000	*2001*	*2002*	*2001/ 2000*	*2002/ 2001*		
Mexico	8.3	8.4	8.9	1.3	5.4	103	86
Australia	8.5	7.6	8.1	– 9.8	6.1	20	414
Thailand	7.5	7.1	7.9	– 5.5	11.7	64	124
Netherlands	7.2	6.7	7.7	– 6.8	14.6	16	480
Switzerland	7.6	7.3	7.6	– 3.5	4.4	7	1045
Belgium	6.6	6.9	6.9	4.7	– 0.2	10	671
Malaysia	4.6	6.4	6.8	39.7	6.4	23	299
Portugal	5.3	5.5	5.9	4.2	7.5	10	587
Denmark	4.0	4.6	5.8	13.9	25.8	5	1078
Indonesia	5.7	5.4		– 5.9		231	24
Republic of Korea	6.8	6.4	5.3	– 6.4	– 17.2	48	110
Singapore	6.0	5.1	4.9	– 15.6	– 2.9	4	1108
Poland	6.1	4.8	4.5	– 21.1	– 6.5	39	117

Source: UNWTO (2004)

First, much tourist expenditure does not remain at the destination, especially in the case of developing countries, where many of the products consumed by tourists have to be imported in the first place. This is called 'import leakage', as opposed to 'export leakage', which refers to money earned by overseas investors, who finance resorts or hotels and take profits back to their country of origin. Tourism leakage can be substantial as in the case of Thailand where an estimated 70% of all money spent by tourists leaves the country (via foreign-owned tour operators, airlines, hotels, imported drinks and food, etc.). Estimates for other Third World countries range from 40% in India to 80% in the Caribbean (Sustainable Living, quoted in UNEP, 2002). This means that it is possible that the net benefit for a destination might be much smaller than indicated by statistics on expenditure or GDP contribution and it might even be negative when wider costs involved in providing tourism are taken into account.

Other negative impacts relate to the cost of infrastructure that is required by tourism and that is usually financed by local taxpayers. Typical examples include airports and roads. Other costs to the country can result through external costs, for example GHG emissions. These costs could be internalised if the country is obliged to purchase carbon credits to meet its Kyoto Protocol obligations. These costs need to be considered carefully when promoting further growth of tourism. Similarly, public sector resources spent on tourism infrastructure or tax breaks granted to businesses potentially reduce government investment in other areas such as education and health (UNEP, 2002). Often, tourism is linked to increases in prices and property values, which then become unaffordable for locals. Overall, there is a risk that countries become overly dependent on tourism as a revenue generator, while at the same time they have little control over demand and development of global tourist flows. Political tensions, downturns in the economies of origin-countries, impacts of natural disasters such as tropical storms, increasing airfares and changing consumption patterns can have a devastating effect on tourism in those countries. These catastrophic events have been described in the chaos model of tourism in Chapter 2 as externalities to tourism. They have the potential to push a destination into a chaotic state that requires substantial reorganisation.

Observed Tourism Trends

Tourism is extremely dynamic and the last few decades have seen major changes in the production and consumption of tourism, in particular against the background of an increasingly globalising world. The trends discussed in this chapter (Table 4.4) are descriptive of First World countries, typically Europe and the USA, and increasingly of Asia.

Table 4.4 Tourism trends

Factor	*Relevance for tourism*	*Relevance for climate change*
Diversification	Niche market development Differing price elasticities for various niche markets	Less carbon-intensive markets Different vulnerabilities to climate change policies, e.g. carbon taxes on aviation
Changes in transport technology; competition in international aviation	Larger aircraft decrease travel time and costs; faster transport system Increasing visitation to far-away developing countries Faster and more accessible railway systems Space travel is a likely niche market in the near future	Larger tourist flows to far destinations Energy-efficient transport options Alternatives to air travel and automobile
Greater availability and use of the Internet for business and leisure	Increased use of the Internet for travel information, planning and purchasing Small companies can access markets to offer individualised and prompt services Increasing demand for flexible, individualised options Access to communication allows tourists to stay connected while travelling	Accessibility of carbon-zero products Substitution of travel by telecommunication Increased demand for travel
Demographic trends	More retired tourists with plenty of free time and relatively wealthy Evening out of seasonality, not bound to school breaks, often second homes	Longer stays and improved carbon efficiency of trips More sensitive to extreme weather conditions More flexible in response to changing climatic conditions

Table 4.4 (*Continued*)

Factor	*Relevance for tourism*	*Relevance for climate change*
Changes in working hours, job conditions and the rise of working mothers	More shorter trips in a year Increased combination of business and recreation Clustering of attractions and experiences to maximise options in a limited time	Higher carbon intensity for trips, more frequent trips Shorter travel distances once at the destination Greater flexibility towards changing climatic conditions
Increased concern over health, well-being, lifestyle	Growth in travel to health spas and resorts Fitness-oriented holidays	Demand for 'slow' travel, including biking and walking holidays
Increased concern about the environment	Continued growth in nature-based tourism Use of environmental product information	Higher demand for less carbon-intensive products

Source: after Faulkner *et al.* (2001)

Few studies have been undertaken on domestic or outbound tourism in developing countries.

Product diversification

A general observation is that while traditional forms of beach-oriented mass tourism are still highly popular, the tourism industry has become much more complex in the last decades. This manifests in the number of players involved in tourism, as well as in the emergence of numerous niche and special interest markets and increasingly diversified products catering for those markets (Table 4.5).

It is difficult to determine sizes or growth rates of the various niche markets. Moreover, niche markets are not distinct or exclusive, and the same tourist may demand different types of products for different holidays or on different occasions. For example, traditional mass tourism to beach resorts is increasingly enriched with additional products, in the form of adventure or ecotourism day trips for example, or other sporting activities. Often, cultural experiences are offered directly at the resort, as in indigenous dance performances, or tourists visit villages as part of a scenic trip from the resort.

Table 4.5 Examples of tourism niche markets

Niche market	*Description*
Adventure tourism	Activities in the natural environment, containing risks in which the outcome is influenced by the participant, setting and management of the experience.
Cultural tourism	Travel directed at an area's arts, heritage, recreational and natural resources.
Ecotourism	Responsible travel to natural areas that conserves the environment and improves the welfare of local people.
Educational tourism	Travel to learn something in a more or less formal way.
Food & wine tourism	Products based around local cuisine and culinary experiences.
Indigenous tourism	Tourism experience that provides insights into the cultural knowledge, lifestyle and beliefs of indigenous people.
Meeting, Incentive, Convention, Event	Business tourism is made up of meetings, incentives, conventions and exhibitions (acronym: MICE)
Nature tourism	Tourism in natural environments
Wellness/health tourism	Products designed to improve the tourist's health, well-being and fitness; often in 'health spas'
Religious tourism	Travel for religious reasons, such as pilgrimage
Sports tourism	Travel to participate in or watch sport events

Similarly, conferences often include recreational packages in the form of day trips or social evenings, or even pre- or post-conference holidays. The Meeting, Incentive, Convention, Event (MICE) is one of the fastest growing segments of the tourism industry. The total industry spending on MICE products in 1999 amounted to US$40.2 billion (Weber & Ladkin, 2003). According to the International Congress and Convention Association, 58% of all conferences are in Europe, 17% are in Asia, 12% in North America, 5% in South & Central America and 4% are in Australia. It is expected that Asia will grow fastest as a destination for MICE tourism. Within Europe, the UK is the most important destination and has between 4000 and 5000 MICE venues, with an estimated income generation of £2 billion (Weber & Ladkin, 2003). The growth of MICE tourism is relevant in the context of growth in air travel as the business-travel market is generally said to be less price-sensitive than leisure-based segments.

Transport developments

Descriptions of historical developments and trends in tourism highlight the close link between tourism and transport advances (Prideaux, 2000). One main cornerstone for tourism at a larger scale was the introduction of packaged trips by Thomas Cook in 1841. These are commonly referred to as the beginning of mass tourism. Other important technological developments that changed tourism fundamentally were the increasing popularity of personal car transport, passenger ships for intercontinental travel and jet aircraft. All of these transport advances had significant impacts on the number of people travelling, the distances travelled and the destinations visited. The development of new aircraft technologies and concepts will not only influence where people travel, but how much they have to pay for their travel and what environmental impacts they cause.

Currently, most international tourism is by road transport (52% in 2001), however, there is an observed growth in air travel. Between 1990 and 2000 air travel grew at an average rate of 5.5% per year, compared with 3.8% for road transport and a negative growth of -1.1% for rail travel (Table 4.6). Air travel is more sensitive to international risks (e.g. terrorism or epidemics such as SARS) than other transport options. After the terrorist attacks on 11 September 2001, air passenger arrivals declined by more than 2% in 2001, and tourists substituted destinations closer to home for those faraway.

The most recent changes in the tourism transport industry relate to the introduction of low-cost airlines. The low-cost carriers – also often referred to as no-frills airlines – are understood to be scheduled airlines that offer very competitive fares as a result of minimised operating costs and a low-service culture. Southwest Airlines in the USA pioneered this concept but it is now widely taken up by other airlines around the world – and many of today's low-cost carriers intensified the original ideas by increasingly aggressive growth strategies (Gillen & Lall, 2004). In Europe there are currently over 34 low-cost airlines, compared with 13 in the USA, 5 in Asia and 3 in Australasia. In contrast to traditional airlines, low-cost airlines show high growth rates and financial prosperity (e.g. easyJet passenger numbers increased by 79% between 2002 and 2003 (easyJet, 2003)). In the UK, about 20% of all passenger-miles flown are on low-cost carriers (Royal Commission on Environmental Pollution, 2002).

Table 4.6 World arrivals by mode of transport

	International tourist arrivals (million)				*Market share (%)*		*Growth rate (%)*		*Average annual growth (%)*
	1990	***1995***	***2000***	***2001***	***1990***	***2001***	***00/99***	***01/00***	***1990–2000***
Air	161.1	207.0	275.9	269.4	35.3	39.4	8.5	– 2.3	5.5
Road	236.5	284.7	342.7	345.9	51.9	50.6	5.9	0.9	3.8
Rail	22.0	16.9	19.7	20.3	4.8	3.0	6.8	3.1	– 1.1
Water	29.3	39.7	46.7	46.2	6.4	6.8	3.7	– 0.9	4.7
Unspec.	7.0	2.0	2.3	2.2	1.5	0.3			
Total	455.9	550.4	687.3	684.1	100	100	6.8	– 0.5	4.2

Source: UNWTO (2003a)

Low-cost carriers often receive incentives from regional governments to stimulate regional economic growth. The airlines themselves apply a variety of strategies and business practices that explain their success (e.g. Gillen & Lall, 2004):

- Use of secondary airports that charge airlines less for using their services; they are also less congested and costs associated with delays are lower.
- Focus on short-haul routes (less than 1500 km) and point-to-point traffic.
- Operate only one aircraft type to reduce maintenance costs and maximise crew flexibility.
- Minimum or no in-flight service offered (less operating costs for food and staff) or charging for service.
- High-density arrangement of seats in one single cabin (i.e. no business class) allowing higher capacity.
- Decreased turnaround times of aircraft (e.g. because no catering is needed or toilets are only emptied on demand).
- Internet booking and electronic ticketing.
- Outsourcing of ground operations staff, making cuts and expansion easier and decreasing the number of staff.
- Low-cost airlines typically operate only one maintenance base.
- Pay lower wages on more flexible employment conditions; many employees are paid on a commission basis.
- No airpoint schemes.

The marketing concept of low fares has reshaped the aviation and wider tourism industry by dramatically improving accessibility and affordability for a large part of society. To a large extent the lower airfares have stimulated new demand (from low-yielding customers), but they have also redirected regular customers away from traditional carriers (Franke, 2004). For example, since the introduction of Ryanair in 1986, demand on the London–Dublin sector has quadrupled. At the same time, however, yield of airlines dropped to one-quarter (Franke, 2004). The expansion of low-cost airlines and increasing competition put pressure on traditional airlines. These adapted, for example, by copying some of the strategies listed above on their short-haul services, while maintaining full service on intercontinental flights. Air New Zealand, for example, introduced its Express Class – minimum service, no business class and electronic booking and check-in – on domestic flights, and because this proved so successful a similar concept was developed for trans-Tasman flights to Australia (Tasman Class) and flights to South Pacific Islands (Pacific class). Other airlines were less successful and either went bankrupt (e.g. Sabena) or were bought by other airlines (e.g. KLM by Air France).

The generation of new demand through substitution of cheap air travel instead of surface travel has important ramifications for GHG emissions (see also Chapter 6). The obvious price-sensitivity of at least a proportion of air travel passengers indicates that current growth trends could be reversed by increased airfares, for example as a result of emission taxes. Potentially, this has significant socioeconomic consequences both for tourists (who have got used to the privilege of increased mobility) and tourist destinations, especially in the case of developing countries where tourism is often considered the most important sector for foreign exchange earnings.

Information technology

The example of low-cost airlines and their extensive use of the Internet for bookings highlights a general trend where tourism is becoming more directly available to consumers, circumventing the involvement of the traditional distribution channels of tour operators and travel agents. The direct and efficient communication line between providers – individual operators as well as tourist destinations – and consumers offers new opportunities for information flow and flexibility in both production and consumption. The Internet, therefore, facilitates speedy communication at a negligible cost. The desire for fast, individualised and flexible information flow forms part of wider societal developments as described further below, and is also increasingly important under climate change scenarios. In this information technology lies a challenge and opportunity for the tourism industry to provide in-depth information to increasingly experienced tourists in a competitive way.

Information on the weather or climate is typically tailored to a particular purpose of the user. For example, winter tourism operators need information on the length of the snow season and the altitude of the snow line. In contrast, individual skiers would require information on the probabilities that a sufficient depth of snow will exist at a particular location and time. The type of weather and climate information that is required by tourists, resort operators and travel companies includes:

- general climate conditions (e.g. climatological averages) used for selling a resort;
- seasonality information (e.g. length of snow cover for skiing conditions);
- real-time weather information (e.g. wind conditions for sailing);
- tailored weather forecasts for tourism businesses (used for short-term planning to cope with fluctuations in numbers as a result of differing weather conditions);
- extreme weather warnings (e.g. tropical storm forecasts);

- information on conditions that may impact upon health (e.g. UV index, heat indices, tourism comfort indices (see Chapter 6));
- climate change information to assess how opportunities or risks may change.

Different types and sources of information on the weather and climatic conditions are needed at different times and levels of urgency. In some situations, tourist operators or destinations require information immediately, for example in the case of an extreme event when evacuation or emergency plans have to be implemented. Similarly, airlines require weather information within 6 h of a flight, for example in relation to take-off temperatures, wind direction and speed and other turbulences on alternative route paths (Altalo *et al.*, 2002). The Internet or warnings distributed by the Meteorological Service are the most immediate sources for this information.

Other information, for example on the weather for the next days or week, is not as urgent, and can be retrieved from other sources, for example newspapers or weather channels. These weather forecasts are used for the planning of outdoor activities or staff required to cater for weather-dependent activities. Beyond this short-term information, weather forecasts or climate predictions for months or years are useful for wider planning, for example the development of more resorts. Airlines require detailed information on climatic conditions at potential new airport hubs, for example relating to the number of foggy days or high wind speeds (Altalo *et al.*, 2002). As shown in Figure 4.3, the uncertainty associated with weather information and forecasts increases with the lead time. Improving the accuracy of climatic forecasts – both short- and long-term – increases the profitability of tourism businesses, simply because disruptions to the business can be managed better.

Given the opportunities for immediate exchange of information, the Internet and other related technologies offer the potential for virtual travel and telecommunication in the business segment. Substituting those technologies for travel is an increasingly discussed pathway for reducing GHG emissions from tourism. Clearly, the International Congress and Convention Association recognised the challenge: 'Virtual meetings will eliminate some meetings. Since 9/11, web meetings and other forms of meeting online have blossomed. Although there will continue to be a strong demand for face-to-face meetings, there are many inefficient meetings that will be eliminated. The cost savings in terms of travel/time out of the office and the shorter turn-around time will be too great to ignore. Fewer meetings also mean fewer planners needed' (Ball, 2004). Conversely, it has been argued that recent technological developments in telecommunications will not be a substitute for personal interaction, rather there will be an increasing need for personal exchange

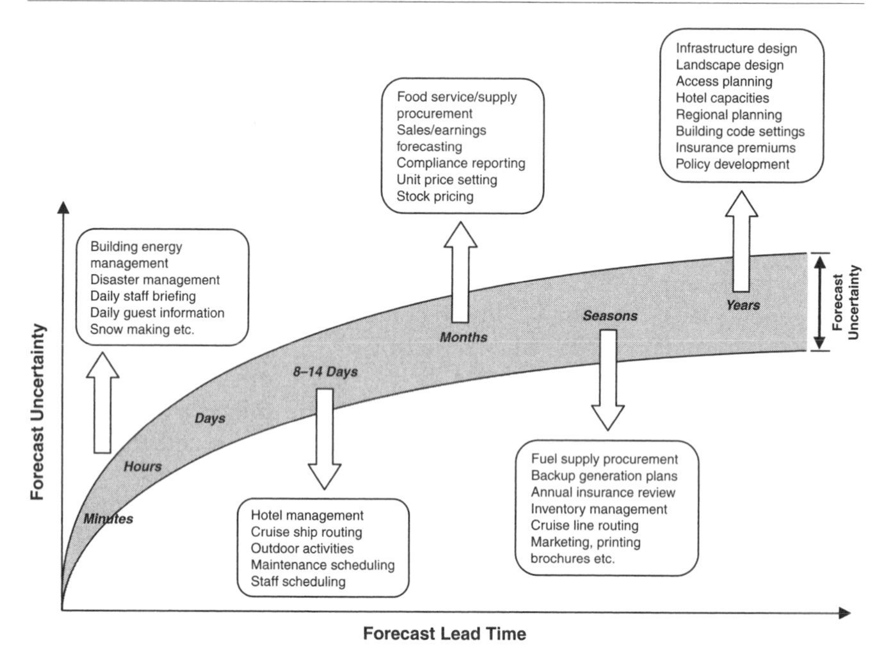

Figure 4.3 Aiding planning of tourism operations by providing information on the weather/climate
Source: Altalo *et al.* (2002)

(Rayman-Bacchus & Molina, 2001). Similarly, the availability of travel information seems to fuel the general public's appetite for travel even further.

Societal trends and changing consumption patterns

Tourism growth and changing demand for tourism products have been explored widely in a socioeconomic context (e.g. Shaw & Williams, 2002). Great differences exist in tourist behaviour across different age groups, gender and ethnicity, and for different socioeconomic groups. An increasing number of people in Western societies have sufficient disposable income and the free time to participate in travel. In Germany, the most important country of origin for international travel, the percentage of the population who took a holiday at least once a year increased from 15% in 1954 to 71% in 2001. The number of days spent on holiday increased as well from 16 days in 1961 to 32 days in 2001. Germans spend their free time on multiple holidays and short breaks, and increasingly in faraway destinations rather than in the traditional destinations in the Mediterranean (Kreisel, 2003).

The trend of ageing societies has major implications for tourism. The OECD (1998) reported that the next 25 years pose a major challenge for OECD countries, given that the number of persons of pensionable age will rise by 70 million, while the working-age population will rise by only 5 million. This means that the workforce as a percentage of the total population will decline dramatically from 2010 onwards. There are contradictory arguments about the impact of ageing societies on tourism. The OECD is concerned that as the workforce decreases relative to retired people it will be difficult to maintain wealth in developed countries. Difficulties in providing adequate pensions and a general decrease in wealth would reduce the ability of people to travel. On the other hand, it is commonly noted that there is an increasing segment of wealthy retired people who take multiple holidays or maintain second homes, for example in seaside resorts. At present, European senior tourists travel one or two times a year, they spend more than non-senior tourists, and preferably stay in middle- or upper-class hotels. Tretheway and Mak (2006) found that both the propensity to travel and spending by Canadian citizens peak for those aged 45–54. High spending rates persist until the age of 75, when travel activity drops markedly. From the age of 55 there is also a shift in spending towards packaged travel.

Time of travel is not restricted for the senior market, and as a result older people travel more often outside the peak season (July and August) than other tourists (Van den Broeke & Korver, 2003). Overall, the senior market offers great potential for tourist destinations in terms of overall visitor arrivals, widening the season and offering products that are less dependent on climatic factors, in particular sunny weather. The senior market, and in particular retiring baby-boomers, will demand products with a focus on convenience and security, for example in relation to transportation, and experiences that contain learning elements and cultural interactions.

Access to tourism is not equal for the different classes of society (see Text Box 4). This relates among others to the number of trips taken per year, number of days spent away and frequency of international trips. Professionals, for example, are taking more holidays (often in the form of multiple short breaks) and are more likely to travel abroad than people from the working class. The trend of an increasing number of women in the workforce has implications for travel behaviour because employment reduces available free time but at the same time increases disposable income. Moreover, the trend whereby older parents have their children later than ever may influence the types of products they demand, for example high-quality accommodation with a wider range of recreational facilities that cater for children.

Often, leisure or tourist activities are influenced by the specific work–leisure relationship: either a tendency for a close link between work

Text Box 4 Hypermobility – high-frequency, long-distance travellers
Contributed by Stefan Gössling, Department of Service Management, Lund University, Sweden

Tourism in industrialised countries has changed substantially in recent years, with a trend towards more frequent, but shorter trips to more distant locations. Within Europe, this development is characterised by the emergence of low-fare carriers, offering, for example, a wide variety of city-break day trips. Globally, an increasing number of people travel to distant or peripheral destinations, often for a considerably short period of time. Clearly, these developments are facilitated by air travel, which, over the past 45 years, has turned from a luxury form of mobility for the wealthy few into a contemporary form of 'hypermobility'. Hypermobility, defined here as mobility that is frequent and often long-distance, is a result of the growing network of airports, perceived cheap fares, better education, higher income and more leisure time, including opportunities to leave work for longer periods. Mobility might generally transform social identities towards cosmopolitan ones, as localities and their characteristics lose importance with respect to space, time and memory (Urry, 1995). Travel, both for work and tourism, thus creates cosmopolitan persons who are increasingly at home anywhere in the world. Ultimately, mobility might thus be seen as a self-reinforcing process finally leading to hypermobile travel patterns (cf. Gössling, 2002b).

As yet, little is known about hypermobile travellers. Statistically, only 4.6% of the world's population participate in international air travel (in 2003; UNWTO, 2005), but there is evidence that a minor share of these accounts for a large part of the overall kilometres travelled. In one case study of international tourists in Zanzibar, Tanzania, Gössling *et al*. (2005a) found that the average distance flown for leisure in 2002/2003 was 34,000 pkm per tourist, excluding the trip to Zanzibar ($n = 252$). The 10 most frequent travellers in this case study had covered almost 180,000 pkm each for leisure travel in 2002/2003, with a maximum of 24 countries visited. Another case study in Mauritius yielded similar results (Gössling *et al*., 2007). In this sample, the average distance flown for leisure in 2003/2004 was 21,436 pkm per tourist, excluding the trip to Mauritius ($n = 242$). Both case studies suggest that there is a minor share of hypermobile leisure travellers, who are both responsible for a major share of the distances travelled as well as for the environmental impacts caused. Demographic characteristics of the samples suggest that these travellers are 20–50 years old, well educated and wealthy, while their awareness of environmental problems caused by energy-intense lifestyles is low.

interests and attitudes and leisure (spill over relationship) or the opposite, where work and leisure are seen as independent and non-complementary (compensatory relationship). The former is often typical of professional or managerial workers, whereas the latter can be associated with manual occupations or mass production. An increasing number of people have a large workload and take less time off. When they take a holiday – often in the form of short breaks – their expectations are therefore high. Many people need to remain in touch with their business while on vacation, which might have implications for the services provided, for example in the arrangements of holiday apartments (e.g. Internet connection in hotel rooms).

In developed countries an increasing number of individuals pursue lifestyles that are no longer need-driven (e.g. satisfying basic needs such as food), but that are motivated by the need for esteem and status (outer-directed) or experiences (inner-directed) (see Mitchell (1983) for a classification by 'values and lifestyles', in Shaw & Williams, 2002). In the UK, for example, it was found that the inner-directed type is growing most rapidly (Shaw & Williams, 2002). This means that an increasing number of consumers become more sophisticated and skilled in their purchasing behaviour in general and in relation to leisure activities in particular. This movement has also been described as post-modernism or post-Fordist forms of consumption, and manifests for example in a move away from mass tourism towards more specialised forms of tourism, as well as an increased focus on activity-based and health-related tourism. Post-modern tourists seek individuality, experience, sophistication, authenticity and possibly more environmentally friendly products. An increasing research body on market segmentation reflects this trend and the desire by researchers, marketers and managers to understand the increasingly diversified market. Examples of such analyses include segmentation based on participation frequencies in different recreational activities (Hsieh *et al*., 1992; Tatham & Dornoff, 1971), the selection of theme attractions (Fodness & Milner, 1992), leisure motivations in tourism (Ryan & Glendon, 1998) and entertainment preferences in the hospitality sector (Jurowski & Reich, 2000).

Responsible tourism

There are a number of terms that describe tourism that is not harmful to the visited environment and communities. Examples include: sustainable tourism (see Figure 4.4 for a definition by the UNWTO), ecotourism (see Ceballos-Lascurain, 1996), responsible tourism, green tourism, ethical tourism and so forth. It is not the purpose of this section to provide a discussion of these different concepts. As Hunter (2002: 11) argued, it is '... yet unclear as to the most basic purpose of ST [Sustainable Tourism] thinking: is ST about creating the conditions

UNWTO (2004) defines that sustainable tourism should:

- *Make optimal use of environmental resources that constitute a key element in tourism development, maintaining essential ecological processes and helping to conserve natural heritage and biodiversity.*
- *Respect the socio-cultural authenticity of host communities, conserve their built and living cultural heritage and traditional values, and contribute to inter-cultural understanding and tolerance.*
- *Ensure viable, long-term economic operations, providing socio-economic benefits to all stakeholders that are fairly distributed, including stable employment and income-earning opportunities and social services to host communities, and contributing to poverty alleviation.*

Figure 4.4 A definition of sustainable tourism provided by the UNWTO

whereby tourism activity can survive over the long-term for its own sake, or is ST about how tourism can best contribute to the more general goals of SD?' Hunter also notes that a tourism-centric view rejects the possibility that another economic sector might contribute better to the goals of sustainable development than tourism.

Table 4.7 provides examples of weaker (light green) or stronger (dark green) advocacies of sustainable tourism. The concept of a spectrum from weak to strong applies to both the supply and the demand side. A number of studies described a growing segment of tourists who wish to

Table 4.7 Exemplified positions for light and dark green variants of sustainable tourism

Light green sustainable tourism	*Dark green sustainable tourism*
Advocacy of new product development	Cautionary approach to new products
Benefits of tourism generally assumed	Benefits of tourism must be demonstrated
Tourism products must evolve according to market need where nature is a commodity	Natural resources must be maintained and tourism products are tailored to minimise impacts
Industry self-regulation (voluntary) as dominant management approach	Wide range of management approaches; legislation and enforcement
Advocates are often involved in tourism	Advocates are often involved in environmental management

Source: after Hunter (2002: 11)

Text Box 5 Are tourists sensitive to GHG emissions?

A New Zealand study indicated relatively low awareness and knowledge of climate-related impacts of air travel on the part of tourists (Becken, 2004b); however, the same study revealed substantial potential for tourists to participate in carbon offsetting schemes that help mitigate their GHG emissions. Tourists indicated that they were interested in participating in tree-planting schemes or similar conservation-related initiatives for reasons such as 'feeling good' or 'giving something back to nature'. Tourists distinguished between their personal responsibility for GHG emissions during their travel (something extraordinary away from home) and their everyday life. At home they are more likely to consider environmental factors in their decision making. Spaargaren's (2003) also found that people apply different principles and rules to various segments of their lifestyle. Gössling *et al.* (2005a) noted tourists' greater awareness of local environmental impacts (e.g. littering) compared with global climate change.

Several studies revealed that people assess risk in terms of 'net benefit', rather than the environmental impacts resulting from an activity (Löfstedt, 1991; McDaniels *et al.*, 1996). For example, people perceive the pollution from vehicle emissions as less severe, because they associate considerable personal benefits with using a vehicle. As holidays are typically undertaken with the goal of deriving personal benefits, it is conceivable that environmental risks associated with holidays are underestimated or denied, which would make any changes in travel behaviour very difficult. The phenomenon of people acting for their personal benefit but against the good of society is commonly referred to as the 'tragedy-of-the-commons' (see also Stoll-Kleemann *et al.*, 2001).

Having more information on environmental impacts could be a precursor for pro-environmental action (O'Connor *et al.*, 1999). What people know about climate change is strongly influenced by its representation and the discourse that surrounds it. As personal experience with the phenomenon is hardly existent, the media play an important role in filling this gap (Corbett *et al.*, 2002). Climate change reporting by journalists often lacks accuracy, is subject to distortions and bias, and tends to misrepresent the (scientific) uncertainty around climate change (Zehr, 2000). In the context of sustainable consumption, and in particular in relation to global problems, however, it has been found that boosting the levels of information and awareness often only increases the level of helplessness and lack of individual control, and therefore could result in reduced individual action (Jackson, 2005).

influence the types of products and services offered on the market by making 'responsible' decisions; i.e. decisions that consider long-term effects of travel (Text Box 5). These consumers belong typically to higher socioeconomic strata with a high disposable income; are well educated, often inner-directed and generally concerned about the environment and the sustainability of the products they consume.

There are many private sector initiatives (e.g. International Hotels Environmental Initiative) that relate to sustainable tourism, including those relating to energy consumption and GHG emissions. Self-regulation and voluntary initiatives seem to be the preferred options by tourism businesses, and command-and-control options are usually not well received by operators.

The Tour Operators Initiative (TOI) is a good example of an initiative relating to a specific subsector. The TOI has been established in cooperation with UNWTO, UNEP and the United Nations Children's Fund (UNICEF). TOI acknowledges that tour operators have a particularly important role in the context of energy use and GHG emissions, because 'they influence consumer demand, destination development patterns, and their supplier's performance, as well as tourists' behaviour' (TOI, 2002). Another sector-specific initiative is the 'Keep climate cool' campaign by the National Ski Area Association (2004), USA (Figure 4.5). This campaign followed the 'Sustainable Slopes' Charter (2000) that identified climate change as a potential threat to the environment and the ski industry. A climate change policy was adopted with the aim of raising awareness of the potential impacts of climate change on the ski business; reducing the industry's GHG emissions; and encouraging others to take action as well.

Despite those environmental initiatives, the consumption of energy and resulting GHG emissions remain a major challenge for sustainable tourism. This is particularly evident for 'ecotourism', which often involves substantial mobility. For ecotourism, a travel component can occur at three distinct scales: first, transport directly associated with the ecotourism experience, for example a boat trip around an ecotourism site; second, travel between various sites or operations; and third, transport from the home location to the destination, where the ecotourism experiences takes place. Ecotourism often occurs in remote areas, in developing countries or in island states (Gössling, 2000), while ecotourists tend to originate from Western countries, typically Europe or North America. This means that this third travel component often requires international long-distance air travel.

The sustainability of ecotourism has to the present been largely assessed on a local level, focusing on directly visible effects. However, as Simmons and Becken (2004) demonstrate for the destination New Zealand, it may be that ecotourism's more menacing effects lie in those

Figure 4.5 Keep winter cool campaign
Source: National Ski Area Association (2004)

that remain invisible. A typical New Zealand ecotourism itinerary covers 3773 km and results in a total transport energy use of 6388 MJ and concomitant release of 430 kg of CO_2 per tourist. These emissions are about the same as the one-way flight for Australian visitors to New Zealand. Visitors from Great Britain emit about six times the emissions of their destination-based travel on their one-way flight to New Zealand. This compares unfavourably to so-called 'mass tourism', for example as in the case of all-inclusive resorts in Santa Lucia, Caribbean (UK CEED, 1998). This study found that one guest-night is equivalent to about (only) 109 MJ (own calculations derived from figures provided by UK CEED). In addition, tourists have minimal transport requirements; typically they are picked up from the airport and dropped off at the end of their holiday and occasionally tourists undertake excursions, typically in collective transport modes.

Notwithstanding the large carbon footprint of ecotourists, they might be the most susceptible to change messages, for example engaging in less frequent, less extensive but longer travel, and supporting green initiatives and technologies. Some destinations develop alternative regional tourist routes (Briedenhann & Wickens, 2004) to encourage tourists to travel slowly (i.e. shorter distances per day) and to spend more time in the regions, hence minimising environmental impacts and maximising economic benefit. Other initiatives of 'greening' a tourist destination have been reported, for example, for Kaikoura, New Zealand, which obtained Green Globe 21 destination benchmarking status in 2002 (McNicols *et al.*, 2002).

It can be argued that consideration of global problems, such as climate change, forms part of what has been described as 'dark' green tourism. When climate change impacts are included as a factor in sustainable tourism, some forms of tourism have to be questioned or simply declared unsustainable. The dependency of most forms of tourism on motorised transport (and resulting GHG emissions) is a major impediment to achieving sustainable tourism. Few forms of transport meet the definition of sustainable mobility as proposed by the World Business Council for Sustainable Development (in Air Transport Action Group, 2002): 'Sustainable mobility can be defined as the ability to meet society's need to move freely, gain access, communicate, trade and establish relationships without sacrificing other essential human or ecological values today and in the future.'

Conclusions

Tourism has become an integral part of many people's lives. It is also a very important economic activity in both industrialised and developing countries. Technological shifts, wider societal trends, changing consumption patterns and the emergence of new tourism destinations and generating markets have major impacts on the nature and distribution of tourism, and accordingly its impacts on the global climate as well as vulnerability to climatic changes. Some of the trends outlined in this chapter will lead to an increase in total tourist mobility (especially if growth in air travel continues), growing hedonism and 'egotourism' (Wheeler, 1993). This results in greater use of resources and an increasing contribution of tourism to global GHG emissions. Current dependencies on tourism for economic development may hamper wider considerations of sustainable development, in particular those that take climate change impacts into account.

However, there is also potential for emerging markets, products and destinations that are more sustainable and better adapted to changes in climatic factors. It is possible that the tourism industry as a whole could adapt to climate change by developing a wide range of specialised products and suitable destinations. To be sustainable new products need to be of a low-carbon nature. The key challenge lies in developing low-carbon transport options or changing current transport behaviour. Highly seasonal mass tourism products could be substituted for more sustainable alternatives, possibly closer to home. Information on weather, climate and climatic change can assist in this process.

The following chapter will explain what climate change is and what it means for tourism.

Chapter 5

Global and Regional Climate Change

Key Points for Policy and Decision Makers, and Tourism Operators

- Human influences on the climate are predominantly through a process known as radiative forcing, resulting in global climate change. The latter is often referred to as 'global warming' despite the changes being much more diverse and complex than a simple increase in the global mean temperature.
- The largest single contributor to radiative forcing is the increased atmospheric concentration of CO_2 as a result of the burning of fossil fuels. The tourism sector is a major user of such fuels, for transport as well as heating, cooling and lighting. Overall, tourism contributes some 5% of the global emissions of gases that result in radiative forcing and hence climate change.
- The tourism sector is already experiencing disruptive changes consistent with many of the anticipated consequences of climate change, including abnormally high air and ocean temperatures and sea levels, and increased destructiveness of tropical cyclones.
- Ozone depletion and global warming are linked, but they result from very different processes, namely chemical destruction of ozone, a highly important atmospheric constituent, and increased atmospheric concentrations of GHG.
- The story lines underpinning climate projections combine futures that range between strong economic values and strong environmental values on the one hand and between increasing globalisation and increasing regionalism on the other.
- Because of the diversity and complexity of climate-determining processes, and despite the considerable and growing intricacy of global climate models, it is still not possible to simulate the full range of interactions between and within the components of the global climate system; the resulting uncertainties regarding future climates are exacerbated by knowledge gaps.
- Increases in temperature are projected to occur at high northern latitudes and over land, with relatively less warming over the southern oceans and the North Atlantic.

- Precipitation will generally increase in the humid tropics and at high latitudes, but decrease in the subtropics, where precipitation amounts are already relatively low.
- Considerable research has been undertaken into the response of El Niño to global warming. Despite this, it is not yet possible to make any conclusive statements about future changes in the frequency, duration and intensity of El Niño events and the weather and climate patterns they influence.
- On the other hand, there is high confidence that the most immediate and more significant consequences of radiative forcing are likely to be changes in the nature of extreme events (e.g. flooding, tropical cyclones, storm surges, heatwaves) and climatic variability (e.g. droughts, and prevailing winds accelerating coastal erosion).
- One manifestation of uncertainty is the ability of the climate system to undergo abrupt and pervasive changes; such changes are termed 'surprises', and present a significant challenge to tourism and other policy makers and planners.

Introduction

This chapter considers the multiple causes of climate change, including those related to tourism. The initial focus of the chapter is on the globally dominant change processes. In addition to 'global warming', the roles of ozone in changing the climate for tourism are discussed briefly. The consequences of all these changes for the tourism sector are then discussed. Emphasis is placed on changes in climate variability and extremes, as such changes are likely to be the main drivers of the increase in climate-related risks to tourism, at least in the near term.

Future large-scale interactions between tourism and climate are fundamentally dependent on how tourism itself evolves in the coming years and decades as well as on the changes in the climate of the major tourism destinations. Thus the chapter includes a review of the scenarios that underlie the evolution of both tourism and climate in the foreseeable future. The projections are used as a basis for describing the anticipated changes in the global climate and their relevance to tourism regions. The chapter concludes with a discussion of two important issues, namely uncertainty in the climate projections and the related possibility that changes in the climate may involve 'surprises' – that is, changes that are substantially different to those that are anticipated in the normative projections.

The Causes of Global Climate Change

This section builds on concepts and information presented in Chapter 2, including the greenhouse effect and the increased concentrations of

GHGs in the atmosphere as a result of human activities. Because most GHGs have long atmospheric lifetimes, and are thus relatively well mixed, concentrations measured at a few sites (e.g. Figure 5.1) reflect the global emissions, and demonstrate how levels have risen dramatically since the Industrial Era began in the mid-18th century.

Human influences on the climate are predominantly through a process known as radiative forcing. Overall, radiative forcing is the process by which human activities modify the natural greenhouse effect, resulting in global climate change. The latter is often referred to as 'global warming'

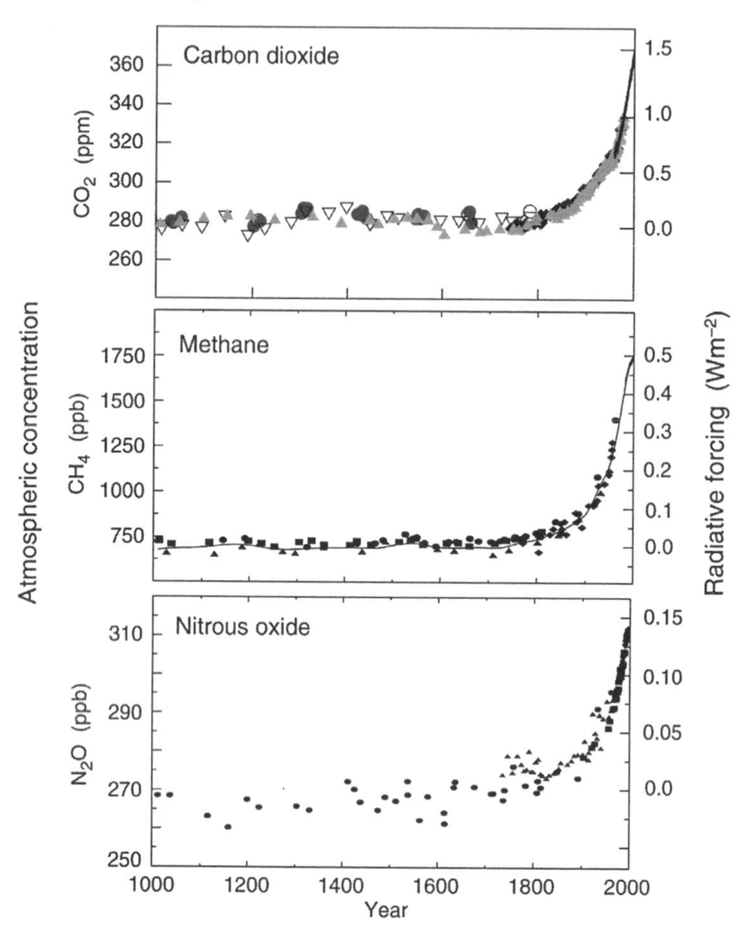

Figure 5.1 Atmospheric concentrations for three GHGs, CO_2, CH_4 and NO_x
Source: IPCC (2001)

despite the changes being much more diverse and complex than a simple increase in the global mean temperature. The largest single contributor to radiative forcing is the increased atmospheric concentration of CO_2 as a result of the burning of fossil fuels. The tourism sector is a major user of such fuels, for transport as well as heating, cooling and lighting, and is thus a significant contributor to the global increase in CO_2 emissions to the atmosphere. Gössling (2002b) estimated that tourism may contribute up to 5.3% of global emissions of gases that cause radiative forcing, with transport accounting for about 90% of this. Aviation alone accounts for 2-3% of the world's total use of fossil fuels and up to 3.5% of all radiative forcing (see also Chapter 7). More than 80% of this is due to civil aviation (Penner *et al.*, 1999). Based on current trends, these impacts are likely to increase significantly – air transport is one of the fastest growing sources of GHG emissions.

The contribution of a specific GHG to radiative forcing of climate change depends on the molecular radiative properties of the gas, the size of the increase in atmospheric concentration and the residence time of the gas in the atmosphere (Table 5.1). This last attribute influences policy options. A GHG with a long atmospheric lifetime will not only be well mixed and thus have global rather than more local consequences, but it also represents a long-term legacy of past and current emissions, the increased radiative forcing, and hence future global warming. A GHG with an extremely long residence time will exist in the atmosphere for millennia, before natural processes remove it.

Policy options are typically assessed using global warming potentials (GWPs) (Table 5.1). These are a measure of the global warming effect of a given substance integrated over a chosen time horizon and expressed relative to that for CO_2. To further facilitate the policy process, the concentrations of GHGs are sometimes converted to CO_2 equivalents (CO_2-e), using the GWP of the relevant gas. Because the atmospheric residence times of some GHGs cannot be determined with reasonable certainty, their GWPs cannot be calculated. The dependency of the GWP on the time horizon reveals how the atmospheric lifetime of a gas influences its contribution to global warming. Moreover, while, for example, N_2O has a high GWP, its contribution to radiative forcing is relatively small. This is due to its comparatively low rate of concentration change (Table 5.1).

Climate Variability and Extremes

Until recently attention has been focused on the gradual, long-term changes in average climatic conditions that will be induced by radiative forcing. However, both observations and modelling studies now indicate that the most immediate and more significant consequences of radiative

Table 5.1 Historical and present concentrations of the main GHGs, their atmospheric lifetimes and their GWPs

	CO_2	CH_4	N_2O	*CFC-11*	*HFC-23*	CF_4
	Carbon dioxide	***Methane***	***Nitrous Oxide***	***Chlorofuoro-carbon-11***	***Hydrofluor-carbon-23***	***Perfluoro-methane***
Pre-industrial concentration (approx.)	280 ppm	700 ppm	270 ppm	zero	zero	40 ppt
Concentration in 1998	365	1745	314	268 ppt	14 ppt	80 ppt
Rate of concentration change (1990–1999)	1.5 ppm/yr	7.0 ppb/yr	0.8 ppb/yr	– 1.4 ppt/yr	0.55 ppt/yr	1 ppt/yr
Atmospheric lifetime	5–200 yr	12 yr	114 yr	45 yr	260 yr	> 50,000 yr
GWP for time horizon of:						
20 years	1	62	275	NA	NA	NA
100 years	1	23	296	NA	NA	NA
500 years	1	7	156	NA	NA	NA

Source: IPCC (2001)

forcing are likely to be changes in the nature of extreme events (e.g. flooding, tropical cyclones, storm surges) and climatic variability (e.g. drought, prevailing winds, accelerating coastal erosion). The tourism sectors of most countries are already experiencing disruptive changes consistent with many of these anticipated consequences of global climate change, including periods of abnormally high sea-levels and sea surface temperatures. The evidence being reviewed by the IPCC for its Fourth Assessment Report, due to be released in 2007, shows important observed changes in climate variability and extremes (Text Box 6).

Text Box 6 Recent changes in the observed climate

(a) Average climate

- The rate and duration of the warming of the 20th century is larger than at any other time during the last 1000 years.
- Global average surface temperature has increased by around 0.8°C since the late 19th century; warming since 1979 has been 0.3°C per decade.
- The 1990s was the warmest decade of the millennium in the Northern Hemisphere; 2005 is one of the warmest two years on record, the other being 1998, when surface temperatures were enhanced by El Niño; 10 of the last 11 years are the warmest on record.
- Minimum and maximum land surface temperatures have increased by 0.2 and 0.1°C per decade, respectively, resulting in more frequent warm nights but fewer cold nights.
- Total cloud amounts over Northern Hemisphere mid- and high-latitude continental regions have increased by about 2% since the beginning of the 20th century.
- Globally, dry areas have tended to become drier and are expanding.
- Continental river run-off has increased in the 20th century, due to both climate warming and the effect of increased CO_2 concentrations.
- Glaciers have retreated in many mountain areas around the world; in recent decades there has been a worldwide decrease in the extent of snow cover and depth in spring, the former by about 10% in the since the late 1960s; less snow at low altitudes.
- While Northern Hemisphere sea-ice amounts are decreasing, no significant trends in Antarctic sea-ice extent are apparent.
- During the 20th century global mean sea level rose by between 1.7 and 1.8 mm/year, with a possible acceleration during the last

decade; this sea-level rise often contributes to observed increases in coastal erosion.

- The Atlantic meridional overturning circulation, which includes the Gulf Stream, which warms Europe, has weakened by around 30% between 1985 and 1995.

(b) Climate variability and extremes

- Higher maximum and minimum temperatures and more hot days; and fewer cold and frost days over nearly all land areas.
- Increased frequency of heavy precipitation events over land, even in areas where the average precipitation has decreased; increased frequency of major floods.
- Increased summer continental drying and associated occurrence of drought.
- Climate warming is having an observable impact on plant and animal populations; warming temperatures have increased alcohol levels and vintage ratings in many wine-growing regions, as well as reducing vintage-to-vintage variations.
- Global increase in extreme high ocean water levels since 1975, related to both mean sea-level rise and to large-scale interdecadal climate variability.
- Rising sea surface temperatures have been linked with observed increases in the frequency of both intense tropical cyclones and coral bleaching.

In many regions, the observed changes in extreme events are associated with shifts in interdecadal and multidecadal climate variability. Tropical cyclones provide one example. Recent work by Emanuel (2005) reveals that an index of the potential destructiveness of tropical cyclones has increased markedly since the mid-1970s, due to both longer storm lifetimes and greater storm intensities. The index is highly correlated with tropical sea surface temperature, reflecting such climate signals as multidecadal oscillations in the North Atlantic and North Pacific, as well as global warming. The annual average number of major hurricanes is proving to be much higher in the current warm phase of the North Atlantic Oscillation than in the previous one – an analysis of global hurricane data from satellites (which is available since 1970) shows that the annual frequency of the strongest hurricanes (categories 4 and 5) has almost doubled over the period 1970–2004. This increase is seen in each of the ocean basins where tropical cyclones occur. The increase in hurricane intensity is coincident with the observed increase in sea surface temperature for the same ocean basins. Significantly, tropical ocean temperatures have increased by more than 0.5°C over the past 50 years, a change unprecedented over at least the last 150 years and perhaps the last several thousand years. The 2005 North Atlantic hurricane season, with 23 storms and 13 hurricanes, broke the 1933 record for 21 tropical storms in one season, as well as the 1969 record for 12 hurricanes in one season. The 2005 hurricane season spawned three of the most intense North Atlantic storms on record, including Hurricane Katrina.

Other extreme events, and the consequences for tourism, are described in Text Box 7.

Text Box 7 Extreme climatic events in the European Alps: a threat for tourism?

Contributed by Martin Beniston, Department of Geosciences, University of Fribourg, Switzerland

As climate continues to change in the 21st century, there is concern that extreme climatic events such as those that have in recent years affected the European Alps (extreme precipitation and floods; heatwaves; unusually mild winters; strong windstorms) may become more frequent and/or more intense. Shifts in weather extremes in a warmer climate may have adverse consequences for tourism (in the case of floods, lack of snow and geomorphologic hazards), but also paradoxically some beneficial effects (notably during heatwaves, when the cooler high elevations of mountain areas could become an attractive alternative to the hot summers in lowland areas).

In recent years, regional climate models have become more sophisticated and now operate at relatively fine resolution. This makes it possible to not only study changes in mean climatic conditions over a given area, but to also examine certain forms of extreme events such as heatwaves or heavy rainfall. There is now a real opportunity to investigate the manner in which extreme events could change in the Alps by the end of the 21st century (Beniston, in press).

Most models agree that by 2100 heatwaves such as the scorching 2003 event that affected much of Western and Central Europe will become commonplace, i.e. one summer in two may be just as extreme as the 2003 event (Beniston, 2004; Schär *et al.*, 2004). The impacts of the 2003 heatwave in the Alps were numerous and essentially linked to the accelerated retreat of many Alpine glaciers as well as to a severe hydrological deficit in many of the rivers that originate in the Alps and supply water to the populated lowlands. The cooler conditions at higher elevations did, however, attract tourists, but there were greater-than-average casualties resulting from a large number of slope instability events that were triggered by the unusually hot conditions, even at higher elevations.

Winter warm spells are a phenomenon that has sharply increased since the early 1970s (Beniston, 2005). Because they are characterised by persistent and very mild conditions, they are accompanied by early snow melt (or late arrival of the mountain snow-pack), leading to potential flood hazards as river basins are often not geared toward large discharge in winter. Such persistent warm spells have very negative consequences for the ski industry for obvious reasons of lack of snow. Many climate models project that by the end of the 21st century there will be an increase in such wintertime events, by about 30% compared to today.

Heavy precipitation is perhaps the most devastating climate-related event in the Alps, not only because of the associated flood hazard but also because heavy rain triggers a wide range of geomorphologic hazards that lead to loss of life and costly damage to infrastructure. On a GDP basis the August 2005 floods in Switzerland were proportionately as costly to the Swiss economy as the Katrina hurricane was to the US economy. Many of the tourism-related attractions in Switzerland, Austria, Italy and France are located in the Alps. Increases in precipitation-related events in the future could lead to heightened awareness of these phenomena and perhaps a loss of tourist revenue for these regions.

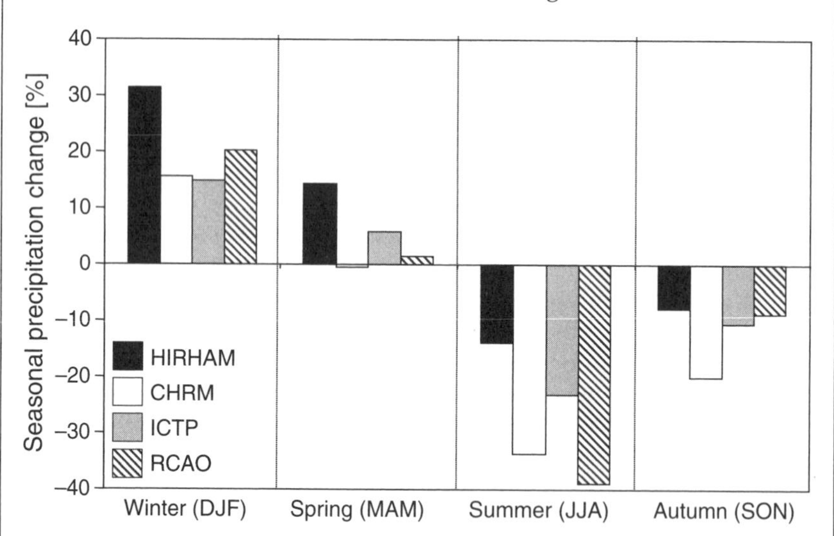

Text Box 7 – Figure 1. Shifts in the seasonal distribution of precipitation in the Northern Swiss Alps between current (1961–1990) and future (2071–2100) climates, according to four regional climate models used in the context of the EU 'PRUDENCE' project (Christensen *et al.*, 2002). All models agree on the sign of change, i.e. increases in winter and spring, decreases in summer and autumn.

Interestingly, while climate models suggest an increase in heavy precipitation in a warmer climate, there is likely to be a seasonal shift in the period when extreme precipitation may occur, namely from late summer currently to spring and autumn in the future (see Figure 1). Because one of the controls on flood events in the Alps is the altitude of the freezing level (and thus the buffering effects of snow versus liquid precipitation run-off), future events may occur in seasons that would still be colder than current summers; under such circumstances, the number of damaging floods would not necessarily increase in the future.

The preceding findings suggest that the future of tourism in the Alps under changing climatic conditions may well be governed by the perception of tourists as they make trade-offs between adverse and more clement weather situations.

Atmospheric Ozone and Climate Change

Another atmospheric constituent, ozone, also plays a key role in climate change. While ozone depletion and global warming are linked, the two are the result of very different processes (respectively, the chemical destruction of ozone, a highly important atmospheric constituent, and radiative forcing due to the increased atmospheric concentrations of potent GHGs).

In the lower atmosphere (the troposphere), ozone is an important GHG. Since the beginning of the Industrial Era ozone levels in the troposphere have risen by some 35% globally. This increase is spatially variable. Importantly, few locations show an increase after the mid-1980s. Ozone is a secondary pollutant, with its formation being influenced by the intensity of sunlight as well as concentrations of precursor gases such as those produced during the burning of fossil fuels. It has a highly variable residence time, due to these same influences. Higher amounts of ozone are observed in cities and other areas where photochemical smog is an issue, but at least in Europe and North America reductions in emissions of precursor gases have reduced this problem. As well as being a GHG, tropospheric ozone has adverse effects on crop production, forest growth and human health. At high concentrations ozone is toxic to many life forms.

The ozone that occurs naturally in the upper atmosphere (the stratosphere) filters out the very-short-wavelength (ultraviolet) energy emitted by the Sun, thus preventing biologically damaging ultraviolet sunlight from reaching the Earth's surface. A number of manufactured gases containing chlorine and bromine are largely inert at the temperatures and pressures prevailing in the lower atmosphere. However, these gases become highly reactive when they reach the upper atmosphere, aiding in the rapid conversion of ozone to oxygen. This depletion of stratospheric ozone, and the resulting increase in ultraviolet energy, increases the incidence of melanoma and non-melanoma skin cancers and eye cataracts, weakens the immune systems of many animals and plants, reduces plant yields, damages ocean ecosystems, reduces fish catches, and damages plastics and many other materials. A decline in stratospheric ozone levels increases risks for sun-based tourist businesses, as well as causing increased health costs and crop losses. For example, in Florida, where tourism is a multibillion-dollar industry, there appears to have been a significant decline in beach tourism in the last 5 years as a result of concerns related to ultraviolet skin damage. Insurance costs of skin cancer treatment are increasing rapidly. Tourist-related sales in the beach areas of Florida amount to over $10 billion per year, not counting large amounts of uncounted real estate business. Thus even a decline of 1% in tourism spending in Florida translates to hundreds of millions of dollars.

Ozone in the upper atmosphere also acts as a GHG. The decline in stratospheric ozone concentrations over recent decades results in a negative radiative forcing and thus a cooling of the Earth's surface and lower atmosphere. But this cooling effect is more than offset by the positive radiative forcing that is occurring due to the increasing concentrations of the long-lived and globally well mixed GHGs. Importantly, from both policy and practical perspectives, over the next few decades stratospheric ozone levels are likely to recover from the present low values. This is due to the Montreal Protocol, an international agreement that is achieving substantial reductions in the production and use of ozone-depleting substances. As a result, radiative forcing associated with stratospheric ozone is projected to become positive in the immediate future, and therefore begin to add to the greenhouse effect.

The issue of ozone depletion highlights a point that is of importance to tourism policy and decision makers: as the atmosphere is a highly complex system with many feedbacks, policy responses may not always achieve the desired outcomes or choices (i.e. tradeoffs) may have to be made (see Chapter 2). The tourism industry has a direct interest in seeing concentrations of stratospheric ozone restored to natural levels. This will reduce health risks for tourists engaged in sun-based and high-altitude activities. On the other hand, restoring ozone levels will accelerate global change, further increasing the climate-related risks to tourism.

Futures for Climate and Tourism

This section presents an overview of the scenarios, projections (in the case of climate) and forecasts (in the case of tourism) that describe the possible future evolution of both tourism and climate. The scenarios underpin the suggested futures for tourism and climate. We talk of tourism *forecasts* as these are the best estimate of how tourist flows will change over time, while climate *projections* indicate that there is a range of climate futures, depending on the story line that is used. Scenarios are typically defined as stories describing different but equally plausible futures that are developed using methods that systematically gather perceptions about certainties and uncertainties. Scenarios are not intended to be truthful, but rather provocative and helpful in strategy formulation and decision-making (Selin, 2006).

Story lines for climate change

As noted earlier in this chapter, climate change is driven predominantly by changes in radiative forcing and, in this respect, principally by the increasing concentration of GHGs in the Earth's atmosphere. Thus the storylines for climate change focus on the social and economic activities that influence the amounts of GHGs emitted to the atmosphere.

In 2000, the IPCC described a set of four GHG emissions scenarios and their associated socioeconomic drivers. They are based on a set of narrative storylines, which are subsequently quantified using different modelling approaches. The storylines combine two sets of divergent tendencies: one set covers the range between strong economic values and strong environmental values, the other set ranges between increasing globalisation and increasing regionalisation. The storylines are summarised in Figure 5.2.

The CO_2 emissions associated with each of the storylines, and the resulting changes in the atmospheric concentrations of CO_2, are shown in Figure 5.3. The ranges evident in both the emissions and concentrations reflect the substantial uncertainties in future emissions of GHGs. This carries through to uncertainties in projecting future climates and sea levels, globally and regionally. It is also clear from Figure 5.3 that even if major and successful efforts are made to reduce GHG emissions, global concentrations of these gases will continue to increase in the coming decades. Thus we are committed to a changing climate, regardless of any realistic steps that might be taken over the next years and decades.

The storylines that underpin the IPCC scenarios can be made more explicit, for example by including descriptions of the contrasting futures for tourism in a global economy that produces the emissions depicted in Figure 5.3. Such tourism-specific storylines were prepared as part of a 'Tourism Futures' workshop for key tourism stakeholders in New Zealand (see Text Box 8).

Storylines for tourism

The preceding storylines describe the economic and related activities that influence future GHG emissions. Another set of storylines underpin forecasts of how tourist flows will change over time into the future. The most comprehensive forecasts of global tourism have been developed by the UNWTO (2001), which has prepared forecasts of international visitor

A1: A future world of very rapid economic growth, cultural and economic convergence, low population growth, rapid introduction of new and more efficient technologies and pursuit of personal wealth rather than environmental quality. This storyline is further differentiated into three groups that describe alternative directions of technological change in the energy system, based on technological emphasis: fossil intensive (A1FI), non-fossil energy sources (A1T), or a balance across all sources (A1B), where balanced is defined as not relying too heavily on one particular energy source, on the assumption that similar improvement rates apply to all energy supply and end-use technologies.

A2: A differentiated world, with high population growth, less concern for rapid economic development and a strengthening of regional cultural identities.

B1: A convergent world with rapid change in economic structures, "dematerialization", introduction of clean technologies, and an emphasis on global solutions to environmental and social sustainability.

B2: A world in which the emphasis is on local solutions to economic, social and environmental sustainability and a heterogeneous world with less rapid, and more diverse technological change, but a strong emphasis on community initiative and social innovation to find local rather than global solutions.

Figure 5.2 Summary of the narrative storylines for the four IPCC GHG emissions scenarios

arrivals for each of its 44 subregional pairs, adjusting historical growth rates up (or down) over the period to 2020. Despite the many limitations, international tourist arrivals are used because this is the most widely reported category of data, incorporating the most standardised definitions.

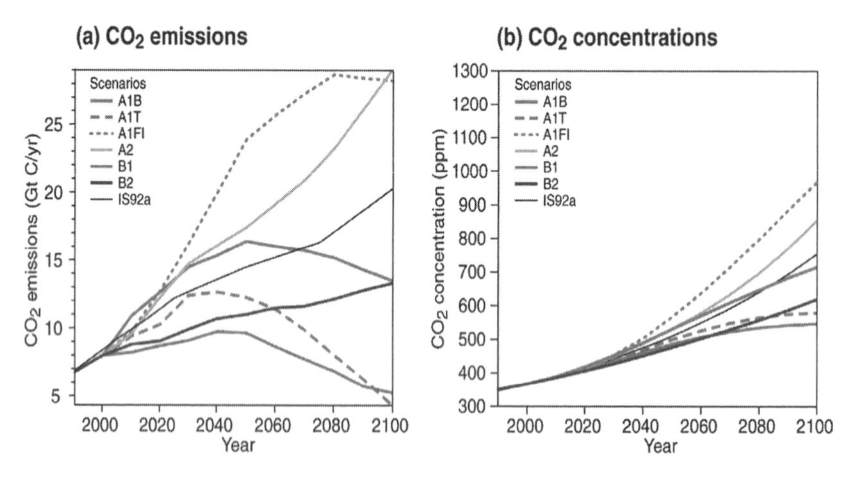

Figure 5.3 Scenarios of CO_2 gas emissions and consequential atmospheric concentrations of CO_2

Source: IPCC (2001)

Text Box 8 Storylines for tourism, based on the IPCC Scenarios
Source: Becken & Frame (2005)

A1: The A1 scenario holds a future of rapid economic growth, and the speedy introduction of new and efficient technologies. Accordingly, tourism will become more sustainable and energy efficient. This is based on technological solutions rather than on environmental awareness. Unless a technological alternative to kerosene is found for aircraft, air travel will continue to contribute to global climate change. The world will be more globally connected and differences between regions will largely disappear; increased cultural and social interactions between regions will take place. Tourist flows will geographically spread more evenly. Globally, there will be more people with a high per capita income, and there will be a bigger market for high-yield tourism. To meet the service and environmental demand of higher-yielding market segments, and to compete internationally, tourism products will need to be of a high quality. It is likely that tourist numbers will grow (despite more expensive air travel). This will lead to skill shortages in the labour market, as well as infrastructural deficiencies. Growth will also have consequences in terms of an increased utilisation of the conservation areas and pressures on natural assets.

A2: The A2 scenario foresees a world in which regions are self-reliant and preserve their local identities. As a result domestic tourism is strengthened while international tourism becomes less significant, especially from distant markets. International tourists will stay longer and often visit for multiple purposes. The reduction of the long-haul and short-break markets will lead to slower economic growth and reduced tourism activity. Regionalisation has implications for tourism products. Each region will develop unique products, and cohesion within regions will increase. Regional Tourism Organisations will be the key organisations for tourism marketing. Technological change will be fragmented and slow, and it is unlikely that alternatives will be available to replace fossil fuels for air travel. Environmental awareness will also be low and not be a reason for people to travel less. As a result of a much more regionalised world, international air travel, especially long-distance, will decrease and passenger transport will be less important in terms of global warming. More transportation of foods will take place, however, to feed a growing global population.

B1: In the B1 scenario the world converges and macro-regional differences become less important. The emphasis is on global solutions to economic, social and environmental sustainability. Global economies will become less material intensive and the focus will shift to service industries. Tourism will be well resourced and sustainability is regarded as a normal aspect of all business operations. Also, tourism will be a leader in terms of new technologies and an advocate for the environment and communities. Cultural and environmental values will be respected. It will become difficult to invest in unsustainable activities. Demanding (possibly older tourists) will put pressure on human resources and there is a risk of skilled labour shortage. The focus on high spenders will increase vulnerability to changes in travel 'fashions'. Increasing travel costs as a result of more expensive fuel will remain a major challenge. Also, unless there is a 'clean solution' for long-distance travel, some environmentally aware tourists will choose not to travel by air.

B2: There will be an emphasis on local solutions to economic, social and environmental sustainability. Economic development is less rapid but diverse. Environmental protection and social equity play an important role. International long-distance travel will become more difficult: transport will be exposed to major external shocks, fuel shortages, carbon charges and national climate policies. As a result, domestic tourism will become increasingly important and there will be a need for more regional airports. Marketing by Regional Tourism Organisations will increase in importance compared with national marketing. A region's unique environments are key assets and will be protected as the major base for tourism. Social structures in communities will be strong and sustainability is a key goal. Tourists will be environmentally conscious and the market might be two-tiered: 'high end travellers versus domestic tourists/budget visitors'.

The storylines that underpin the UNWTO forecasts described in the following section are summarised in the following:

- Economic factors will remain positive for tourism as moderate-to-good rates of global economic growth will continue. Information and communication technologies will further transform both supply and demand. Transport technologies will continue to advance, with ever larger capacity aircraft and improved availability of aircraft that permit better 'matching' of aircraft types according to demand.
- There will be some success in addressing problems related to congestion, delays, pollution and safety. Barriers to international travel will be reduced by fewer countries requiring visas and replacement of passports by technology-driven systems of personal identification.
- Effective protection of tourists from conflict, disasters and risks to their health and security will be in place. Tourism will continue to

show that it is very resilient and manifests a strong capacity for recovery from civil turmoil, war, disasters and tourists' health or security being compromised.

- The trends of ageing populations and growth of contracting workforces will persist in the industrialised countries, as will continued growth in tourism by older market segments, and increased travel between the north and the south by the large volume of migrants in industrialised countries, for the purpose of visiting family and friends. Further erosion of the traditional Western household through rising divorce rates and later marriage and family building will result in a diversification of travel segments. The growing power of international economic and market forces will further reduce the ability of individual states to control their economies, as well as the ability of private corporations operating in limited geographic spheres of operation to dominate domestic markets. However, at the local scale, and particularly in developing countries, the conflict will intensify between identity and modernity.
- The public awareness of sociocultural and environmental issues is likely to grow in the coming years, in part as a result of increased media reporting on major problems such as threat to rainforests, pollution, global warming, coral reef bleaching and issues like the dwindling water supplies worldwide.
- Growing urban congestion in both the industrialised and developing worlds will increase the desire to engage in discretionary tourism to escape and/or to indulge. Changing work practices will result in more but shorter holidays, and less distinction between work and leisure time.

Most of the factors described above operate in combination, and collectively will produce a polarisation of tourist tastes and supply. Over the coming decades large-scale 'mainstream' tourism, involving the movement of large volumes to extensively developed destinations, will prosper alongside smaller-scale, 'individualised' tourism – alternative or new tourism (refer also to Chapter 4).

The UNWTO forecasts 1.6 billion international tourists by the year 2020, approximately 2.5 times the arrivals recorded in the late 1990s. Some 1.2 billion of these arrivals will be intraregional, with the remainder being long-haul (Table 5.2).

With respect to outbound tourism, Europe will remain the world's greatest source of international tourists, being responsible for almost one-half of all tourist arrivals worldwide (46.7% in 2020). East Asia and the Pacific will become the second largest region for outbound travel (405 million in 2020), relegating the Americas to third place. Africa, the Middle East and South Asia will all show above average growth rates. At

Table 5.2 Forecast international tourist arrivals for 2010 and 2020

	Base year 1995	*Forecast 2010 (million)*	*Forecast 2020 (million)*	*Market share 1995 (%)*	*Market share 2020 (%)*	*Average annual growth (%)*
World	565	1006	1561	100	100	4.1
Africa	20	47	77	3.6	5.0	5.5
Americas	110	190	282	19.3	18.1	3.8
East Asia/ Pacific	81	195	397	14.4	25.4	6.5
Europe	336	527	717	59.8	45.9	3.1
Middle East	14	36	69	2.2	4.4	6.7
South Asia	4	11	19	0.7	1.2	6.2

Source: UNWTO (2001)

national level, the main sources of international tourists will not change substantially over the coming decades. The major industrialised countries – Germany, Japan, the USA, the UK and France – will continue to dominate. However, there will be two important additions – China will be at fourth place, producing an estimated 100 million international tourists by 2020, travelling predominantly to close-by destinations. The Russian Federation will generate over 30 million international tourists.

Climate projections

The complex interactions between and within the various components of the global climate system (Chapter 2) make it difficult to predict the response of the global climate system to increased concentrations of GHGs in the Earth's atmosphere. Sophisticated computer-based global climate models are used to estimate winds and ocean currents, the exchange of heat and gases between land, air and ocean, the transport of heat and gases around the Earth, and cloud cover, sea ice and land cover. Such models are computationally expensive – it can take up to two months of computing to simulate one century! Climate models, with their associated scenarios, are used to answer 'what if?' questions. As noted above, we therefore speak of climate *projections* as opposed to climate *predictions*. Divergences between projected climates as a result of using different models (there are at least 23 research centres worldwide that have each developed a global climate model, with its own set of assumptions and other distinguishing features) and scenarios can shed light on the uncertainties that are a consequence of modelling a complex system where there is also imprecise knowledge of GHG emissions and other future conditions, such as land use.

To simplify interpretation of the resulting climate projections it is common practice to analyse and present results for a mid-range emissions scenario and the mid-range estimate determined from the output of several (often 20 or more) global climate models runs for the same input conditions, that is, a multimodel ensemble. This approach will be adopted here, while the last section of this chapter will examine uncertainties and deviations from such normative conditions.

Global changes in mean conditions

In the coming decades, climate change will manifest largely as changes in the frequency and consequences of extreme events (e.g. rainfall and temperature extremes) and of interannual variations (e.g. ENSO), rather than as long-term trends in average conditions. Nevertheless, it is instructive to review the nature of the long-term trends, on which the extremes and variability will be superimposed. The largest increases in temperature are projected to occur at high northern latitudes and over land, with less warming over the southern oceans and North

Atlantic (Figure 5.4). Night-time minimum temperatures are likely to rise more rapidly than daytime maximum temperatures, reducing the average diurnal range in temperature in many areas.

As shown in Figure 5.5, precipitation will generally increase in the humid tropics and at high latitudes, but will typically decrease in the subtropics, where conditions over land are already dry on average. Patterns of change in cloud cover tend to match those of precipitation change, with greatest increases in cloud in the humid tropics and at high latitudes (Williams *et al.*, 2005). As the Earth warms, Northern Hemisphere snow cover and sea-ice extent will decrease and glaciers and ice caps will lose mass. Arctic sea ice cover may become seasonal in the second half of the 21st century (Gregory *et al.*, 2002).

Over the period 1990–2100, global average sea-level is projected to rise by between 0.09 and 0.88 m (IPCC, 2001). This is primarily a consequence of the expansion of sea water as a result of the increase in oceanic temperatures, and of the continued melting of mountain glaciers and small ice caps. The range reflects differences in projections by seven ocean-atmosphere models and in anticipated GHG emissions. The central value of 0.48 m is 2.2–4.4 times the rate of sea-level rise during the 20th century. Due to local changes in heat content, salinity and wind stress, the change in sea-level relative to adjacent land masses is not uniform. Changes in this *relative* sea-level (i.e. ignoring changes in land elevation) are likely to be larger than average in the Arctic and in the Southern Atlantic, Indian and Pacific oceans, and smaller than average in the Southern Ocean.

Global changes in climate variability and extremes

As noted above, and importantly for the tourism sector, in the coming decades climate change will be dominated by changes in the frequency and severity of climate extremes and variations. As ENSO is the most distinctive and significant form of climate variability, affecting weather, ecosystems, economies and societies in many parts of the world (see Text Box 9), there is great interest in how it will respond to a stronger greenhouse effect. ENSO is a result of ocean–atmosphere interactions internal to the tropical Pacific Ocean and the overlying atmosphere. Unusually warm temperatures in the Eastern equatorial Pacific (termed an El Niño, or warm, event) reduce the normally large sea surface temperature difference between the Eastern and Western sides of the tropical Pacific. The resulting eastward shift in the organised convection and rainfall that generally prevails in the Western tropical Pacific – the 'Western Pacific Warm Pool' – and the associated release of energy into the atmosphere resulting from the related condensation, alters the heating patterns of the atmosphere and hence wind patterns, including the mid-latitude jet stream, storm tracks and monsoon circulations. This causes weather anomalies around the world, resulting in societal and economic impacts.

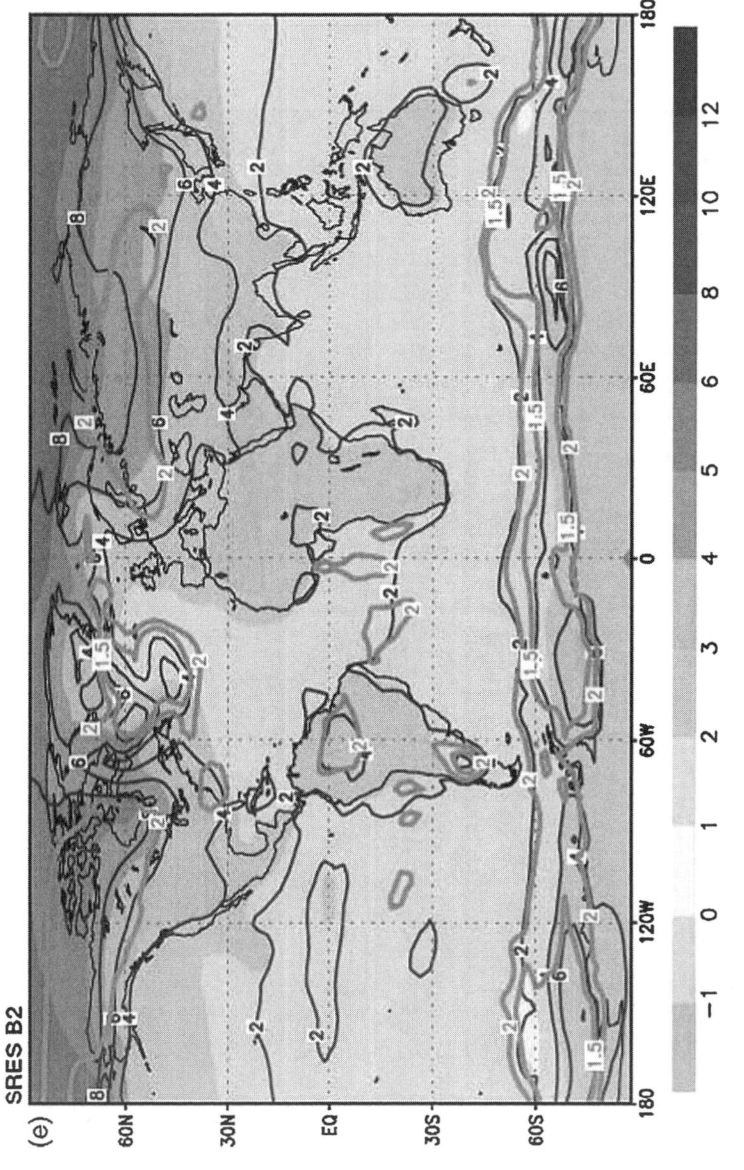

Figure 5.4 Multimodel ensemble annual mean change of temperature for the period 2071–2100, relative to the period 1961–1990, for the B2 emissions scenario

Source: IPCC (2001)

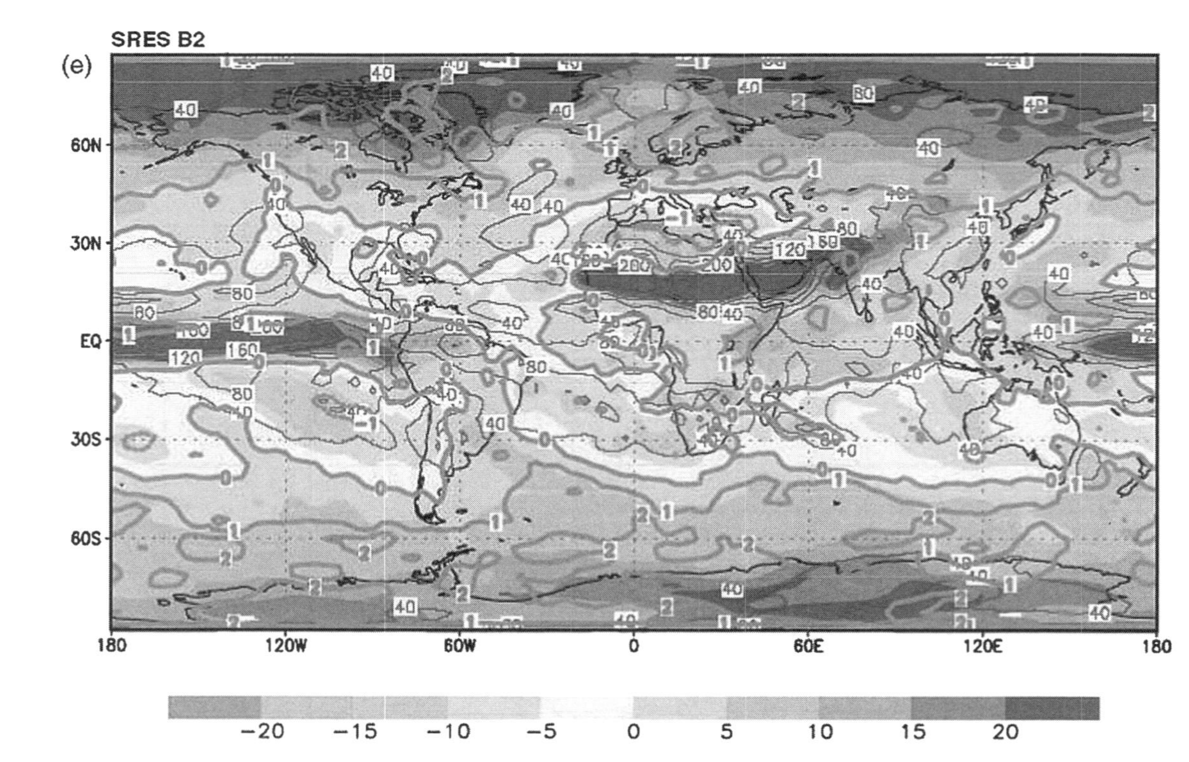

Figure 5.5 Multimodel ensemble annual mean change of precipitation for the period 2071–2100, relative to the period 1961–1990, for the B2 emissions scenario

Source: IPCC (2001)

Text Box 9 El Niño and tourism

One of the first apparent consequences of an El Niño is an increase in precipitation in the Eastern Pacific and adjacent parts of South America and a decrease in precipitation for areas in the Western Pacific. But as the El Niño develops, the extent of its influence on global weather increases, producing anomalous weather and climate conditions worldwide, such as a significant decrease in tropical storm activity in the Atlantic Ocean. This is associated with drought in the Caribbean and Central America. On the other hand, tropical storm activity increases in the Eastern Pacific, while unusually wet conditions prevail in the Southern USA and Eastern Africa. Sea levels are unusually high in the Central and Eastern Pacific.

The direct and indirect consequences for tourism of such anomalous conditions have been identified in several studies, including Glantz (2000) and National Drought Mitigation Center (1998). For example, increased tropical storms in coastal Mexico, along with mudslides and outbreaks of malaria, cholera and dengue fever, deter tourists from visiting areas adjacent to the Gulf coast. The drought, floods and fires that inflict Southeast Asia during an El Niño have a major impact on the region's tourism industry, due in part to the haze resulting from the fires reducing visibility, impeding air traffic and increasing health risks. Water availability is another issue that affects tourism, as other economic sectors.

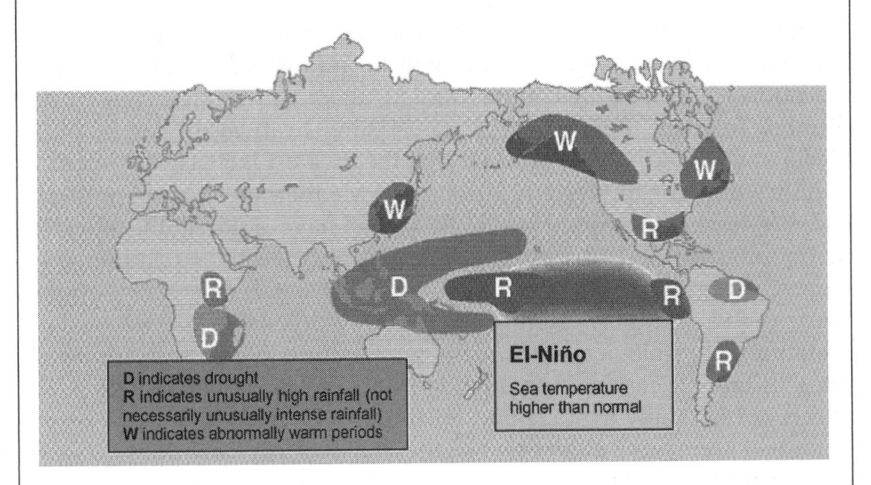

Text Box 9 – Figure 1. Global effects of El Niño (Glantz, 2000)

In Ecuador elevated sea levels and increased wave activity result in accelerated coastal erosion, often damaging tourism and related facilities. Tourists are also deterred by the more frequent heavy rains. In Palau, near surface temperatures in the coastal waters of Palau climb to over 30°C, causing massive coral bleaching events. In the 1997–98 El Niño event one-third of Palau's reefs were killed, with the number of some species declining to 99% below pre-bleaching levels. The associated economic loss to Palau was estimated at US$91 million, in part due to a 9% decline in annual tourism revenues.

Most recent modelling studies indicate that, in a warmer world, the pattern of tropical Pacific sea surface temperatures becomes more El Niño-like – sea surface temperatures in the Eastern tropical Pacific increase faster than those in the West, with an associated eastward migration in the tropical Pacific rainfall pattern (e.g. Yamaguchi & Noda, 2006). But for the six (out of 19 studied) models that were best at simulating present-day ENSO conditions, van Oldenborgh *et al.* (2005) found no significant changes toward El Niño-like conditions in the latter part of the current century. This, and similar findings in other studies, means that it is not yet possible to make any predictions about the future nature of El Niño events, or of the opposite cool event, the La Niña.

Several studies (e.g. Weisheimer & Palmer, 2005) signal the likelihood of increased incidence of extreme high temperatures in the future. Consistent with this, Vavrus *et al.* (2006) found that the frequency of cold air outbreaks during winter will decline in most areas of the Northern Hemisphere. Due in part to the 2003 European heatwave there is now high interest in how such events will be influenced by the enhanced greenhouse effect. In the coming decades there will be an increased risk of more frequent, intense and longer-lasting heatwaves. For example, Meehl and Tebaldi (2004) indicate that the greatest increases in intensity of heat will occur in Western Europe, the Mediterranean and South-east and Western USA. Morabito *et al.* (2004) have studied the effect of thermophysiological discomfort due to hot weather conditions in Florence, Italy, on tourist admissions to hospital emergency rooms. They found a highly significant linear relationship between daily minimum physiological equivalent temperature (PET) and admission rates for tourists coming from high latitudes in Europe and America. The effects of high daily average and maximum PET on hospital admissions of tourists were more evident when these discomfort conditions occurred on consecutive days – that is when heatwave conditions occurred (Figure 5.6). Even though people usually living in colder countries at high northern latitudes of Europe and America were more affected than people coming from other countries, tourists coming from Mediterranean

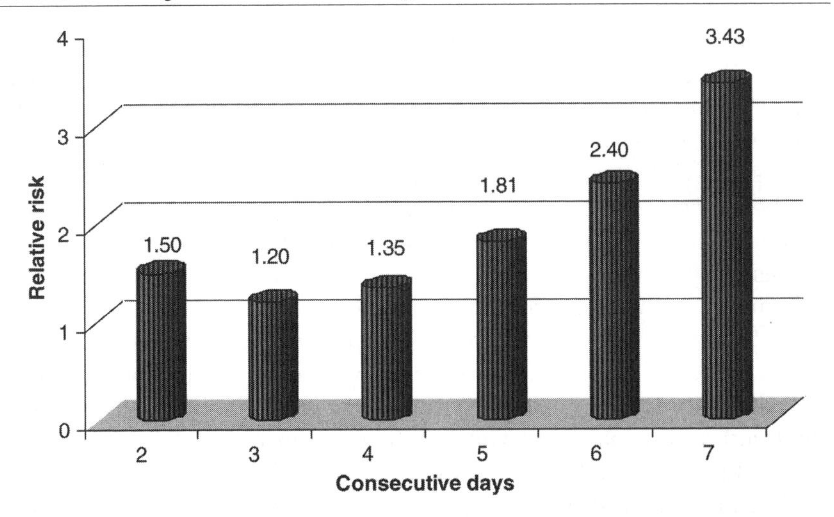

Figure 5.6 Relative risks of hospital admissions for Central and North European tourists on consecutive days with high daily maximum PET assessed during summer 2003
Source: after Morabito *et al.* (2004)

countries, such as Spain and Greece, also showed a great susceptibility when daily minimum PET values increased. On the other hand, people coming from Central and South America did not show any vulnerability to such thermophysiological discomfort conditions.

A growing number of studies are suggesting that observed increases in precipitation intensity, and associated flooding, during the 20th century will likely continue through the 21st century, along with a somewhat weaker and less consistent trend for increasing dry periods between rainfall events – that is, precipitation will occur in more intense events, with longer periods of little precipitation in between (e.g. Allen & Ingram, 2002; Pal *et al.*, 2004; Tebaldi *et al.*, 2006). Benestad (2006) reports that extremes in monthly precipitation will be more common in most mid- and high-latitude areas, as well as in the tropics. This is consistent with the results of Räisänen (2005), who found that not only do wet extremes become more severe in many of the areas where mean precipitation increases, but drought will also become more commonplace in areas where the mean precipitation decreases. A future warmer climate will be associated with increased summer dryness and winter wetness in most parts of northern middle and high latitudes (Wang, 2005).

In the future, tropical cyclones are likely to become more severe, with greater wind speeds and more intense precipitation (Knutson & Tuleya, 2004). There is less certainty as to how the frequency of tropical cyclones may change, with findings being somewhat dependent on the spatial resolution of the climate models. The most robust result appears to be that of Oouchi *et al.* (2006), who used a high-resolution global climate model. Compared to the present day, by the end of the 21st century the number of strongest tropical cyclones with extreme surface winds had increased, while weaker storms decreased. Globally, tropical cyclone numbers decreased by 30%, but increased in the North Atlantic. The cyclone tracks remained consistent, while maximum wind speeds increased by about 10%.

Lambert and Fyfe (2005) found that increased greenhouse warming is likely to reduce the total number of extratropical cyclones, but increase the frequency of intense systems. There was no apparent change in mid-latitude storm tracks, a finding inconsistent with that of Yin (2005), who identified a systematic poleward shift of storm tracks. Consistent with the likely increased frequency of intense extratropical systems is the finding that for most mid-latitude ocean regions an increase of extreme wave height is likely to occur in a future warmer climate (e.g. Wang & Swail, 2006).

Uncertainties,[1] Abrupt Climate Change and 'Surprises'

Because of the diversity and complexity of climate-determining processes in the Earth–atmosphere system, and the high resolution and long time periods over which calculations must be made, computer-based models are used to estimate how global, regional, national and even more local climates may evolve in the future. The high resolution and detailed representations of processes in coupled atmosphere–ocean global climate models mean that they are the only modelling tools capable of producing realistic simulations of internal variability, extreme events and the complex interactions that manifest as global and regional climate change feedbacks (Soden & Held, 2006). Despite the considerable, and growing, intricacy of these global climate models, they often have difficulty simulating the full range of interactions between and within the atmosphere, oceans and terrestrial systems due to inadequate representation of the relevant processes. This includes internal climate variability, as well as not resolving some smaller-scale features that affect local climatic conditions.

The resulting *uncertainties* regarding future climates are exacerbated by imprecise knowledge of such important considerations as future, and even current, GHG emissions, concentrations of aerosols and their precursors, and changes in land cover. Attempts to reduce the uncertainty due to scale and interaction effects often involve 'nesting' a more spatially detailed regional climate model within a lower-resolution global model. This is now a major area of research, as are other efforts to reduce uncertainties in climate change estimates.

Given the current uncertainties, it is now common practice to aggregate a large number of estimates generated using different but plausible GHG emission scenarios and several different climate models – that is, a multimodel ensemble. The assumption is that, by using a wide variety of different climate models, the results can cover the entire range of uncertainty, while close agreement between models in estimates of a future climate indicates the outcome is more likely. Figure 5.7 illustrates this approach, for an arbitrarily selected area and future time. It is clear that all models simulate higher temperatures, and regardless of the emission

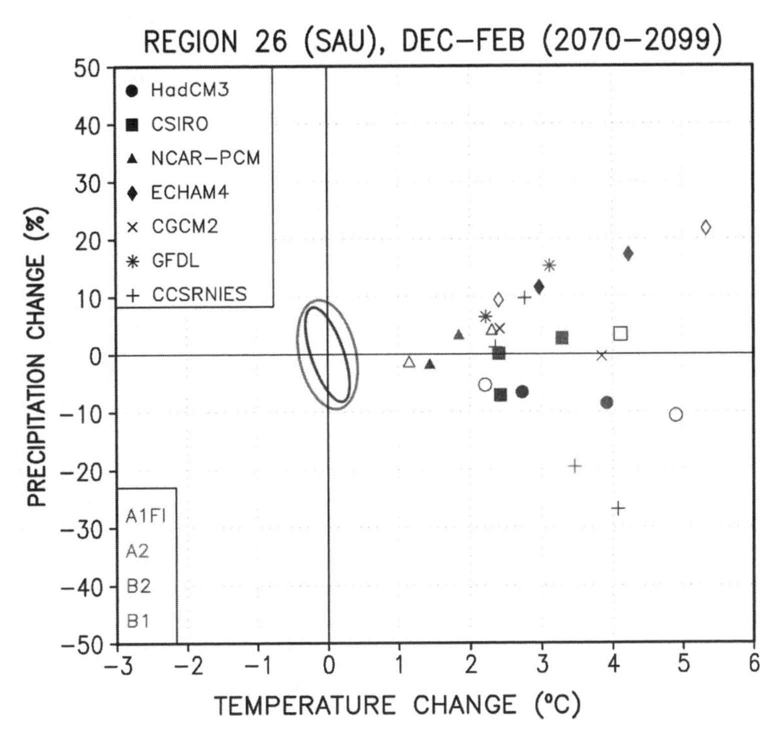

Figure 5.7 Estimates of changes in temperature (x-axis) and precipitation (y-axis). Each scatter point represents the temperature and precipitation response for one of the greenhouse emission scenarios simulated by a global climate model. The models used have been developed in Australia, Canada, Germany, Japan, the USA and the UK. The scenario used is depicted by the colour of the point – see legend at the bottom-left. The shape of the symbol defines the model – legend at the top-left. Ovals centred on the origin indicate the natural tri-decadal variability of temperature and precipitation, derived using two global climate models run without any changes in the atmospheric concentrations of GHGs.

Source: Ruosteenoja *et al*. (2003)

scenario. Thus there is high confidence that the temperature will increase, but the differences between both models and scenarios means there is less confidence in specifying by how much. The situation is quite different for precipitation, as there is not even agreement between models as to the direction of change. Nevertheless, even for precipitation it is apparent that the models suggest the changes will be greater than the levels of natural variability, shown by the ellipses centred on the origin.

As already noted in Chapter 2, one manifestation of uncertainty is the capability of the climate system to undergo abrupt and pervasive changes. This is only now being understood, albeit in a preliminary manner. Examples of past and potential climate-induced abrupt changes, including surprises, are ocean circulation changes and the resulting anomalies in air temperatures and precipitation (Schaeffer *et al.*, 2004), disappearance of permanent Arctic sea ice, loss of glaciers (Bradley *et al.*, 2004), disappearance of the Greenland ice sheet (Gregory *et al.*, 2004), and changes in vegetation cover and composition and soil cover. Conventional understanding has been that the soil will help offset the increased concentrations of atmospheric CO_2 by assimilating additional carbon through increased growth of plants and soil organisms. But the levels of temperature and precipitation increase anticipated for some regions by about 2050 will increase respiration and the soil could change rapidly from a moderate carbon sink to a strong source of atmospheric carbon, thereby enhancing the radiative forcing (Cox *et al.*, 2004). It is estimated that northern ecosystems have accumulated 25–33% of the world's soil carbon. In a warming climate, CO_2 and CH_4 trapped in permafrost have a high potential for release into the atmosphere, increasing radiative forcing still further. This can result in a (positive) feedback loop and more permafrost thaw. Western Siberia has already warmed faster than almost any other region, with an increase in average temperatures of some 3°C in the last 40 years. The warming is believed to be a combination of human-induced climate change, a cyclical change in atmospheric circulation known as the Arctic oscillation, and feedbacks caused by melting ice, which exposes bare ground and ocean. These absorb more solar heat than white ice and snow. Thus a featureless expanse of frozen peat is turning into a watery landscape of lakes, some more than a kilometre across. It is suspected that an unknown critical threshold has been crossed, triggering the melting. Sudden melting on such a wide scale could unleash billions of tonnes of CH_4, a potent GHG, into the atmosphere. West Siberian wetlands alone contain some 70 billion tonnes of CH_4, a quarter of all the CH_4 stored on the land surface worldwide. If the bogs dry out as they warm, CH_4 will oxidise and escape into the air as CO_2. If the wetlands remain wet, as is the case in Western Siberia today, then CH_4 will be released straight into the atmosphere. CH_4 is 20 times more potent as a GHG than is CO_2. Gas has already been observed bubbling from thawing permafrost so fast it was preventing the surface from freezing, even in the midst of winter (*New Scientist*, 11 August 2005).

But the increasing sophistication of climate change science also brings good news. For example, previous estimates of the upper limit of climate sensitivity (global mean temperature change due to a doubling in atmospheric CO_2 concentrations) might exceed 9°C. Past changes in climate allow estimates of climate sensitivity. A recent study of Northern Hemisphere temperatures in the pre-industrial period 1270–1850 suggests that it is unlikely that the climate sensitivity exceeds 6.2°C (Hegerl *et al.*, 2006).

Conclusions

Tourism is one of the many human activities contributing to climate change. The changes will not only exacerbate the climate-related risks currently faced by the sector but will also give rise to new risks, including the possibility of climate change 'surprises'. Everyone engaged in tourism, be they providers, consumers, regulators or other players, will be affected by climate change. While the greatest consequences are likely to be experienced in the 'climate–tourism hotspots' (see Chapter 2), no tourism activity is isolated from the inevitable increases in the level of risk.

Chapters 6 and 7 will go into further detail regarding the contribution of tourism to climate change, and how this might be reduced in the future, through initiatives that reduce the sector's emissions of GHGs. Chapter 8 will consider the issue from an alternative perspective, namely that of identifying and addressing the consequences of climate change for the tourism sector. Unacceptable risks need to be reduced, while also making the most of any potentially beneficial consequences.

Note

1. This section focuses on uncertainties in relation to climate change and not tourism forecasts.

Chapter 6

Methodologies for Greenhouse Gas Accounting

Key Points for Policy and Decision Makers, and Tourism Operators

- There are no systematic methodologies for accounting of tourism's GHG emissions; when inventories are carried out, tourism is typically implicit in traditional sectors such as transport or energy.
- Deriving CO_2 emissions from energy use is done by applying fuel-specific emission factors. In contrast, emissions of non-CO_2 GHG emissions are more difficult to determine, as these depend on a variety of factors related to the combustion process.
- It is important to distinguish between direct and indirect (or embodied) energy impacts (and concomitant GHG emissions) of tourism activity. Direct impacts are those that result directly from tourist activities, while indirect impacts are associated with intermediate inputs from second- or third- (or *n*th) round processes.
- Accounting for tourism energy use or GHG emissions can be undertaken through a bottom-up or a top-down analysis. Bottom-up analyses generally require data collection from individual units of analyses, for example tourism businesses or tourists, and provide a great level of detail. In contrast, a top-down analysis attempts to estimate tourism's share of a larger system, for example a nation or a destination. This means top-down analyses provide an abstract but holistic overview of tourism embedded in the economy.
- Transport databases or companies often provide information on the energy intensity of specific transport modes, for example in the form of 'energy use per passenger kilometre (MJ/pkm)' or 'energy use per vehicle-kilometre (MJ/vkm)'. This energy intensity can be used to determine a tourist's energy requirement to travel for a certain distance, and following from this, GHG emissions can be estimated.
- Energy use data for different types of accommodation and tourist attractions or activities are not usually readily available, and therefore this information has to be collected from businesses. Information on electricity, fossil and other solid fuel should be collected in the most commonly used unit (e.g. kWh, litres) or whatever unit is most convenient for the tourism business.
- Information to be collected for the energy-relevant analysis of tourist behaviour includes: (a) tourist itineraries, (b) transport

modes used, (c) vehicle occupancies, (d) accommodation types and number of nights spent, and (e) attractions visited or activities undertaken.

- Many countries have International Visitor Surveys and National Travel Surveys in place to collect information on tourist travel; these surveys are usually not comparable between countries. Information reported by tourists can be converted into energy use and GHG emissions.
- Input–output analysis can be used in a top-down approach to GHG accounting for tourism; this method allows estimating the direct and indirect effects of tourism as a separate sector in the economy.

Introduction

Tourism activity requires the use of fossil fuels, resulting mainly in emissions of CO_2. Sources for CO_2 emissions are both stationary (e.g. a hotel) and mobile (e.g. a car). Tourism also contributes at a small scale to non-energy related emissions, for example CH_4 emissions from waste. Tourism has become an important economic activity in many countries, but there are no systematic methodologies that help to account for tourism's GHG emissions. When GHG inventories are carried out – for example following the IPCC (1996, 2000) guidelines – tourism is typically implicit in traditional sectors such as transport. It is the purpose of this chapter to provide basic information on how to account for tourism GHG emissions at a country or destination level. Accounting for GHG emissions allows for targeted mitigation initiatives and effective policy development in the tourism sector (see also Chapters 7 and 9). Two approaches will be discussed in this chapter: bottom-up and top-down. Emissions that are not related to energy use are not discussed.

Energy Use and Greenhouse Gas Emissions

Energy comes in many forms (e.g. chemical, thermal, kinetic or potential), and one form can be converted into another. In conversion processes energy cannot be 'used up', i.e. the total quantity of energy remains constant (First Law of Thermodynamics). Conversion or transformation is often necessary to generate a form of energy that can be used for a specific purpose, for example for mechanical work. However, parts of the original quantity of energy may simultaneously be transformed into a form of energy that is not wanted (e.g. heat) and therefore considered 'wasted' or 'lost'. The ratio of 'useful energy output' to 'energy input' is generally defined as energy efficiency. The efficiency varies greatly for different conversion processes (Table 6.1).

Table 6.1 Conversion from one form of energy into another and typical efficiencies

Converter	*Form of input energy*	*Form of output energy*	*Typical efficiency (%)*
Electric motor	Electrical	Mechanical	80–95
Petrol engine	Chemical	Mechanical	20–25
Hydroturbine	Kinetic	Mechanical	10–99
Electric lamp	Electrical	Light	5
Solar cell	Light	Electrical	8–15
Solar collector	Light	Thermal	25–65
Wood stove	Chemical	Thermal	12–30
LPG stove	Chemical	Thermal	60–79

Source: after RWEDP (1997)

A common unit of measure in energy audits is based on the heat content of the energy source or on fuel equivalence values, such as coal or oil equivalents (see Text Box 10). The heat content measures a maximum available quantity of energy and does not consider the energy ultimately used for the desired output. This approach therefore ignores the different quality of energy sources, which results in a different efficiency of energy output compared with energy input (for more detail see Patterson, 1996).

In energy analyses information is typically collected initially in the physical measures of kilogram (kg), litres (l), cubic metres (m^3), cords of wood (3.6 m^3) and kWh of electricity, and transformed into the common unit of joules (J) or megajoules (MJ) by using fuel-specific conversion factors (see Text Box 10). These factors reflect the heat content of the respective energy source. Fuel properties and thus conversion factors vary depending on the quality of the primary energy source and the refining process. Accordingly the factors may differ for different countries or regions.

Except for electricity derived from renewable energy sources, such as solar, hydropower, wind or biomass,[1] the consumption of energy is based on the combustion of fossil fuels. Burning fuels results in the emission of various gases, as well as particles (e.g. soot). These gases contribute to local air pollution (e.g. sulphur dioxide) or act as GHGs.[2] Overall, CO_2 is the most important GHG resulting from human activities. About 80% of anthropogenic CO_2 emissions are due to the burning of fuels (IPCC, 1995); the remainder is linked to non-energy activities, such as cement production. The CO_2 emissions from a combustion process depend on the carbon content of the fuel source. Emissions can be readily calculated by using fuel-specific factors[3] (Table 6.2). As these factors assume complete oxidisation, it is valid to apply these factors without further consideration of the combustion process. Most countries provide averaged CO_2 emission factors for the liquid, solid and gaseous fuel sources available in that specific country.

Text Box 10 Basic information needed for energy and GHG accounting.

The most commonly used energy measure is the joule (SI unit); other energy measures include calories (e.g. for food) and kilowatt-hours (e.g. for electricity). The following conversion factors apply:

Energy measures	*Conversion factors*	*SI unit*
British thermal units (BTU)	1055.055	Joule
Calories (cal)	4.1868	Joule
Kilowatt-hours (kWh)	3.6×10^6	Joule
Tons of oil equivalent (toe)	4.187×10^{10}	Joule
Tons of coal equivalent (tce)	2.929×10^{10}	Joule

SI, Système International d'unités.

To account for large amounts of energy, it is common practice to use prefixes such as mega (one million) or giga, up to exa.

SI prefixes commonly used in energy calculations

Factor	*Prefix*	*Symbol*
10^{18}	Exa	E
10^{15}	Peta	P
10^{12}	Tera	T
10^9	Giga	G
10^6	Mega	M
10^3	Kilo	K

Once the consumption of energy from a specific fuel source has been determined, it is possible to derive CO_2 emissions based on the carbon content of the fuel and assuming fuel combustion (i.e. full oxidisation of carbon into CO_2) of the fuel. CO_2 emission factors vary for different countries; those used in New Zealand are presented below:

Fuel type	Energy content (gross calorific value)	CO_2 emissions (g/MJ)
Petrol	34.5 MJ/L	66.6
Diesel	38.1 MJ/L	68.7
Marine diesel oil	38.3 MJ/L	68.8
Aviation fuels	36.9 MJ/L	68.7
Kerosene/JetA1	36.8 MJ/L	68.7
Wood	5,820–10,910 MJ/m^3	No net emissions
LPG	50 MJ/kg	60.4
Natural gas	39–47 MJ/m^3	56.2–52.7
Coal	25.1 MJ/kg	90.4
Electricity	3.6 MJ/kWh	42

Source: Baines (1993)

Notes: Some energy statistics use gross calorific value, which describes the maximum available energy in the form of heat content. This includes energy released by the complete combustion and heat recovered from water vapour once it is completely condensed. The IPCC (1996) recommends use of net calorific values, which provides the amount of useable energy; that is, considering energy in the form of water vapour to be 'lost' for further use. Wood heat content depends on the species and the humidity; the above values reflect wood in air dry condition; the lower value relates to *Salix alba* (willlow) and the higher one to *Chamaecytisus palmensis* (tree lucerne).

The emission factor for electricity depends on the generation mix in a specific country; and it may also vary across seasons and different weather conditions. The proportion of hydropower in a given country, for example, depends on water availability and levels in hydro lakes. In Europe, factors could range between 500 and 740 g/kWh of CO_2 (Van den Vate, 1997). The average emission factor in the USA is 606 g/kWh of CO_2 (Energy Information Administration, 2002). The emission factor for nuclear energy is zero; nuclear energy poses other environmental risks. In New Zealand, as a result of a large share of hydropower, the average factor is 152 g/kWh (or 42 g/MJ). As electricity is an important energy source in the tourism sector, it is necessary to obtain country- or site-specific emission factors for electricity consumption. In the case of diesel generation at tourist resorts it is possible to derive emissions from electricity consumption through the amount

Table 6.2 Estimated GHG emission factors for European passenger cars

Vehicle type	NO_x	CH_4	*NMVOC*	*CO*	N_2O	CO_2
Passenger car: uncontrolled emissions, assumed petrol economy of 11.2 l/100 km						
Total g/km	2.2	0.07	5.3	46	0.005	270
g/kg fuel	27	0.8	63	550	0.06	3180
g/MJ	0.6	0.02	1.5	13	0.001	73
Passenger car: oxidation catalyst, assumed petrol economy of 8.1 l/100 km						
Total g/km	1.4	0.07	1.4	7.5	0.005	190
g/kg fuel	22	1.2	24	125	0.08	3180
g/MJ	0.5	0.03	0.6	2.9	0.002	73
Passenger car: oxidation catalyst, assumed diesel economy of 7.3 l/100 km						
Total g/km	0.7	0.005	0.2	0.7	0.01	190
g/kg fuel	11	0.08	3.0	12	0.2	3140
g/MJ	0.3	0.002	0.007	0.3	0.004	74

Source: IPCC (1996)

of diesel used by the generator (typically at an efficiency of between 30 and 35%).

Deriving emissions of non-CO_2 GHG emissions is somewhat more difficult. The procedure of simply applying averaged emission factors to the amount of energy does not work in the same way as for CO_2, because the quantity of emissions from a specific activity depends on a number of factors. For example, in the case of tourist transport, emission factors depend on the vehicle model (age, accumulated mileage and technology), typical driving behaviour, fuel type and maintenance of vehicles.

The IPCC (1996) provides default emission factors for various vehicle types in their Greenhouse Gas Inventory Workbook. They suggest using the 'most disaggregated technology-specific and country-specific emission factors available' (IPCC, 2000). Emissions factors are typically expressed as gram per kilometre, gram per kilogram of fuel or gram per megajoule. It can be seen in Table 6.2 that the emission factors for CO_2 are constant across vehicle types and technologies, whereas non-CO_2 emission factors differ. The IPCC workbook also provides emission factors for air travel, whereby different types of aircraft are distinguished, as well as emissions associated with landing-and-take-off cycles (LTO) and travel at cruising altitude.

Accounting Framework

In Chapter 2 tourism has been described as an open and complex system. Those characteristics provide a challenge for developing GHG inventories. Accounting for GHG emissions requires defining system boundaries and decision rules about what to include and what to exclude. For example, one could follow the UNWTO concept of tourism defined by consumption and include all the activities that are involved in tourists' activities. The emissions produced to provide services and goods to tourists could be included as well. The tourism industry purchases products from other industries, for example furniture, appliances or vehicles, all of which require energy input to be produced or delivered. Emissions from those activities are more indirect in nature and could be accounted for but kept separate from the direct GHG emissions (for more detail on indirect or embodied energy see Biesiot & Norman, 1999; Enquête-Kommission, 1994; Maibach *et al.*, 1995; Van den Vate, 1997; Weber & Perrels, 2000). The methodology for the top-down analysis described later in this chapter is conducive to the calculation of multipliers and thereby the estimation of indirect effects.

Acknowledging that tourism in an open system and very dynamic, it is pragmatic and practical to draw boundaries according to the UNWTO definition and attempt to account for tourism's direct energy use and emissions. The result of the inventory has to be interpreted against the background of the somewhat arbitrary system boundary and the assumption of tourism being a stable and rather static system. The inventory is a snapshot of one point in time.

Types of approaches

Accounting for tourism energy use or GHG emissions can be undertaken through a bottom-up or a top-down analysis (Becken & Patterson, 2006, see also the IPCC guidelines, 2000). Bottom-up analyses generally require data collection from individual units of analyses, for example tourism businesses or tourists. This means that bottom-up models involve a detailed analysis of a system in its disaggregated components. The individual analyses are then aggregated to obtain a picture for the whole tourism industry.

An industry-based bottom-up analysis provides an indication of the industry's total contribution to national GHG emissions, as well as a breakdown into different subsectors. However, little is revealed about tourists' demand and consumption patterns. For example, it would be very useful to understand what kinds of tourists are using what types of transport modes and how far they travel on average. For this reason, it may be useful to follow a bottom-up approach that involves both industry and tourist behaviour analyses. Energy use and CO_2 emissions from tourism can then be presented as a result of the combined effects of energy efficiency of different tourism industries and tourists' travel choices.

Bottom-up models recognise that markets are imperfect and that there are barriers for taking up theoretically plausible pathways. They assist investigating technologies that make up the supply and demand components of tourism, in order to identify low-cost GHG reduction opportunities. The analysis of tourist behaviour as part of the bottom-up approach is critical for formulating energy efficiency and GHG policies that target human behaviour. As data on energy use and/or tourist behaviour are generally not available, this approach potentially requires extensive collection of primary data. This is a very time-consuming and expensive task, given the diversity of the tourism sector, but the resulting energy database offers a great level of detail for energy use, fuel sources and emissions broken down into different transport modes, accommodation categories and tourist attractions, and has potentially much greater accuracy than a top-down approach.

A top-down analysis attempts to estimate tourism's share of a larger system, for example a nation or a destination. This means top-down analyses provide an abstract but holistic overview of tourism embedded in the economy. To this end it employs a macroeconomic approach using aggregate data on economic activity and associated energy consumption. This allows for intersector comparisons and wider analyses of environmental and economic interactions. Top-down models are, for example, used for policy analysis and allow assessment of macroeconomic instruments such as a carbon charge. Economic-wide feedbacks on prices and commodity and factor substitution effects can be explored as well (Wing, 2004). A usual basic assumption in these studies is that the market is perfect and that demand responses can be described by price-elasticities, i.e. how demand follows changes in price. Top-down models are deterministic in nature and do not allow for non-linear or chaotic changes in the tourism system or wider economy. They generally do not seek to address uncertainty.

Top-down integrated economic–environment accounts (formalised in an IO model) provide a direct platform for the application of a number of analytical methods, for example determining eco-efficiencies or ecological footprints (refer to Patterson & McDonald (2004) for further details). While economic data are usually available for the traditional sectors and increasingly for tourism in the form of a satellite account, environmental data are often not readily available and require at least an aggregated collection of primary data or modification of aggregated national energy data.

There are increasing calls to combine the two methodologies in integrated models in which the top-down data complement the (reduced) bottom-up data. Using the same reference years, the data generated in the bottom-up analysis can be incorporated in the environmental accounts of the top-down analysis. This refines the top-down approach and generates more accurate results. Ideally, the results of the combined analysis could enable policy formulation to take into consideration various outcomes, such as (a) what macroeconomic instruments result in what effects, (b) how does a specific measure affect tourism relative to other sectors, (c) what industries are the main contributors to CO_2 emissions, (d) what categories within each industry should be targeted and (e) how are specific tourist types affected?

Bottom-up Analysis of Tourism Industries

The Tourism Satellite Account distinguishes between tourism-characteristic, tourism-related and non-tourism-related industries in terms of final demand. This differentiation takes into account that some industries typically cater for tourists (e.g. accommodation), while others produce products or services that are purchased by both tourists and non-tourists (e.g. banking and insurance). When accounting for tourism's environmental impact it is pragmatic to focus on tourism-characteristic industries. The tourism-characteristic industries as defined in the TSA are usually summarised as 'passenger transport, accommodation and tourist attractions' (Collier, 1999). The methodology presented here does not include restaurants and other catering services, because of their heterogeneous nature, which would increase the sampling frame considerably.

Destination-based tourist transport

Tourists use a wide range of transport modes while on holiday. Common modes may differ for different tourist destinations. For example, tourism in Europe is largely based on the private car, and to some extent on air and rail travel. Some destinations are characterised by very specific forms of tourism transport, for example cruise ship tourism in the Caribbean or Alaska, safari tourism in Africa (often requiring purpose-built transport), rail travel in Switzerland, India or China, and trekking tourism in the Himalayas. Some destinations are characterised by tourism that involves extensive multidestination travel, for example for sightseeing or ecotourism holidays. Often, such touring destinations offer a broad mix of transport options, so they are in a position to cater for a wide range of both packaged and unpackaged travellers. Examples of touring destinations include Chile and Argentina, Iceland, South Africa, Australia and New Zealand. Large countries often require the use of domestic air travel as the main mode of tourist transport.

Most countries collect and provide national data on the energy intensity of commonly used transport modes. This information is typically provided in the form of 'energy use per passenger kilometre (MJ/pkm)' or 'energy use per vehicle-kilometre (MJ/vkm)'. These energy intensities can be applied directly in tourism analyses. Often, however, some adaptation to existing data is required to fit the tourism context. Adaptation is necessary, for example, when tourism-specific load factors may differ from non-tourism ones. This may be the case for rental cars compared with private cars or tourist coaches compared with intercity buses or school buses. Also, it is possible that tourist vehicles differ systematically from those used for non-tourist purposes (e.g. more modern and energy-efficient vehicles in rental car fleets).

Many airlines provide information on fuel efficiency (e.g. 'energy use per passenger revenue kilometre') of domestic flights in their annual reports. Airlines also release information on routings and the mileage of specific travel segments. Based on this, the total mileage for an itinerary can be estimated. The distance travelled by a tourist can be converted into energy use by multiplying it with the airline's specific fuel consumption. An airline's energy intensity is usually expressed in the form of fuel consumption (in litres or MJ) per passenger-revenue kilometre (i.e. excluding crew) or more simply passenger-kilometre (pkm). The energy intensity of a flight depends on several factors, namely the aircraft type (including engine and airframe), the ratio of LTO time to cruising time[4] and the load factor. Lenzen (1999) estimates an energy intensity of 1.75 MJ/pkm for long-distance flights, taking into account an average

Table 6.3 Fuel use and average sector distance for representative types of aircraft

	Aircraft			
	A310	*A320*	*B727*	*B737 400*
Average sector distance in nautical miles (nm)				
Total flight	1228	663	583	531
Total climb	81	159	117	100
Cruise	1034	393	384	339
Descent	113	111	82	92
Fuel use (kg)				
Total flight	12,160	4342	6269	3750
LTO (flight < 3000 ft)	1541	802	1413	825
Flight minus LTO (flight > 3000 ft)	10,620	3539	4856	2925
Fuel use (kg per nm)				
Flight minus LTO (flight > 3000 ft)	8.65	5.34	8.33	5.51

The IPCC advises these data be treated with caution as national circumstances might apply
Source: after IPCC (2000)

number of LTO cycles. To convert energy consumption into CO_2 emissions, a factor of 69 g/MJ (Baines, 1993) to 74.9 g/MJ (Australian Institute of Energy, no date) for kerosene needs to be applied.[5] More refined approaches could distinguish between emissions below 3000 feet (914 m) and above, and also count the number of LTOs per aircraft type (IPCC, 2000). This approach would allow estimating the effect of GHGs other than CO_2. As mentioned earlier, the emission factors for non-CO_2 emissions depend on the aircraft type and technology, phase of the flight (LTO versus cruising) and speed (Table 6.3).

Land-based transport modes that are specific to tourism, such as campervans, cable cars or some ferries, are likely to require additional collection of energy data and typical load factors to derive energy intensities. The data collection is simplified when there are only a limited number of transport providers who ideally collect emission-relevant information in their data system (e.g. rental car companies). The following example of campervan travel indicates the aspects that require consideration when attempting to derive energy intensity:

- Fleet composition: how many companies in the country rent campervans, and what vehicle types are represented; what is the age of vehicles and what type of fuel is consumed, i.e. petrol (and what kind of petrol) or diesel?
- Vehicle-specific fuel consumption: how much fuel (petrol or diesel in l/100 km) is consumed on average by different types of campervans?
- Other factors: what is the typical passenger load factor per campervan, what is the driving behaviour (aggressive versus considerate) and what percentage of vehicle-km are travelled on different road types (e.g. motorways, urban roads, mountain roads)?

The information provided by companies is not always detailed enough to account for all the factors described above. It is more likely that a sample of companies will provide average consumption figures for their fleet and an approximate ratio of petrol to diesel vehicles, as shown in Table 6.4. Often, this kind of information lacks accuracy.

The energy intensity (MJ/vkm) of a specific transport mode (e.g. campervan) is calculated by multiplying each vehicle type's share within the fleet (e.g. number of diesel and petrol cars) with the specific fuel consumption (l/vkm) and the energy content of the fuel (e.g. 38.1 MJ/l for diesel or 34.5 MJ/l for petrol). To derive average energy intensity (MJ/vkm), the weighted energy intensities of the different vehicle types must be added up (Equation 6.1).

$$\text{Average energy use/vkm} = \sum_j \text{specific fuel consumption}_j \text{ (l/vkm)} \times \text{energy content}_j \text{ (MJ/l)} \times \text{proportion of } j \qquad (6.1)$$

where j is vehicle type.

Equation 6.1 Average energy intensity of a transport mode

As explained earlier, it is straightforward to translate energy use into CO_2 emissions, but it is more challenging to determine emissions of non-CO_2 gases. Attempts have been made to integrate non-CO_2 emissions into one single equivalence factor, which has to be multiplied with the emission factor for CO_2 to obtain *CO_2-equivalents* (CO_2-e) (see Chapter 5 for an explanation of radiative forcing). For example, the total of GHG emissions or CO_2-e for car travel is about 1.05 times bigger than CO_2 emissions alone. For air travel, the equivalence factor is 2.7 (Table 6.5). These factors are only indicative and are likely to vary for different fleets and driving behaviours. Gössling *et al.* (2005c) also estimated so-called detour factors (DF), taking into account that tourists do not necessarily travel the shortest route between two points (Table 6.5).

Table 6.4 Fleet and fuel consumption data for New Zealand campervan companies

Company	*Information obtained*	*Average MJ/ vkm*	*Average load factor*	*Average MJ/ pkm*
1	290 vehicles in fleet (230 diesel, 60 petrol); avg. fuel consumption of vehicles: 12–15 l/100 km (diesel) and 12–14 l/100 km (petrol)	5.01	2.2	2.3
2	1200 vehicles in fleet (984 diesel, 216 petrol), avg. fuel consumption of vehicles: 14 l/100 km (diesel) and 15 l/100 km (Toyota Hiace, petrol)	5.31	2.2	2.4
3	All campervans use diesel, no information on consumption rates	NA	NA	NA

The gross calorific values used in this example were: 34.5 MJ/l for petrol and 38.1 MJ/l for diesel. Average occupancy factors need to be estimated (by the company) or data need to be collected (via a tourist survey)

Table 6.5 Emission factors for different transport modes (Gössling *et al.*, 2005c)

Transport mode	*Emission factor β_m for CO_2 (g/pkm)*	*CO_2 equivalence factor o_m (g CO_2-e/g CO_2)*	*Detour factor (km actual/km direct)*
Air (EU)	140	2.7	1.05
Air (ICA)	120	2.7	1.05
Rail	25	1.05	1.15
Car	75	1.05	1.15
Coach	18	1.05	1.15
Ferry	70	1.05	1.05
Cruise ship	70	1.05	1.3
Cycle/moped	10	1.05	1.15
Other	75	1.05	1.15

EU, European Union, i.e. flights with a maximum range of 2000 km; ICA, Intercontinental Air Transport, i.e. flights with a range greater than 2000 km.
Occupancy rates: air, 70% (EU), 75% (ICA); cars, 50%; long-distance rail, 60%; coach, 75%

Accommodation

Energy use data for different types of accommodation are often not readily available, and therefore this information has to be collected from businesses. To control sample size, the need to disaggregate the industry into a sufficient number of categories (to address the great variety in energy profiles) has to be balanced with the minimum requirements for sample sizes within each category. The total sample should be representative of the total of tourist accommodation businesses within a destination, whereby factors such as climatic conditions, size of the business, age of building stock are among the most important ones to be considered.

Warnken *et al.* (2004) explored several methodologies for energy accounting in the accommodation industry in Australia. In particular they suggested:

1. **The floor area method**. The idea of this method is to calculate an average energy use per square metre of floor area for different types of businesses.
2. **Regression analysis**. This method builds on identification of key influencing variables, such as building material and shell, floor area, number of employees, etc.; these independent variables are predictors of energy use.

3. **Mandatory reporting**. Accommodation providers would be required to submit data on energy use to a central agency that compiles this information alongside already collected data on visitor nights and occupancy rates.

Warnken *et al.* (2004) concluded that the floor area method is too coarse to capture the differences among businesses, and it fails to include external factors such as climate. The regression model is more sophisticated, but would need to be updated and refined continuously and is therefore time-consuming and costly. To achieve the greatest benefit in terms of energy accounting and GHG reduction measures, the mandatory method seems preferable over the other two methods.

An energy survey of accommodation businesses would need to collect two types of information: first, data on present energy use broken down by different energy sources or fuel types, and second, information on the building and the business operation. The latter is necessary to derive measures such as energy use per visitor-night and also to assess energy-saving potentials. Data to be obtained could include:

- consumption of fossil fuels and other energy carriers,
- building use and services,
- equipment and appliances, and business vehicles,
- building fabric and shell,
- capacity and occupancy rates, seasonality of guest nights and
- geophysical parameters (e.g. frost days).

Information on electricity, fossil and other solid fuel should be collected in the most commonly used unit (e.g. kWh, litres) or whatever unit is most convenient for the survey respondent (Figure 6.1). Consumption in physical units can then be converted into energy content (MJ) using calorific values.

The total annual energy consumption for each business is calculated by summing the energy consumed for each fuel type in one year (Equation 6.2). Energy consumption for each accommodation business can then also be expressed on a per visitor-night basis to facilitate comparison.

$$\text{Total energy} = \sum_j \text{fuel consumption}_j \text{ (physical unit)} \times \text{energy content}_j \text{ (MJ/physical unit)} \qquad (6.2)$$

where j is the fuel source.

Equation 6.2. Total energy use of an accommodation business

CO_2 emissions can be derived by applying fuel-specific emission factors, as already discussed for transport. Again, for GHGs other than

Q. What was the total number of visitor nights from ___ to ___?
(Note: one visitor night = one guest for one night. If you had two guests staying for three nights, you would have (2 x 3) = 6 visitor nights). ________________________________

Q. Electricity usage: Please refer to the electricity bills for the period __to ___ for actual *units* consumed ______________________kWh or ______________________$
You could use this table to help calculate the total yearly consumption for electricity.

	Jan	Feb	Mar	Apr	Ma y	June	July	Aug	Sep	Oc t	Nov	Dec	Total
Units kWh or $													

Q. How much other energy did your business use for the period __to ___? *Please provide figures.*

Petrol/diesel for business vehicles ______________litre or ______________$
Petrol/diesel for other operations ______________litre or ______________$
Liquid Petroleum Gas (LPG) ______________litre or ________kg or ________$
Natural gas ______________litre or ________m^3 or ________$
Coal ______________kg or ______________$
Other sources, please detail: ______________________________

Figure 6.1 Example of energy-related questions for accommodation businesses

CO_2 it is more difficult to estimate emissions as they depend on the combustion process. Non-CO_2 emissions from accommodation businesses are likely to be negligible compared with CO_2 emissions.

The largest problem in analysing energy use associated with tourist accommodation is the availability and the quality of data, especially for smaller businesses. Many operators are not interested in measures of energy consumption and do not collect this information systematically. It is also difficult to obtain long-term, reliable data; among other reasons, because of poor recording practices and high management turnover. Large accommodation providers often employ an engineer who looks after energy supply and demand and who monitors energy use at different functions. For these reasons, Warnken *et al.* (2004) suggested that in order to ensure buy-in from the industry, it would be desirable to develop standardised spreadsheets in an electronic form. These would assist managers to provide the data more easily and transform them into some useable output for their own use, for example monitoring of energy use at various functions of their business. These auditing tools, however, would have to take into account the varying situations for different types of accommodation, for example backpacker hostels versus five-star hotels.

Attractions and activities

Most tourism textbooks state that attractions are the initial pull factor to a destination for holiday visitors. Naturally, attractions and the mix of activities differ for tourist destinations and for this reason the first step in auditing this subsector is to define what kinds of tourist attraction or activity businesses exist, and which ones should be part of an energy or GHG inventory. The attraction subsector is extremely heterogeneous and could include anything from built attractions, such as museums or theme parks, outdoor-oriented activities and events (e.g. sport events or concerts), to shopping, for example in the form of craft or souvenir shops. Most tourist destinations maintain business directories; alternatively, relevant businesses can be identified through yellow pages or guidebooks.

Data on energy use or emissions for tourist attractions or activities are rarely readily available, except maybe for larger attractions (possibly run by the public sector) that monitor and publish data on energy use as part of their operational costs. For this reason, it is likely that a survey of the attraction subsector needs to be undertaken. Pragmatically, the sampling would follow a stratified approach in which a sufficient number of tourism businesses within each 'category of attractions' are surveyed. For some destinations, there are probably only a few categories, for example in the case of beach destinations these might be 'water sports', 'water transport', 'scenic tours' and 'activities in built environments'. Other destinations, for example cities, may offer a much wider range of possible tourist attractions.

Emission factors for CO_2 can be applied as described above for accommodation (Figure 6.2). The energy accounting may sometimes prove difficult because of complex business situations (various business activities by one operator), insufficient accounting or lack of data, high business and staff turnover, lack of interest by operators and a general diversity in the industry, which impede achieving a reliable and representative sample. Energy efficiencies in the attraction industry are often 'guesstimates'.

There are many tourist activities that do not consume energy and produce emissions, for example, a walk on the beach. These kinds of activities usually take place outside a business operation and are therefore

Example: *Steam boat scenic river cruise with on-board restaurant and bar*

Visitor numbers: 36,000 per year
The tourist attraction uses electricity for running the office and the ticket counter, and consumes gas for cooking and diesel for the boat. Annual consumption rates are:

Electricity: 83,410 kWh * 3.6 MJ/kWh => 300,276 MJ * 0.042 kg CO_2/MJ[1] => 12,612 kg CO_2
LPG: 6156 kg * 50 MJ/kg => 307,800 MJ * 0.064 kg CO_2/MJ => 19,699 kg CO_2
Diesel: 76,000 L * 38.1 MJ/L => 2,895,600 MJ * 0.069 kg CO_2/MJ => 199,796 kg CO_2

TOTAL annual energy use: 3,503,676 MJ
TOTAL annual CO_2 emissions: 232 t CO_2

Energy use per visitor: 97 MJ
CO_2 emissions per visitor: 6 kg

1) The emission factor for electricity applies to the New Zealand context for 2002.

Figure 6.2 Deriving energy use and CO_2 emissions for a tourist attraction

not part of the industry analysis or an emission inventory of the attraction industry.

Bottom-up Analysis of Tourist Travel Behaviour

The tourist can be seen to be at the centre of the bottom-up analysis, as all energy use can ultimately be related to choices that tourists make in relation to their travel. These 'travel choices' largely concern transport, accommodation and recreational activities. Data on travel choices are typically collected from tourists by means of a survey. Ideally, the information obtained includes:

- tourist itineraries to and at the destination,
- modes of transport used on different travel segments,
- vehicle occupancies for different transport modes,
- accommodation types and number of nights and
- attractions visited or activities undertaken.

Many countries undertake extensive tourist surveys for both international and domestic tourists. Surveys differ concerning the detail to which travel itineraries are recorded, the description and aggregation of transport modes or accommodation categories, the reporting of recreational activities and the target population.

International tourist data

International tourists are typically intercepted at border crossings or in the departure lounges of international airports. Some countries combine inbound surveys with outbound surveys. The administration of a survey involving international tourists is relatively straightforward and less expensive in the case of island destinations where tourists arrive by air or ship at a selected number of gateways.

In New Zealand, Australia and Fiji, for example, the *International Visitor Surveys* (IVS) are flight-based, which means that specific international flights are selected and departing passengers are approached in such a way that a representative sample is achieved. The surveys sample about 1–2% of the tourist population. In all three countries information is collected on destinations visited, transport modes used and accommodation used. The spatial resolution of destinations recorded is less fine in Australia and Fiji, compared with New Zealand where destinations are recorded on a location level and not an aggregated regional level.

The *International Passenger Survey* (IPS) in the UK covers both inbound and outbound tourism in one survey. Tourists are surveyed as they enter or leave the country via the main air or sea routes and the Channel Tunnel. In 2002, 55,100 interviews were carried out with international tourists departing from the UK and 61,000 interviews with UK residents returning from abroad. In the inbound survey, tourists report the towns where they stayed overnight and based on this it is theoretically possible to analyse international tourists' itineraries in the UK. Transport modes are also recorded, as well as the number of people travelling in one vehicle.

International tourism to and from the USA is recorded by the *Survey of International Air Travelers*. Information is collected on travel patterns, characteristics and spending patterns of international travellers to and from the USA. Data on destination-based travel include countries, states and major cities or attractions visited, main destination, port-of-entry, nights away from home or spent in the USA, recreational activities undertaken, types of transportation used during the trip, accommodation used and spending behaviour. In addition to the air traveller survey, road transport surveys are in place for traffic to neighbouring countries.

Similarly to the USA, Canada administers a combination of several surveys to account for air and road travel. The main components of Canada's *International Travel Survey Program* (ITS) are a mail-back questionnaire distributed at 150 designated ports of entry, and the *Air Exit Survey of Overseas Visitors* (AES) undertaken at major airports. The ITS collects information on tourists' expenditure, activities undertaken and places visited (overnight stops and places that did not involve an overnight stay). The survey covers both Canadian residents returning from an international trip and international tourists to Canada. In total, the number of usable questionnaires is approximately 40,000 for American tourists to Canada and 47,000 for tourists from overseas countries, from which almost 6000 came from the AES.

International tourist surveys in countries with multiple border crossings are more complex and costly. Sweden, for example, maintains an *Incoming Visitors to Sweden* (IBIS) survey at 11 selected border crossing points where foreign tourists leave Sweden. The survey is undertaken at land crossings, ferry ports and international airports. The situation is even more complex in France, where international tourism is recorded through *Enquêtes aux Frontières*. Tourists are intercepted at main entry points (e.g. highways, airports and train stations), and the survey is undertaken every 2–3 years (although the last useable data set dates back to 1996 with around 100,000 questionnaires). Again, the information collected allows assumptions on tourists travel (and in particular transport) behaviour, and therefore energy consumption.

Domestic tourist data

Some countries undertake surveys specifically for domestic tourism, for example Australia (*National Visitor Survey*) and New Zealand (*Domestic Travel Survey*). Most countries, however, account for domestic tourism as part of their National Travel Surveys (NTS). These NTS are often large-scale, multipurpose surveys that measure travel behaviour from a transport perspective. The main purposes are to improve transport planning and forecasting, infrastructure needs, accident rates, taxation policies and specific transport needs of various socioeconomic groups. Surveys are often based on households that are interviewed by mail, telephone or face-to-face. Interviewees are typically asked to recall all travel for a given period of time, or they are asked to record their travel activities in diaries during a randomly assigned period of time. The scope of these national surveys differs substantially for different countries and comparability is therefore difficult. Main differences among NTS are:

- main purpose of the survey (e.g. transport or tourism),
- definitions of travel and trips,
- interviewing methods, sampling frames and sample sizes,
- geographical resolution (national versus local data),
- the mix of overnight and day trips,
- eligibility (e.g. age; household versus individual) and
- data analysis and availability.

The size of NTS varies substantially. *Mobility in Germany*, for example, draws a national stratified random sample of 25,000 households based on population registers. The 2002 *UK National Travel Survey* interviewed 5796 tourists face-to-face on their personal travel. Key questions include purpose of trips, distance, length of stay, transport modes and number of companions. The US *National Household Travel Survey* (NHTS) covers American day travel and long-distance trips. Information collected includes transport modes used, travel times, trip duration and purpose. All together, a quarter of a million day trips and about 45,000 long-distance trips were recorded in the 2001 NHTS. In France, two instruments monitor domestic travel: *Suivi de la demande touristique* and *Enquêtes vacances*. The former surveys 20,000 people per month, including both domestic tourists and excursionists, i.e. day visitors. The survey is very detailed and includes destinations and transport modes. *Enquêtes Vacances* covers trips longer than 3 days, surveying 50,000 tourists every 5 years (G. Dubois, personal communication, January 2005). Even though some of these travel surveys are not necessarily tourism-focused, they still provide information that could be useful for estimating tourists' energy use and GHG emissions when travelling domestically.

Analysing tourist surveys

Information reported by tourists can be converted into energy use and GHG emissions. Different analytical steps are required for the different components of the tourism product, namely transport, accommodation and attractions. The first step in the transport analysis involves estimating travel distances in kilometres from tourist itineraries. The minimum information needed to construct a tourist itinerary is the location of origin and main destination; however, the more detailed the itinerary (e.g. the sequence of overnight stops) is recorded, the more accurate the estimate of the total travel distance. In the case of large samples and a large number of possible locations visited by tourists, it is useful to code the multitude of possible tourist destinations into a manageable number of key locations and then build a database of travel distances between those destinations. Road travel distances can be retrieved from maps or automobile associations. In some countries the road network is digitalised and can be linked with coordinates of key tourist locations. Air travel should be treated separately, because travel distances could differ for different types of air travel and land-based travel, and combinations of both (see Figure 6.3). Information on mileage is available from airlines.

The tourist itinerary data need to be combined with energy data for different transport modes, which are ideally recorded separately for different travel segments. The result of this integration is that every tourist can be described by the distances travelled using different transport modes during their trip. There are a number of possible sources of error that tend to result in an underestimate of travel distance and therefore CO_2 emissions. These include, for example, the tourists' ability to remember their activities and routings, their interpretation of transport modes (e.g. ('coach tour', 'backpacker bus' or 'scheduled coach'), detours to the shortest route and under-reporting for long-staying tourists.

Deriving energy use and emissions for accommodation and attractions is more straightforward than for transport. Tourists report what kind of accommodation they used either on a night-by-night basis or at least in the form of 'main accommodation type' stayed at. In combination with the length of stay it is possible to estimate the number of nights spent in different accommodation types. As accommodation types are characterised by their energy intensity (MJ/visitor-night), the total energy use (and emissions) can be estimated for each tourist by multiplying energy intensity with number of nights. The same procedure is applied

Example 1: Trip from Manchester (UK) to Valencia (Spain)
Scheduled airline via London: 1,817 km
Car and low-cost airline: 283 km (car) + 1,530 km (air)
Car and channel train: 2,160 km (car) + 58 km (train)

Example 2: Trip from Cologne (Germany) to Valencia (Spain)
Scheduled airline via Munich: 2,103 km
Charter airline, direct flight: 1,610 km
Car: 1,724 km

Figure 6.3 Comparison of travel distances for different transport options using two examples

to tourist activities and attractions, where energy intensities (MJ/visit) are multiplied by the number of visits a tourist made to a specific attraction or attraction category.

Ultimately, every tourist could be characterised by the distance travelled by different transport modes, the number of nights spent in different types of accommodation and the number of visits to a number of tourist attractions – and the energy use and the CO_2 emissions resulting from these travel choices. The behaviour of an *average tourist* can be derived by calculating average values, for example for energy use per trip or per day. This can be extrapolated to the total population of tourists to a destination. A more refined approach to estimating total energy use and emissions is to derive a number of tourist types and depending on their representation in the total population of tourists extrapolate energy use and emissions from those proportionally. Such a model of tourist types provides a deeper understanding in typical behavioural patterns and profiling those types allows more directed strategies for reducing emissions. Such energy-relevant tourist types based on travel behaviour were derived for the New Zealand context using cluster analysis (for a detailed methodology, see Becken *et al.* (2003)).

Top-down Analysis

Contributed by James Lennox, Sustainability & Society Group, Landcare Research, New Zealand.

Energy use and GHG emissions of the tourism sector can also be accounted for using a top-down approach. Top-down methods are most frequently applied at the national level, but can also be applied on smaller or larger scales. Increasingly, energy and GHG accounts are constructed following internationally standardised conventions specified in the System of Economic and Environmental Accounts (SEEA) (United Nations, 2003). GHG accounting is also influenced by accounting rules under the Kyoto Protocol (see Chapter 9). Standardisation and compatibility with national economic accounting standards (SNA '93 United Nations, 1993) facilitates integrated environmental and economic analyses. Integrating environmental data with TSAs (see Chapter 2) is of particular interest and is discussed later in this section.

The simplest approach is to construct GHG accounts for tourism-characteristic and tourism-related conventional industry sectors. If industry sectoral GHG accounts exist in sufficient detail, these only require presenting the data in a different way. If GHG accounts do not already exist, then they must first be constructed in the usual way (which is beyond the scope of the present work). More commonly, GHG accounts may exist for industry sectors at a relatively aggregated level. In this case, it may be desirable to disaggregate existing accounts so as to better distinguish key tourism (and GHG-emitting) sectors such as accommodation and passenger transport. Tabatchnaia-Tamirisa *et al.* (1997) took this approach to construct energy accounts for tourism in Hawai'i. Meaningful disaggregation requires the use of supplementary data that can characterise the energy use of industries to be disaggregated. These data may be used in an *ad hoc* way or within more formal statistical estimation frameworks (with respect to the latter, see for example Golan *et al*. (1996)). Having identified total emissions from the relevant industry sectors, TPRs (from the TSA) may be applied to scale the accounts of each sector.[6]

Emissions by households should also be included in a conventional GHG account, although commonly only the most significant emission-generating activities are considered – fuel use in private motor vehicles and possibly, for domestic space heating. Private motor vehicles are often used extensively by tourists, especially in the case of domestic tourism in developed countries. Private vehicle GHG emissions may be disaggregated into tourism and non-tourism components using vehicle-kilometre data as a proxy. Tourism is unlikely to contribute significantly to domestic fuel consumption for space heating or other purposes, but holiday homes and caravans are potential sources of such emissions.

GHG accounts as described above include emissions from industries supplying directly to tourists and emissions generated by tourists own activities. For the purpose of managing emissions from tourism though, it may be more useful to account for all emissions relating directly or indirectly to tourism; that is, emissions may be considered along entire supply chains rather than just their penultimate and ultimate steps. Comprehensive estimates of both direct and indirect emissions can be made with the aid of environmental–economic input-output (IO) models. These models also permit detailed analyses of intersectoral dependencies in relation to tourism and GHG emissions.

Economic IO analysis was first developed by Wassily Leontief in the 1930s. A central concern of Leontief (1947, 1951) was the organisation of productive sectors in producing commodities (goods and services) for final consumption. IO accounts describe the interindustry transactions within the productive sector, the final consumption of households, government and overseas purchasers, and purchases of capital goods by producers. In addition, capital depreciation, compensation of employees and other elements of value added are accounted for. From transactions accounts recording the supply and intermediate and final uses of commodities, IO models can be derived. A basic IO model is simply a large matrix of coefficients that describes the proportions of inputs each industry requires to produce one unit of its output.[7] Using this matrix, the activities of industry sectors can be related to the final demands for commodities (e.g. Miller & Blair, 1985). Economic IO accounts can be supplemented with rows of environmental inputs (natural resources) and/or outputs (wastes

Table 6.6 A simplified environmental accounting matrix

	Commodities	*Industries*	*Government/ households*	*Rest of world*	*Nature*
Commodities		Intermediate consumption	Domestic final demands	Commodity exports	Residuals from consumption
Industries	Domestic production				Residuals from production
Rest of world	Commodity imports				
Value added		Value added			
Natural resources		Physical inputs from nature			

and emissions) as illustrated in Table 6.6. The rows of an IO table correspond to supplies and the columns correspond to uses. For example, in row one, commodities are supplied to industries, government, households and overseas purchasers. Column one shows that the outputs of domestic and overseas industries generate the supply of commodities.

Leontief, Hite and Laurent, and others pioneered the use of IO models as a tool for integrated environmental–economic analyses (Leontief, 1970, 1972; Leontief *et al.*, 1977; Hite and Laurent, 1971). Subsequently, a considerable body of literature has emerged concerning the relationships between economic structure and pressures on the environment and on energy resources. Environmental–economic IO models derived from these accounts (e.g. Miller & Blair, 1985) may serve a variety of purposes, including:

- analysing direct and indirect environmental impacts associated with final demands or resulting from changed demands;
- analysing interactions between productive sectors and their consequent environmental impacts; or
- deriving IO-based eco-efficiency metrics.

In recent years, IO models have frequently been used to analyse connections between economic activities and GHG emissions (e.g. Hetherington, 1996; Lenzen, 1998; Lenzen & Dey, 2002; Foran *et al.*, 2005). IO analysis has also been used to analyse economic impacts of tourism in both academic studies and by governments or the industry (e.g. Dwyer *et al.*, 2004). However, we are aware of only one major IO study that links environmental impacts (including GHG) and tourism; that of Patterson and McDonald (2004) in New Zealand. Also, most IO studies of tourism focus on consumption and/or tourism-characteristic sectors within a conventional IO framework, which does not distinguish a 'tourism sector' as such.

The first steps in integrating TSA and IO accounts data is to reconcile tourism expenditure by category of tourist (consumer) with tourism purchasing sector (which may be either the consumer, firm or other organisation). Almost all expenditure by international visitors will be considered as exports. Domestic tourist expenditure is more complicated, as it is necessary to distinguish at least the business, government and household sectors. Ideally, individual business sectors' spending on tourism should be separately identified, but such data are unlikely to exist. Instead, proportional or other allocation methods may be used (Patterson & McDonald, 2004). Unless business tourism is of major importance or interest, this should not be a critical limitation. The same applies to government tourism.[8] A further practical difficulty is that tourism expenditure data may only be recorded in terms of the primary purpose of travel. Evidently, business and public-sector tourists are likely to make some private purchases during their travel too.

Leontief multipliers show the impacts of demand changes on productive activities, and consequently, on GHG emissions. They can be used to estimate the direct and indirect impacts associated with any vector of tourism final consumption (e.g. domestic leisure tourism or international tourism). Any direct impacts arising from tourism final consumption must still be added. Total impacts can be decomposed into contributions accruing along each stage of supply chains at the level of industries. It is also possible to estimate other economic impacts of tourism consumption (e.g. value-added or employment) and thereby to derive a variety of 'eco-efficiency' indicators, such as value added per tonne CO_2-eq for different tourist types. It should be noted that the vector (or in fact, matrix) of domestic business tourism purchases should not be analysed in the same way – it is an intermediate and not a final demand.

Tourism can also be considered from the supply side and in terms of the supply–demand interactions. In this case, a better approach is to modify the conventional IO framework such that tourism activities are disaggregated out of the conventional industry sectors and reaggregated into one or more tourism sectors. New sectors/subsectors are defined in terms of tourism and non-tourism intermediate and final consumption (Table 6.7). In general terms, the tourism sector/s incorporates significant activity from tourism-characteristic industries, some activity from tourism-related industries, and little or no activity of non-tourism industries. This approach was taken by Patterson and McDonald (2004), disaggregating and reaggregating tourism-directed activity into a single tourism sector, in an IO model with 25 productive sectors.

A tourism- and GHG-specific IO model would permit analyses including:

- GHG associated with various tourism final demands (e.g. domestic leisure and international tourism);
- GHG emissions from the tourism sector;
- GHG emissions from the non-tourism sector indirectly caused by tourism demands;
- eco-efficiency and other ratios; and
- analysis of backward, forward and intrasectoral linkages in relation to GHG emissions from industries and consumers (including emissions of tourism consumers).

Inclusion of a tourism sector in an IO model permits many different analyses of current conditions (or of past conditions, using historical IO tables). However, such a model is less suited to prospective analyses. IO models become more reliable when both productive activities and commodities are classified so as to be as homogeneous as possible.

Table 6.7 A tourism sector distinguished in an IO model

	Other industries	*Tourism industry*	*Government, households*	*Exports*
Other industries	Non-tourism interindustry flows	Intermediate consumption by tourism sector	Domestic final demands for non-tourism commodities	Other commodities
Tourism industry	Domestic business tourism (incl. self-consumption by tourism sector)		Non-business domestic tourism	Incoming international tourism
Rest of world	Imports by non-tourism sector	Imports by tourism sector		

This is because IO coefficients of less homogenous sectors are more likely to be affected by changes in the relative importance of activities within a sector (e.g. hotels versus restaurants). The problem may be reduced by retaining more of the original structure. For example, existing tourism-characteristic sectors could be used to reaggregate tourism activity disaggregated out of other sectors. Conversely, non-tourism activity in tourism-characteristic industries could be disaggregated and reaggregated into appropriate non-tourism sectors. Some heterogeneity would inevitably remain – tourism consumption being heterogeneous – but at least key parts of the tourism sector could be distinguished and their relative change considered in prospective studies.

Conclusions

Most countries undertake GHG inventories as part of their commitment to mitigating global climate change. The IPCC produced guidelines and default emission factors for deriving GHG emissions from a wide range of activities. In those traditional inventories, tourism is largely represented in the transport and energy categories; however, it is not possible to extract tourism and understand its contribution to national emissions, let alone the nature and distribution of those emissions across tourism subsectors. This chapter discussed two approaches for accounting for tourism's energy use and resulting GHG emissions, under the assumption that boundaries can be drawn for the tourism system.

A bottom-up methodology was presented that contained both an industry and a tourist behaviour analysis. Here, the tourism industry was disaggregated into air travel, other transport, accommodation and activities. Energy and emission data gained from such analyses are usefully combined with information on tourist behaviour, ideally broken down into different types of tourists. This provides a very detailed picture about who is using what kinds of energy for what activity and how much this contributes to the total amount of GHG emissions from tourism in a particular destination. A top-down approach has been discussed as well. This involves, for example, IO analysis to extract tourism as a separate sector in national accounts. This methodology gives a macro-perspective on tourism compared with other sectors; it also allows estimating indirect effects.

Both methods allow different questions to be answered. It seems that many destinations – especially those where tourism has grown to be one of the most important economic contributors – would benefit from more detailed inventories of their tourism sector's contribution to national climate change goals. Emissions resulting from various tourist activities will be discussed in the following chapter.

Notes

1. Burning biomass (e.g. wood) involves oxidation of organic material and therefore releases CO_2. Emissions can be considered climate-neutral, because the carbon released will be re-sequestered by regrowth within short periods of time.
2. Environmental impacts from energy use are various, ranging from anthropogenic climate change, to acid rain and other air pollution, ozone generation in the lower troposphere and ozone depletion in the stratosphere, habitat destruction and the emission of radioactivity (Dincer, 1999; Stern *et al.*, 1997).
3. A small proportion of carbon is released as CO, CH_4 or NMVOCs. All of those gases oxidise to CO_2 within a few days or up to 12 years. It is pragmatic to account for emissions as if all carbon was oxidised fully.
4. Each landing and take-off cycle generally increases the energy consumption by about 1000 MJ per passenger (Hofstetter, 1992).
5. Aviation fuel is divided into two basic categories: aviation gasoline (avgas) and jet fuel (kerosene). Avgas is used in aircraft that use internal combustion engines (e.g. scenic flight operators), while jet aircraft and turboprops use kerosene jet fuel. Jet fuel is a refined petroleum distillate intermediate in volatility between diesel and gasoline (petrol). Kerosene Jet A1 is used worldwide in civil aviation. Although avgas and jet fuel differ in things such as volatility and freezing points, their energy density, or heat content, is just about the same.
6. Note that the TPRs should exclude supply met by imports if the GHG account is being constructed for emissions emanating from the national territory.
7. The coefficients can be expressed in terms of physical quantities or values. Quantity and value models are directly related by prices.
8. In IO accounts, the public sector may be separated into 'productive' and 'unproductive' parts. The former are those that recover a significant proportion of their costs from users.

Chapter 7

Climate Change Mitigation Measures

Key Points for Policy and Decision Makers, and Tourism Operators

- It is interesting to observe that while the oil crises compelled an immediate response in research activity, the increasing concern about global warming appears to have generated only a limited and delayed research activity thus far.
- The carbon footprint of air travel can be substantial for some tourist destinations. Improvements in the energy efficiency of air travel are sought in the areas of airframe and engine design, and air traffic management.
- Much tourist travel to destinations is undertaken by private car; air travel ranks second in importance, especially for international tourism. Tourist transport by train is less significant.
- Alternative technologies for road transport include: electric, hybrid, hydrogen vehicles; and biofuel or gas as less carbon-intensive fuel sources. Similar options are available for water transport.
- The often negative social representations of transport alternatives constitute a challenge when addressing transport problems in tourism. Climate-friendly transport for tourists needs to be attractive and also represented as such in the wider discourse.
- Specific transport initiatives range from destination-wide transport management (e.g. car-free resorts) to restrictions on certain routes, encouragement of public transport usage (e.g. by offering attractive travel tickets in urban areas), the establishment of cycle paths or networks, and benefits offered at tourist attractions or accommodation for non-car users.
- Energy use in the accommodation sector is usually due to heating and cooling, hot water supply, cooling for fridges and freezers, and lighting. In warm holiday destinations, the single largest energy end-use is air conditioning.
- The energy intensity of tourist activities or attractions depends largely on the relative importance of a transport component; for example motorised water activities are comparatively energy and emissions intensive.
- A number of renewable energy sources are relevant for tourism. These are wind, photovoltaic (PV), solar thermal, geothermal,

biomass and waste. For most tourist operations it would be most practical to exploit several renewable energy sources, for example hybrid systems.
- Carbon-compensation projects are an option for 'neutralising' emissions that cannot be avoided. There are already a number of schemes that offer tourists the opportunity to offset their GHG emissions by investing in energy efficiency, renewable energy sources or carbon sink projects. Carbon sinks are typically forest sinks; i.e. trees absorb the same amount of CO_2 from the atmosphere that tourists emitted through their activities.

Introduction

Research in the area of energy use and leisure was sparked by the 1970s' energy crises with a clear focus on security of energy supply. In particular, researchers investigated how vacation travel would be affected by a limited energy supply and higher oil prices (e.g. Corsi & Milton, 1979; Kamp *et al.*, 1979; Williams *et al.*, 1979). A number of studies observed various shifts in travel behaviour in response to the oil crises, for example a general reduction in leisure travel (McCool *et al.*, 1974), especially with the private car (Williams *et al.*, 1979), trip chaining (linking several trips into one) (Jones, 1983), the purchase of more economic, smaller vehicles and an increase in shorter holidays based on camping, mainly within the holidaymaker's own country (Knikkink, 1982). There were also some studies that concluded energy crises will not alter travel behaviour, or only to a small degree (e.g. Moncrief *et al.*, 1977).

In more recent years research on the consumption of energy, in particular fossil fuels, has mainly been undertaken in the context of GHG emissions and climate change. However, in the growing body of climate change research in general, tourism's role as contributor has largely been neglected. Environmental impacts from tourism have so far been studied from a more local perspective and often in a business context. For example, energy-efficiency analyses for tourism businesses have been undertaken with the aim of reducing costs, minimising local air pollution and gaining a marketing advantage with 'green consumers'. The interface of tourism and resource management remains under-researched (Carter *et al.*, 2001), and as a result little is known about the energy intensity of various tourism activities, as well as associated GHG emissions. Some research has been undertaken concerning tourist transport (Høyer, 2000; Müller, 1992; Müller & Mezzasalma, 1992), in particular air travel (Gössling, 2000; Knisch & Reichmuth, 1996; Olsthoorn, 2001). The accommodation industry has been addressed in a small number of studies (Buckley & Aranjo, 1997; Deng & Burnett, 2000), and even fewer studies have specifically explored tourists' recreational activities (Byrnes & Warnken, 2006; Stettler, 1997). It is interesting to observe that while the oil crises compelled an immediate response in research activity, the increasing concern and knowledge about global warming appears to have generated only limited and delayed research activity thus far.

The total contribution of tourism to global energy use and GHG emissions is unknown, but likely to be substantial. Gössling (2002b) estimated that tourism transport, accommodation and activities alone contribute up to 5.3% of global GHG emissions from energy use. Other components, for example restaurants and retailers, have not been considered by Gössling, but are likely to be comparatively minor (Patterson & McDonald, 2004). The most important share of tourism's GHG emissions (14081 PJ and 1399 Mt of CO_2-e) is transport, at about 90% of total emissions (Figure 7.1). Tourism's role in the context of global GHG emissions and climate change will increase given that the industry is growing worldwide and that long-distance travel is increasing in popularity. The diversification from traditional mass tourism into an array of special-interest, nature- and activity-based tourism segments also potentially increases the demand for fossil fuels, and therefore GHG emissions (Simmons & Becken, 2004).

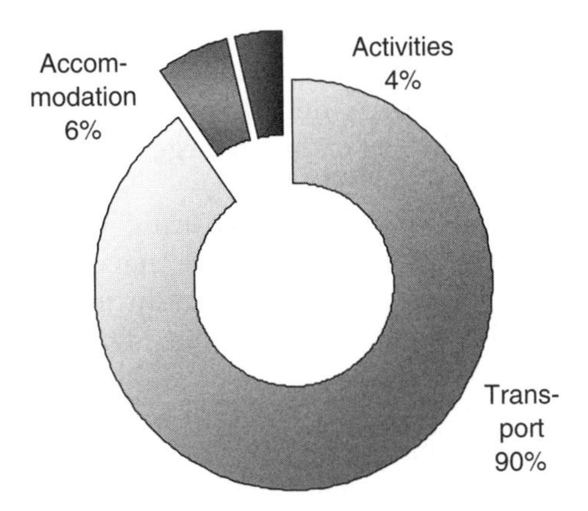

Figure 7.1 Global CO_2-e emissions from tourism
Source: after Gössling (2002b)

Reducing the human impact on the global climate is commonly referred to as climate change mitigation. Broadly, there are four mitigation strategies for addressing GHG emissions from fossil fuel consumption:

- Reducing the need for energy: for example by changing management practices or behaviours, such as walking instead of using motorised forms of transport.

- Improving energy efficiency: the ratio between the energy input (e.g. electricity consumed by a light bulb) and the useful energy output (light energy provided by the bulb).
- Increasing the use of renewable energy sources: replacing the consumption of fossil fuels with energy sources that are not finite, such as hydro, wind and solar energy.
- Sequestering CO_2 through carbon sinks: CO_2 can be stored in biomass (e.g. as forest), in aquifers and in geological sinks (e.g. in depleted gas fields).

This chapter describes tourism's consumption of fossil fuel and production of GHG emissions, and discusses options for reducing both. In particular, it examines tourist transport, accommodation and activities.

Transport

Transport accounts for about one-quarter of global CO_2 emissions, and it probably poses the greatest challenge in terms of GHG mitigation strategies in industrialised countries. In a recent scenario analysis of CO_2 emissions from car travel in the UK, Kwon (2005) found that only under the most optimistic assumptions both in terms of travel behaviour (e.g. reduced number of trips, trip distance, significant use of telecommunications technology) and technological advances (e.g. a rapid transition to alternative vehicle technologies, decrease of average engine size) is it possible to reach a hypothetical emission target of 20% reduction of CO_2 emissions by 2030 relative to 1990 levels.

There are three generic areas for reducing GHG emissions from transport: improving energy efficiency of the transport task, changing to less carbon-intensive energy sources and reducing the transport task (Ministry for the Environment, 2005). Actions underpinning those areas are either technological or behaviour-based (Table 7.1).

Leisure-related travel accounts for about 50% of all transport in industrialised countries. This percentage has been found, for example, in travel surveys in Norway, Germany and Austria in relation to travel distance (Heinze, 2000; Høyer, 2000; Knoflacher, 2000) and in Sweden for travel time, travel distance and travel frequency (Carlsson-Kanyama & Linden, 1999). As explained in Chapter 2, leisure travel includes tourism travel (i.e. for at least one night away from the usual environment), day trips and other recreational travel, for example mobility related to sports activities (Stettler, 1997). Mobility is an essential prerequisite for tourism. Tourists not only travel from their place of residence to the destination, but they also require transportation systems that ensure mobility within a destination, either to visit tourist attractions or to travel along recreational routes. An increasing number of trips are no longer to one

Table 7.1 Framework for considering transport GHG reductions

Change area	*Key actions*	*Policy measures*	
		Technological	*Behavioural*
Improve energy efficiency	Improve technical efficiency of the vehicle	Regular maintenance to manufacturer's standards	Choose fuel-efficient and design-efficient vehicles
	Increase use of more energy-efficient modes		Mode shift to walking, public transport, cycling
	Improve operational efficiency of the transport task	Roading measures to reduce congestion; increase occupancy rates	Improve driver techniques
Change to lower-carbon energy sources	Use of lower-carbon fossil fuels (e.g. gas)		Install and use LPG systems
	Use of carbon-neutral fuels	Develop and make available fuel cells	
	Shift to low-energy propulsion systems	Electric propulsion for rail, buses, battery vehicles	
Reduce transport task	Reduce quantity of trips		Telecommunication for business trips
	Reduce trip lengths		Locate closer to main location

Source: after New Zealand Ministry for the Environment (2005)

single destination (e.g. as in traditional beach tourism), but to more than one destination. The most extreme example of multidestination travel is round-the-world travel, where airlines offer attractive combinations of flights that allow tourists to circumnavigate the globe with a multitude of stop-overs in various continents.

The most important transport modes for tourist travel are the private car and air travel. Other land-based transport, such as rail and bus, and water transport are less important in terms of passenger volume, and even less so with respect to GHG emissions. In the following, air transport, land-based transport and water transport are discussed with regards to current emissions and potential to reduce energy use and GHG emissions.

Air travel

Energy use and emissions of international air travel

The environmental impacts and external costs (e.g. air pollution, noise, accidents, infrastructure requirements and congestion) of air travel have been discussed in various publications (e.g. Janic, 1999; Price & Probert, 1995). The largest external costs of aviation relate to GHG emissions and climate change impacts. Other external costs, for example land use, congestion, capital costs and noise, are comparatively favourable (Maibach & Schneider, 2002). The most comprehensive analysis of impacts on the atmosphere is the Intergovernmental Panel of Climate Change (IPCC) *Special Report on Aviation and the Global Atmosphere* (Penner *et al.*, 1999). In this report, it has been estimated that aviation accounts for 2–3% of the world's total use of fossil fuels and up to 3.5% of the anthropogenic greenhouse effect. More than 80% of this is due to civil aviation.

Air travel can be usefully disaggregated into domestic and international travel, or in the case of Europe into intra-EU travel and travel outside the EU (Figure 7.2). The mix of different kinds of passengers varies for different countries of origin, tourist destinations and airports. In the case of New Zealand, for example, almost all international tourists arrive by air. The energy consumed by all international tourists travelling to New Zealand (one way) in 2000 amounted to 27.8 PJ or 1900 kilotonnes of emitted CO_2 (Becken, 2002). When adding emissions from international arrivals to New Zealand's national CO_2 emissions for the same year, these would increase the total by 6%. This increase does not take into account emissions from New Zealand's considerable outbound tourism. As can be seen from Table 7.2, the energy use and emissions vary considerably for different markets. Tourists from European countries emit almost six times as much as tourists from Australia. As a result of the high per-passenger emissions and the size of the market, the UK is the largest contributor to air travel emissions from New Zealand tourism.

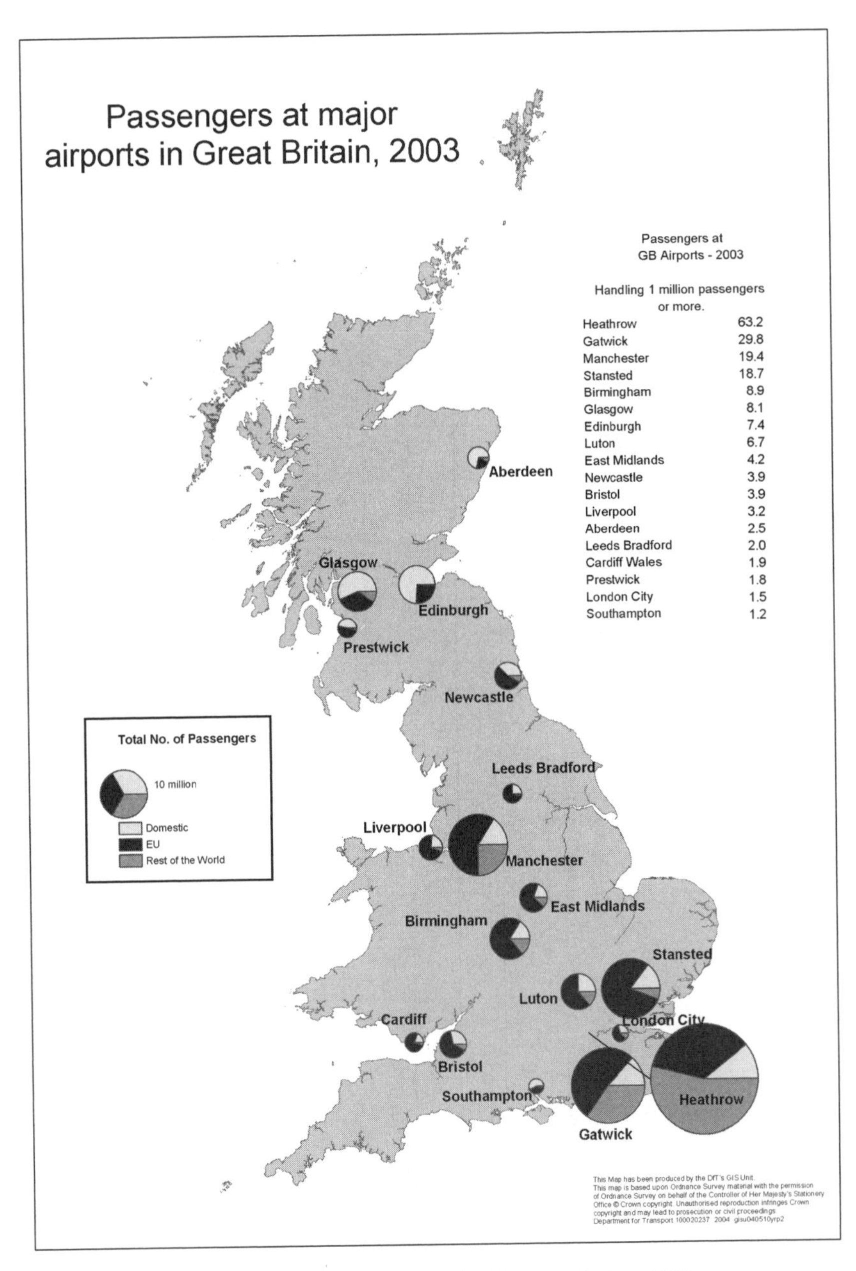

Figure 7.2 Passengers at major airports in Great Britain, 2003
Source: Department for Transport (2004). This map is based upon Ordnance Survey material with permission of Ordnance Survey on behalf of the Controller of Her Majesty's Stationery Office © Crown copyright.

Table 7.2 Tourist arrivals in New Zealand for main countries of origin, average flying distance, energy use and CO_2 emissions

Country of origin	*Total air arrivals*	*One-way distance (km)*	*Energy use per visitor (GJ)*	*CO_2 per visitor (t)*	*CO_2 per country (kt)*
Australia	521,912	3446	6.0	0.42	210
USA	173,182	11,146	19.5	1.4	230
UK	167,202	19,955	34.9	2.4	400
Japan	146,953	9931	17.4	1.2	180
Germany	45,603	20,701	36.2	2.5	110
Korea	43,386	10,684	18.7	1.3	56

Source: Becken (2002)

Gössling *et al.* (2002) calculated the ecological footprint of tourism in Seychelles. In Seychelles, more than 80% of all visitors come from Europe, mostly from France, Italy, Germany and the UK. Almost all tourists arrive by air transport. As in the New Zealand study, energy use (and area of forest required to sequester carbon) was calculated by estimating flight distances and multiplying by an energy intensity of 2 MJ per passenger-kilometre (pkm). Most likely routings were assumed and included national connecting flights in the source countries, international connections (via hubs such as Nairobi or Dubai), and national flights within Seychelles by helicopter and aircraft. The average visitor to Seychelles consumed 25.5 GJ for air transport, compared with 1.2 GJ for other transport and 0.6 GJ for accommodation. Overall, the footprint analysis showed that the major environmental impact of travel is a result of transportation to and from the destination. Gössling pointed out that current efforts to make destinations more sustainable (e.g. energy conservation in hotels) could only contribute marginally compared with the large amounts of energy consumed for air travel.

It is more difficult to allocate emissions from travel to a destination in the case of multidestinational travel, as is, for example, the case for many tourism destinations in Europe. Peeters and Schouten (2006) estimated the ecological footprint of travel to Amsterdam by allocating travel emissions proportionally to length of stay in different destinations during one trip. Similar to Gössling's study, they found that transport to Amsterdam is responsible for the largest contribution to the overall environmental pressure (about 70%), compared with minor contributors, such as accommodation (21%), visits to attractions and other leisure activities (8%) and local transportation (1%). They also noted that long-haul tourists accounted for less than 25% of tourism revenues but were responsible for 70% of the environmental footprint of inbound tourism to Amsterdam, hence implying low eco-efficiency for long-haul travel.

Emission reduction potentials

Fuel is a major cost for airlines at between 15 and 25% (Hanlon, 2007) and reducing energy use is in the commercial interest of airlines (Table 7.3). Increasing fuel costs – be it a result of increasing global oil prices or carbon charges – provide an incentive for airlines to accelerate efforts to reduce fuel consumption. The world's airlines improved their fuel efficiency considerably by around 50–70% between 1950 and 1997 (Penner *et al.*, 1999; Peeters *et al.*, 2005), and they reached an average energy use of 4.8 l of kerosene per 100 passenger kilometres in 1998 (ATAG, 2002). Given the major improvements in the past, it is becoming increasingly difficult to achieve efficiency gains (Penner *et al.*, 1999).

Major aircraft manufacturers pursue different concepts to improve service and fuel efficiency. Boeing is focusing on fast and efficient

Table 7.3 Energy intensities for various airlines from airlines' environmental reports

Airline	*Reference year*	*MJ/passenger-revenue km*
Lufthansa (total fleet)	2003	1.7
Singapore Airlines passenger services	2003/2004	2.0
Cathay Pacific	2003	2.1
British Airways	1999	2.3
Scandinavian Airlines	1999	2.9

point-to-point connections; i.e. longer non-stop flights with medium-sized aircraft (200–250 seats). The use of advanced technologies means that Boeing's Sonic Cruiser could be 15–20% faster at improved fuel efficiencies compared with today's commercial jets. The Sonic cruiser would fly at altitudes of about 13,000 m at a range of up to 7700 nautical miles (ca. 14,300 km). In contrast, Airbus continues to build on the hub-and-spoke concept. They have developed the Superjumbo A380 (Figure 7.3), a double-decker plane with a capacity of 555 seats and a non-stop range of 8000 nautical miles (ca. 14,800 km) to connect major hubs. The effects of aviation's emissions into the atmosphere have been discussed in Chapter 3.

Much of the fuel is consumed during take-off and climbing phases. For short flights, the landing and take-off (LTO) cycle forms a substantial part of the flight. As a result the per-passenger-kilometre fuel use is very high and the emissions impact of short-haul flights is disproportionate.

Figure 7.3 Airbus 380

The longer the flight, the less important becomes the take-off phase relative to total fuel consumption. For long-distance flight, however, the weight of the fuel required is substantial (up to two times the weight of the payload) and to lift the heavy aircraft to cruise altitudes requires extra engine thrust, hence resulting in more fuel consumption. It has been estimated that the most fuel-efficient flight distance is around 5000 km depending on aircraft type (Peeters, 2000).

Efforts to reduce emissions should focus on the operation of aircraft. Improvements are sought in the areas of airframe and engine design, and air traffic management – a discussion of which now follows.

Fuel consumption depends on the design of the airframe and the weight of the aircraft, both of which influence the thrust force. Research into improving airframe designs is targeted at improving the lift-to-drag ratio[1] (L/D). Aircraft drag consists of three main components:

- **Skin friction** drag due to the flow over the airframe skin. Skin friction drag can be reduced by avoiding protruding elements on the airframe (even an object as small as a nail head increases the drag measurably) and enhancing the streamline of the design. An important opportunity applied to all modern sailplanes is to use laminar flow on wings instead of turbulent flow. Laminar flow means that the air close to the airframe skin (think in millimetres) is stable and smooth. All airflows start laminar but become turbulent within a few percent of flowing along the wing chord. A laminar wing is designed such that laminar flow is retained on up to 70% of the wing chord. As the drag from turbulent flow is several times larger than for laminar flow, this potentially reduces friction drag. However, in practice the construction of the wing must be extremely precise and smooth, and the wing must be kept extremely clean (even small dead insects disturb the laminar flow). Therefore, this kind of laminar flow has not been achieved practically for commercial aircraft.
- **Induced** drag due to generating lift causing lift-vortexes by the wing.The induced drag (at optimum cruise flight 50% of the total drag) can be theoretically reduced to zero if the wingspan is increased to infinity (which, of course, is impossible). Induced drag is inversely proportionate to the wing aspect ratio (AR is a measure of how long and slender a wing is; i.e. AR is the wingspan-to-chord ratio). Boeing aircraft have ARs of around 7 or 8 compared with those of Airbus, which range between 8 and 9. A higher AR wing generates a lower induced drag and a slightly higher lift than a wing with a lower AR. A higher AR comes at the cost of increased wing construction weight and also tends to increase mach drag.

- **Mach**[2] drag due to compressibility effects of the air at speeds near the speed of sound (significant for modern subsonic aircraft).Mach drag can be reduced by flying slower and by choosing a well designed wing section (the form of the cross-section of the wing). Designs are trade-offs between high-speed (cruising) and low-speed (landing and take-off) characteristics of the wing. For high speeds a swept-back wing with a low thickness ratio is preferable, while for landing and take-off a non-swept-back wing and a higher thickness ratio is the best solution.

One new concept for airframe design is the blended wing-body (so-called 'Flying Wing'), where the aircraft's body is contained within the wing. This design potentially has a lower drag with a fuel consumption of possibly 30% less than traditional aircraft with the swept-wing-fuselage airframe design (Royal Commission on Environmental Pollution, 2002). This kind of aircraft, however, is not expected to enter the market within the next decade. More short-term improvements can be obtained from higher-aspect-ratio wings, advanced aerodynamic design and reducing weight by new materials, such as aluminium alloys, titanium and composite materials.

Much improvement has been achieved in engine technology (about 40% over the last 40 years, Penner *et al*., 1999), with the most efficient engines being gas turbine engines with a high-bypass to high-pressure ratio. Improved technologies for fuel efficiency (e.g. staged combustion) involve higher pressures and engine temperatures. These, however, result in increased NO_x emissions. Thus, the development of new technologies often constitutes a trade-off between reducing fuel use and thereby CO_2 and water vapour emissions and NO_x emissions. The IPCC expects further reduction potentials from improved engine technology in the order of 10% by 2015 and 20% by 2050 (Penner *et al.*, 1999).

Another opportunity for advanced zero-climate-impact aircraft might be the fuel cell aircraft (Peeters, 2000; Snyder, 1998). Fuel cells run on hydrogen. Hydrogen is produced in an electrolyser, where an electric current splits water into hydrogen and oxygen. The reverse process happens in a fuel cell: electricity is generated through a chemical reaction of hydrogen and oxygen to water. This is a highly efficient means to store and generate power, compared with traditional combustion engines. Hydrogen requires different systems for storage, as hydrogen would be transported in a liquid form under cryogenic temperatures in high-pressure and insulated storage tanks. These systems would be heavier than those for kerosene and therefore increase fuel consumption. Hydrogen takes four times the volume per unit of heat it contains (see Brewer, 1991), and hydrogen storage tanks would need to be much larger than current fuel tanks, thereby adding extra weight and drag to the aircraft. On the other hand, per unit of energy the weight of hydrogen is 2.8 times lower than for kerosene, saving a lot of the design take-off weight. The overall impact on energy consumption seems generally to be advantageous: the aircraft would be 40 – 50% more energy efficient, but fly at somewhat lower cruising speeds and altitudes.

Using liquid hydrogen would eliminate CO_2 and particle emissions but not reduce the problem of NO_x, and it would also lead to production of more water vapour (about 2.6 times). This could result in the increased build-up of contrails and cirrus clouds, resulting in increased radiative forcing (Royal Commission on Environmental Pollution, 2002). Moreover, currently hydrogen is produced based on fossil fuels, which means that emissions are only reduced at the point of use but not in general. Unless hydrogen is produced from renewable primary energy sources or from fossil fuels, including large-scale storage of CO_2, there will be no global reduction in GHG emissions. Other alternative fuels, such as ethanol, have a lower calorific value and are also associated with undesired side-effects, such as the formation of formaldehyde (Royal Commission on Environmental Pollution, 2002).

Substantial fuel reductions can be expected from improved operations and air traffic management. The IPCC (Penner *et al.*, 1999) estimated that up to 10% could be saved in fuel by reducing congestion and optimising flight paths (Williams & Noland, 2006). In Europe, Reduced Vertical Separation Minima (RVSM) was introduced in 41 states, which reduced emissions by 5% in the upper atmosphere. Optimisation of air traffic management will be facilitated through new navigation systems such as the Galileo satellite navigation system developed in Europe and possibly entering the market in 2010. Also, achieving higher load factors would decrease the emissions per passenger kilometre. Currently, load factors typically vary between 70 and 75% on international routes. A reduction in cruise altitude could also diminish aviation's impact on the climate as a result of reduced formation of contrails. However, flying at lower altitudes is usually associated with a fuel penalty (due to lower energy efficiency) and there is a trade-off between increased CO_2 emissions and reduced radiative forcing from contrails (Williams *et al.*, 2003). Williams *et al.* found that for inter-European flights a decrease of flight altitude and therefore contrail formation results in a positive effect (i.e. reduced radiative forcing) despite an increase of 3.9% in fuel burn.

The emissions associated with the production of an aircraft are negligible compared with those resulting from flying. Airbus estimated that the energy use in service makes up 99% of the total energy used throughout the life cycle of an aircraft. Moreover, it has been reported that emissions from aircraft manufacturing of an airbus A-320 are equivalent to emissions from one week of service (Royal Commission on Environmental Pollution, 2002).

Infrastructure required for aviation is comparatively minor relative to other transport modes. The energy requirements for airport operations are small compared with those for the operation of aircraft. Notwithstanding this, emission reductions can be made and can be a substantial contribution to reducing GHG emissions from tourism at a national scale. This is demonstrated by the case study of Auckland International Airport, New Zealand (Text Box 11).

Text Box 11 Auckland International Airport
Source: Energy Efficiency and Conservation Authority (2004).

About 9.4 million passengers passed through Auckland International Airport in 2002 and capacities could cope with a growth of 30%. The airport's electrical energy efficiency almost doubled between 2000 and 2003 (3.0 to 1.7 kWh per passenger). Each passenger is responsible for the emission of 260 kg of CO_2 as a result of electricity consumption and 150 kg of CO_2 from gas usage; total annual CO_2 emissions amount to 3854 tonnes. Most of the energy is used for climate control (heating and cooling) and lighting.

To reduce energy use, recent building extensions and renovations have taken into account lighting and thermal features of building materials, lights are metal halides and fluorescent with daylight sensors or timer systems, and lights are also connected to motion detectors (e.g. in toilets). A building management system adapts temperatures to arrival and departure volumes, as well as to external weather conditions. Heating is met by a gas-fired boiler whose enclosure is well insulated to preheat air and reduce the cost of heating.

Land and water transport

Energy use and emissions of different land transport types

Much tourist travel to destinations is undertaken by private car. The relative importance of the car and the aeroplane differ for domestic and international tourism. In the case of domestic tourism, typically at least three-quarters of all journeys are based on the private car (Table 7.4), whereas the importance of air travel increases for international travel (up to 100% for island countries). The use of different transport modes is further detailed in the case study of European tourism shown (Text Box 12).

The energy and carbon intensity of a transport mode depend on the fuel source and engine technology, fuel consumption per vehicle kilometre (among others dependent on driving behaviour) and the

Table 7.4 Breakdown of transport modes for tourism travel in different countries

Country	*Transport modes*	*Comment*	*Reference*
USA	Cars/personal vehicles 89%; air travel 7%; bus 2%; train 1%, ship/ferry 0.1%	Domestic travel (more than 50 miles) in 2001; 44% are overnight trips	Bureau of Transportation Statistics, 2004
UK	Cars 73%; train 12%; air travel 5%; bus/coach 6%; caravan/campervan 1%	Domestic tourism by UK residents	United Kingdom Tourism Survey (UKTS), 2003
Germany	Cars 49%; air travel 35%; bus 9%; train 6%	For holiday trips longer than 5 days	FUR, 2001
New Zealand	Cars 77%; air travel 13%; bus 9%; company car 5%; ferry 3%; rental car 2%; bus 2%	Domestic tourism	Forsyth Research, 2000
France	Cars 72%, train 14%, plane 8%, ferry 3%, others 3%	French outbound tourism (more than one night and 100 km from home)	'Suivi de la Demande touristique' (SDT) Survey, 2002*

*Personal communication, G. Dubois, September 2005

Text Box 12 European MUSST Project
Contributed by Paul Peeters, NHTV, Breda University, The Netherlands

The European Commission is undertaking steps towards achieving sustainable transport. A first step was a feasibility study regarding a multistakeholder European targeted action for sustainable tourism and transport (MuSTT, for more details, see Peeters *et al*., 2004). The MuSTT study includes domestic, international and intercontinental tourism (ICA). The focus of this excerpt is outbound tourism, which includes all European tourists visiting any place outside their normal place of residence. The focus of the study is on Origin/Destination (O/D) transport, which is transportation between the normal place of residence (origin) and the tourism destination. The most common tourist transport modes have been distinguished: car, airplane, train, coach and ferry. The following data refer to all 25 members of the European Union (EU25), unless stated otherwise.

Methods

For MuSST the European model used for TEN-STAC (scenarios, traffic forecasts and analysis of corridors on the Trans-European Network) has been used. This model covers all interregional passenger transport by EU member states citizens and distinguishes 'holidays' as one travel purpose. However, TEN-STAC's definition of holiday as a leisure trip of at least two nights does not match the definition applied by WTO. A range of adjustments and assumptions were necessary to overcome those data issues. In general, transport properties like modal split were taken from TEN-STAC, while numbers of international tourists were based on WTO data. The number of domestic tourists has been based on TEN-STAC, but corrected for the WTO definition of tourism. The resulting MuSTT model contains information on tourist transport by a range of transport modes in relation to number of trips, passenger-kilometres (pkm) and emissions of CO_2, CO_2-es, NO_x and PM, for a total of 915 O/D-pairs.

Tourism transport in Europe

In 2000, the total number of outbound trips by EU25 citizens has been calculated to be 875 million trips (61% domestic, 29% intra-EU25, 4% to European countries outside the EU25 and 6% to other continents). The prognoses for 2020 show an overall increase for outbound tourism trips by 2.3% per year up to 1371 million trips. However the growth is not evenly distributed across travel distance classes (Figure 7.1). Large growth rates occur for trips with a large O/D return distance; however, long-distance markets currently make up only 6% of all outbound tourism. Long-distance markets are growing faster and shares are expected to increase to 13% in

2020.Mobility, i.e. distance travelled, is more meaningful than number of trips. In 2000, ICA represented 37% of outbound mobility (measured in pkm). By 2020, this will rise to 56%. In contrast, the domestic market – 63% of all trips – generates only one-third of all tourism mobility in 2000. This mobility share will further decrease to 20% in 2020. Total mobility is expected to increase from 2020 billion pkm in 2000 to 4480 billion in 2020. Europe's outbound tourism mobility in 2000 represented about 7% of world mobility.

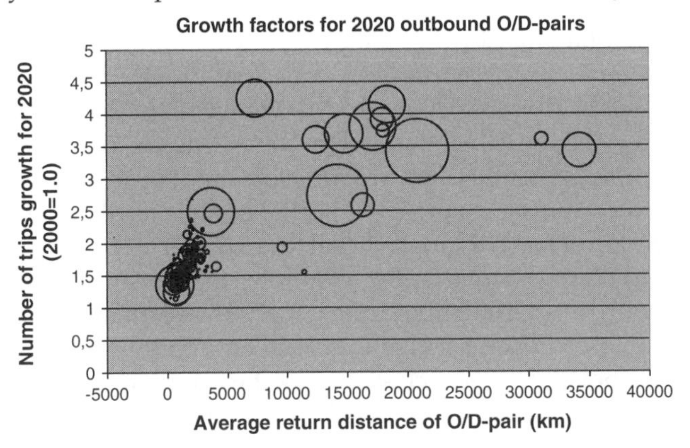

Text Box 12 – Figure 1. Increase in trip numbers by distance classes. The area of the bubble represents the total of pkm for each O/D category in 2020.

Greenhouse gas emissions

The impacts of mobility on GHG emissions depend broadly on two parameters: the modal split and the technological and operational efficiency. Most tourism trips depend on ground transportation, but air travel makes up the main share of mobility and GHG emissions (**Figure 2**).

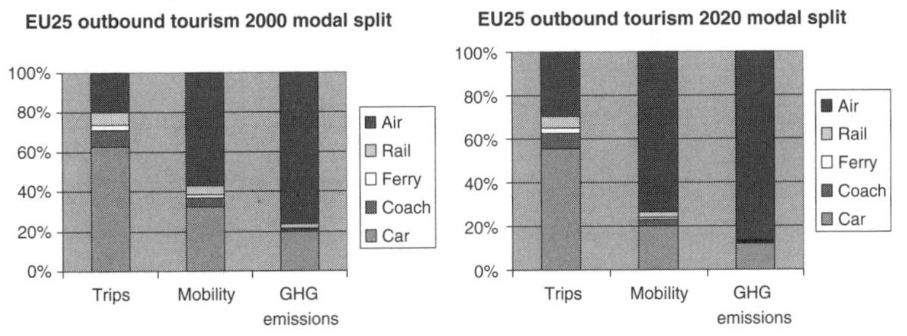

Text Box 12 – Figure 2. Modal split broken down by number of trips, mobility (pkm) and GHG emissions of EU25 outbound tourism transport (including ICA) in 2000 and 2020.

Air transport will heavily influence any overall reductions in GHG emissions. The IPCC (Penner *et al*., 1999: 5) report an annual overall efficiency (MJ/pkm) increase of 1.1–1.9%. Assuming a rate of 1.4% over the period 2000–2020, future GHG emissions from European tourist transport could be estimated. The same technological gains were assumed for the other transport modes. In 2000, total GHG emissions were 474 million tonnes of CO_2-e; intra-EU25 and domestic tourism contributed 210 million tonnes. Emissions are projected to increase to 878 million tonnes by 2020. Tourist transport makes up about 9% of total emissions in the EU25 (5052 million tonnes CO_2-e in 2000). Since EU25 tourism transport emissions are expected to grow by 90% between 2000 and 2020, and all other EU25 GHG emissions are expected to decline, this share may increase to 16–18%. Within tourism it is customary to divide the market by geographical attributes, for example domestic, international, interregional or intercontinental tourism. This means that the size of a particular region influences mobility statistics. For example, the distance travelled for domestic tourism in Luxembourg is generally less than 100 km, while for China this could be as much as 5000 km. To evaluate impacts of tourist transport, distance classes may be more relevant than the crossing of borders. The large number of O/D-pairs available within the MuSTT database allows some insights into distance-related travel behaviour. Table 7.1 shows, for example, that the two highest return distance classes represent only 13% of all trips, but cause 50% of mobility and 63% of GHG emissions.

Text Box 12 – Table 7.1. Distribution of trips, passenger-kilometres and GHG emissions over some distance classes for all EU25 outbound tourism

Return distance class (km)	*Number of trips (%)*	*Mobility pkm (%)*	*GHG emissions (%)*
0–1000	71	31	18
1000–2000	16	19	19
2000–4000	7	13	16
> 4000	6	37	47

Concluding remarks

Within the European Union tourism is sustained by the domestic and intra-EU25 markets, which make up more than 90% of trips. Intercontinental travel contributes only about 5% to all EU25 tourism, though this share may double by 2020. Ground transportation dominates, but air travel is becoming more and more important. Air transport is responsible for almost 80% of GHG emissions from European tourism transport. Tourism transport contributed 9% of total GHG emissions in the EU25 in 2000. This share could increase when one considers current trends of GHG reductions to comply with the Kyoto target of an 8% reduction compared with 1990 levels. The MuSTT study shows the importance of tourist transport in general and air transport in particular in the context of climate change mitigation. The uneven distribution of emissions over tourism distance markets offers an opportunity to reduce emissions significantly, while affecting only a relatively small part of all tourism.

typical load factor. As can be seen from the New Zealand situation (Table 7.5), energy use and emissions differ substantially for different transport modes. Air and water transport are comparatively energy-intensive, while collective modes of transport such as buses and trains are more energy efficient. Examples of emission factors have already been presented in Chapter 6.

Table 7.5 Transport energy efficiencies of New Zealand transport modes used by tourists (reference year 2001)

Transport mode	*Energy efficiency (MJ/pkm)*	*CO_2 emissions (g/pkm)*
Domestic air	2.54	175
Rental car*	0.94	63
Private car*	1.03	69
Coach (tour bus)	0.32	22
Scheduled bus	0.51	35
Ferry	2.63	181
Camper van	2.39	165
Train	0.38	26
Shuttle bus, van	0.56	39
Motorcycle	0.87	60
Backpacker bus	0.39	27

*Average occupancy rates for rental and private cars were 2.5 and 3.2 tourists per vehicle, respectively
Source: Becken & Cavanagh (2003)

Text Box 13 East Japan Railway Company
Source: East Japan Railway Company (2005)

The East Japan Railway Company operates over 13,000 rolling stock and 1700 stations, and carries 16 million passengers. Business activities go far beyond transportation services, and include hotels, retail, travel agencies, advertising, real estate management and financial services. In 1992, JR East Group developed three basic policies:

- To contribute to customers' lives and local communities by providing a comfortable environment.
- To develop and provide the technology needed to protect the global environment.
- To maintain an awareness of environmental protection and raise the environmental awareness of our employees.

An annual Sustainability Report is produced. In particular, the following goals (established in 1996) were achieved by 2002.

- A 20% reduction of CO_2 emissions in general business activities (to 2.32 million tons of CO_2 in 2002).
- Realisation of an energy-saving railcar representation of 80%. JR developed an energy-saving railcar that consumes about half the electricity of conventional railcars. The introduction of those railcars and other facilities has reduced CO_2 emissions by 430,000 tons.
- A 15% reduction in energy consumption for train operations in proportion to unit transportation volume.
- A 60% reduction of NO_x emissions at a company-run thermal power plant.
- The promotion of environmentally friendly driving practices has lowered the instances of quick acceleration. This has reduced energy use and also accident rates by 38% over three years.
- Implementation of specific environmental conservation activities (including tree planting) on an annual basis.
- The carbon efficiency of JR has improved from 94.5 to 71.5 tonnes of CO_2 per billion yen of operating profit.

JR is also developing the NE Train – a hybrid type of railcar. This new train is expected to reduce energy consumption by about 20%. Fuel-cell-powered NE trains are part of JR's future development plans.

Trains are generally more efficient than cars (see also Text Box 13). The actual operational energy consumption for trains depends among others on the speed, the gradient and the number of accelerations (Jørgensen & Sorenson, 1997). Occupancies of trains are usually lower than those of aeroplanes. In Denmark, for example, occupancy factors range from 28% for S-trains to 56% for InterCity trains. Similarly, the German ICE trains have an occupancy levels between 39% (Frankfurt to Munich line) and 61% (Hamburg to Frankfurt line) (Jorgensen & Sorenson, 1997). Taking into account average travel conditions, Jorgensen and Sorenson derived emission factors for different types of passenger trains in Denmark (Table 7.6).

Table 7.6 Energy consumption and emissions per passenger-kilometre (g/pkm) for passenger trains in Denmark

	Energy use (MJ/pkm)	*CO_2*	*SO_2*	*NO_x*	*HC*	*CO*	*Particles*
InterCity	0.52	36	0.01	0.6	0.03	0.1	0.02
International	0.74	54	0.02	1.0	0.04	0.2	0.05
Interregional	0.89	66	0.02	1.2	0.05	0.2	0.06

Source: Jorgensen and Sorenson (1997)

The Deutsche Bahn AG offers a Web-based environmental mobility check for energy intensity and GHG emissions associated with individual transport choices within Germany. The calculations are based on average occupancy rates for public transport (train and bus), and car properties can be selected to best reflect the customer's alternative (e.g. mid-class car, petrol engine with a catalytic converter from 1997 and an occupancy of 1.4 passengers). The output shows energy use for the vehicle, use of primary energy (i.e. including embodied energy) and CO_2 emissions. For example, on the Frankfurt-to-Munich route, train travel is equivalent to 3.2 l of petrol of consumer energy, 10.0 l of primary energy and the emission of 17.8 kg of CO_2 (Figure 7.4). This compares very favourably to car travel. Emissions from electric trains stem from power plants and to derive primary energy-use a factor of about three has to be applied in terms of energy consumption relative to vehicle energy use.

Many countries envisage substitution of trains for air travel through the increasing network of high-speed trains (e.g. in France on the Paris-to-Marseille route), because not only are high-speed trains more energy-efficient than aeroplanes, they are also powered by electricity and therefore contribute to a lower level of local air pollution.

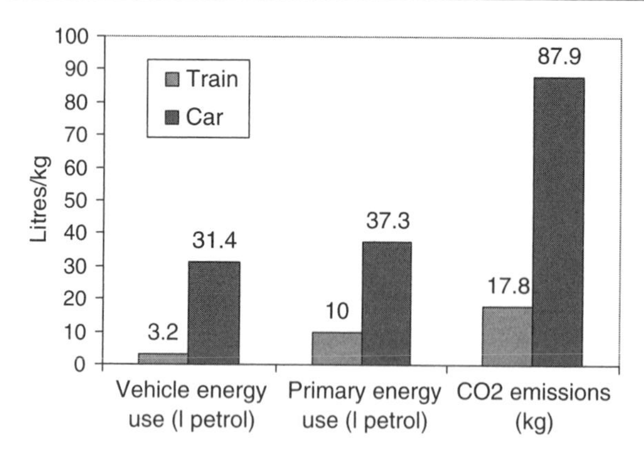

Figure 7.4 Comparison of the energy use and CO_2 emissions for a journey from Frankfurt to Munich by rail and car based on the Environment Mobility Check provided by the Deutsche Bahn AG (2004) (www.reiseauskunft.bahn.de)

Energy use and emissions of water transport

Water transport (both freight and passenger) is a significant contributor to global GHG emissions. International shipping alone contributes about 2% to global fossil fuel consumption, and its contribution to anthropogenic CO_2 emissions is in a similar order (Corbett & Fischbeck, 1997). Several forms of water transport are relevant to tourism. The most important categories are ocean cruise ships, river cruise ships and passenger ferries. The tour boat sector is another important subsector, which will be discussed in the section on tourist activities.

Worldwide, about 11 million tourists travel on the more than 260 cruise ships that are in operation (Gössling, 2002b; Peeters *et al.*, 2004). This represents about 1.6% of all global international tourist trips. The largest cruise ship tourism market is North America, with about 75% of the total. Major destinations for cruise ships are the Caribbean, Alaska, the Mediterranean and the South Pacific. Traditionally, ocean cruise tourism has attracted older and rather wealthy market segments, but the fastest growth rates are now in the 25–39 years age group. Also, products are increasingly diversified, catering for a wide range of customers from adventure travellers to business tourists. Many of the cruise ship products are offered in the form of 'fly and cruise' packages. Typically, cruises offer land-transport options when in the harbour. These land trips have been subject to research in the areas of economic (e.g. Dwyer & Forsyth, 1996), social and environmental impacts (Johnson, 2002).

Little has been reported about the energy use and GHG emissions of cruise ships. There are several reasons why it is difficult to analyse the energy use of cruise ships:

- Cruising companies are often large international companies that are difficult to approach, in particular for information that is not market-related.
- Fuel consumption varies considerably for different types of cruise ships and it is difficult to obtain information on a per-passenger-day basis (distances covered is less relevant, so the consumption per day is a more appropriate measure).
- Diesel consumed by the ship is for transport and on-board electricity generation necessary to meet catering and hospitality functions of the vessel. A cruise ship fulfils the functions of both transport and accommodation.
- Cruise ships accommodate a large number of staff, which adds to energy demand and the energy intensity of the ships.

Some indication of the energy use of a cruise ship could be obtained through a search on the Web. In combination with information on the maximum passenger capacity, the energy use per passenger was estimated. For the four Web sources[3] identified, energy use ranged between 1500 and 4000 MJ per passenger-day. These figures show that cruise ships are extremely energy-intensive, even when up to 500 MJ per day is allocated to accommodation and entertainment (e.g. swimming pools, restaurant, bars and cinemas). This would leave at least 1000 MJ per passenger-day for transport, which is equivalent to about 500 km of air travel every day of the cruise trip. These estimate confirm those by Peeters *et al.* (2004), who reported CO_2 emissions of between 155 and 700 kg/day, the equivalent of between 2000 and 10,000 MJ per passenger-day. Gössling (2002b) estimated an energy use for the coastal steamers of the Norwegian company 'Hurtigruten' in the order of 7.2 MJ/pkm (including accommodation). Gössling concluded that the cruise ship sector could use up to 1.6 Mt of fuel (68 PJ of energy use) per year, and emit about 5 Mt of CO_2.

Apart from cruise ship tourism, a large number of other boat operations cater for tourists. River cruises, for example, are an important niche market with about 1 million passengers per year (Peeters *et al.*, 2004). Among tourists from the UK, for example, the most important rivers for cruising are the Rhine (30,000), the Nile (25,500), Asian rivers (18,000) and the Danube (12,200). Little research has been done on river tourism in general and environmental issues, such as fuel consumption or GHG emissions. More general research is available on passenger ferries, used by both tourists and non-tourists. According to Gössling (2002b) there are about 2150 short sea ferries in operation and their fuel consumption could be in the order of 1.9 Mt per year. Gössling assumed that one-third of this can be attributed to leisure; this would result in approximately 26 PJ of energy use and 1.9 Mt of CO_2 per year for leisure-based ferry transport.

Emission reduction potentials

Technologies and alternative fuel sources

Several alternative technologies for road transport are already in use or are being developed. Alternatives are sought that result in lower emissions compared with current petrol- or diesel-based combustion engines. Most of the new technologies are also associated with a number of shortcomings (Table 7.7). This is particularly true when comparing the complete life-cycle, as some fuels may emit little or nothing during vehicle use, but more during vehicle manufacture or other upstream activities, for example emissions from electricity generation.

Electricity-powered vehicles have been in use for a number of years, but have made way now to hybrid vehicles that have two power sources, namely electricity and petrol or diesel. The hybrid vehicles differ from the original battery vehicles in that they recharge during driving. Emission reductions of CO_2 are in the order of 20–25% for hybrid cars compared with internal combustion engines (Transportation Association of Canada, 1999). Hybrid cars are seen as an intermediate step to the mass production of fuel cells, which run on hydrogen. At present, hydrogen is produced using energy from renewable sources or from petroleum fuels and therefore the use of hydrogen cars still results in emissions. Notwithstanding this, hydrogen is a long-term prospect for transportation.[4] Hydrogen fuel cells based on renewable energy sources are favourable with regard to all parameters, while those based on fossil fuels are still cleaner with regard to acidification and photo-smog than conventional engines, but not necessarily in terms of CO_2 emissions.

Bio-diesel (e.g. ethanol or methanol) can be extracted from corn and blended in gasoline with different results in emission reductions for different mix ratios. E85 (ethanol blended in petrol up to 85%) achieves CO_2 reductions of about 40%. The advantage of lower mix ratios (e.g. 10%) is that cars do not require changes in the fuelling system (Transportation Association of Canada, 1999).

The use of compressed natural gas (CNG) and liquefied petroleum gas (LPG) reduces CO_2 emissions by about 25%. The gases contain less carbon, the recovery and processing are less energy intense, and other emissions (e.g. carbohydrates) are less toxic (Transportation Association of Canada, 1999). Fuel switching from diesel to CNG can reduce CO_2 emissions, but may lead to an increase in CH_4 and NO_x emissions, therefore, decreasing the overall benefit in decreased GHG emissions. Another drawback is the inconvenience of accommodating the heavy high-pressure tanks.

Table 7.7 Comparison of alternative fuel and vehicle technologies for vehicles

Technology/ fuel source	*Description*	*Benefits*	*Shortcomings*
Electric	Battery powered	• Zero emissions at vehicle • Low noise • Low operating costs	• Limited driving range • Lengthy charging times • High cost of the battery • Little maintenance • Electricity generation potentially produces emissions elsewhere
Hybrid	Combination of electric motor with internal combustion engine	• Very fuel-efficient • Low emissions • Established infrastructure (no recharging of battery required)	• High cost • Current models have relatively small engines • Some emissions remain
Hydrogen	Fuel cell: generates electricity by oxidation of hydrogen	• Twice as efficient as internal combustion technology • Zero emissions at vehicle	• Short durability of fuel cell • High cost • Lack of infrastructure for refuelling • Electricity generation potentially produces emissions elsewhere
Ethanol and methanol	Blend can be used in existing combustion engines	• Low emissions • Renewable energy source	• Larger tanks are required • Little infrastructure for refuelling
CNG or LPG	Used in existing combustion engines	• Low emissions • Medium network for refuelling • Available in dual-fuel, i.e. gas and petrol	• Shorter driving range • Storage space for tank needed

Research has also been undertaken on various technologies (e.g. advanced diesel engines; operations management) and fuel sources to improve fuel efficiency of vessels and reduce emissions (CALSTART, 2002; Farrell *et al.*, 2002). Also, the potential of renewable energy sources for water-going vessels is explored, for example the use of wave energy, wind or solar energy to produce hydrogen that is stored in fuel cells on board. Currently, those alternatives are too costly to enter the market in the short-term. Theoretically it is also possible to run vessels on renewable fuel sources such as copra (coconut oil), which could be an interesting option for tropical small island destinations.

Individual transport behaviour

The most efficient use of an independent vehicle is one that is adapted to the actual needs of the traveller. Often these needs are 'over-matched', as for example in the case of hired four-wheel drive vehicles, because of fashion and a flexibility that most tourists will not exploit. Not only type or make can be chosen wisely, but also the size; this means matching capacity with likely occupancy. Driving behaviour can affect the fuel consumption considerably. For example, 'aggressive driving' as compared with 'restrained driving' increases the specific fuel consumption by about 30%. The use of air-conditioning increases the fuel bill by 10–15%, and an extra load of 100 kg increases fuel consumption by another 7–8% (Van den Brink & Van Wee, 2001). Changing driving behaviour was found to be among the most promising measures to reduce passenger transportation emissions in Canada (Transportation Association of Canada, 1999).

It could be argued that time efficiency should not be the major issue in travel. Often, the tourist does not gain anything by travelling fast to a destination and missing out points of interest en route. By considering travel as an attraction in its own right, slower transport modes offer a greater chance to relax and enjoy the environment. For example, cyclists travel at a very slow pace, strongly motivated by exploring the area in more detail and experiencing nature and the scenery (Ritchie, 1998). In fact, cycle tourists spend 75% of their time on cycling, and consequently have a lower demand for other, potentially energy-consuming, activities. In contrast, air travellers reach their destination quickly at the expense of a large amount of energy, and have additional time to participate in other activities.

Dickinson and Dickinson (2006) note that, despite a variety of examples of how to address tourist transport, it is still unclear what works well and for what reasons. They argue that transport planning is often based on measures of people's attitudes that predict how people will respond to a transport initiative. Many studies of transport behaviour assume people operate rationally, but this may not be the case. Often people do not behave according to their attitudes, especially concerning travel, where

Table 7.8 Examples of various transport management initiatives

Type of initiative	*Examples*
Destination-wide tourism traffic management initiatives	GAST (Switzerland), where nine communities joined to promote car-free tourist resorts and use of train to the resort; e.g. in Saas Fee with train access from major cities with convenient luggage service, use of public transport at the destination, and cheaper rental car rates. NETS (Network for Soft Mobility) is an EU-funded project to promote the use of environmentally sound transport modes for tourism, including a number of pilot projects. Strasbourg, France, operates tramlines to reduce inner-city traffic, pedestrian zones, restricted use of private vehicles, limited parking space at high fees, park and ride system with free shuttle service, bike rental, electric tourist train for sightseeing in the inner city.
Containment/restriction and pricing strategies	Road charging system to manage congestion in the Upper Derwent Valley in the Peak District National Park, UK (2 million visitors). Car-free resorts and islands in the North and Baltic seas, e.g. Hiddensee, Baltrum, Zingst where travel by train is encouraged and walking and cycling holidays are promoted.
Encouraging use of public transport routes	Wayfarer ticket in the greater Manchester Area including travel on bus, train and Metrolink to encourage public transport use throughout the day and specifically on the weekend through a weekend ticket (extending as far as the Peak District National Park). France maintains a network of high-speed trains (TGV), for example TGV from Paris to Marseille (3.5 h) and the *train de neige*, to reduce air travel.

(*Continued*)

Table 7.8 (*Continued*)

Type of initiative	*Examples*
Improving routes for cyclists and walkers	UK's National Cycling Network offers 5000 miles of connected cycling routes on traffic-free trails, traffic-calm roads and minor roads. La Rochelle and Ile de Re, France, offer 300 bicycles to rent and luggage service associated with bike hire, cycle paths, day-pass tickets for bicycles and public transport.
Initiatives at visitor attractions and accommodation providers	Initiatives by the British National Trust and Royal Society for the Protection of Birds to offer discount entry and other benefits for non-car-users. Fahrtziel Natur is a cooperation between the Deutsche Bahn and the Naturschutzbund, Verkehrsclub and WWF Deutschland to encourage train travel to National Parks by offering attractive packages and information on transport and the destination, including nature activities.

Source: after Dickinson & Dickinson (2006)

a multiplicity of social realities underpin attitudes and subsequent behaviour. Social representations of reality (Moscovici, 1981) include tacit, widely accepted knowledge and common beliefs that are widely shared in society. Social representations are often created by the media or powerful community members, rather than by individual experience. Typical social representations for tourist transport include 'if public transport was improved people would use it more', 'transport problems are caused by others' and 'car use cannot be restricted'. In those representations and their inertia lies a challenge when addressing transport problems, and a more thorough investigation of those representations and their origins is required for a successful implementation of environmentally friendly transport alternatives.

Sustainable transport initiatives

Much research has been undertaken to investigate ways in which individual car travel can be reduced and how a shift to more environmentally sound forms of transport could be accomplished. Such transport alternatives have to be perceived as attractive for tourists, day visitors and local residents. In many cases, locals resist new transport initiatives, because they fear that tourists will choose another destination instead of adapting to public transport systems (Eaton & Holding, 1996).

Lumsdon *et al.* (2006) note that there has been a change in approach from an emphasis on investment in physical infrastructure to more 'soft' measures to encourage a modal shift towards sustainable forms of transport. The new approach is referred to as travel demand or mobility management. Initiatives range from destination-wide transport management (e.g. car-free resorts) to restrictions in certain routes, encouragement of public transport usage (e.g. by offering attractive travel tickets in urban areas), the establishment of cycle paths or networks and benefits offered at tourist attractions or accommodation for non-car users (Table 7.8). Those measures also involve improving information systems, increasing reliability and personal security, and improving interchange facilities (see also Text Box 14).

Text Box 14 Bad Hofgastein/Werfenweng, Austria
Source: Thaler (2004), Werfenweng (www.werfenweng.org.au)

A 'Sustainable Mobility – Car-Free Tourism' project has been implemented in two major tourist centres in the Austrian province of Salzburg: Bad Hofgastein and Werfenweng. There are 7500 inhabitants and over one million overnight stays per year in those case studies. Bad Hofgastein is among the 10 most tourism-intensive communities in Austria. In both case studies, tourists are encouraged to travel by train, and free pick-ups and luggage services are offered. Tourists can participate in the Sustainability Mobility programme when they:

- arrive by train;
- leave their car key at the reception of the tourism office; or
- stay at one of the 24 Soft Mobility vacation hotels and get a Soft Mobility Card (€5 administration fee).

These tourists are then eligible for a wide range of benefits:

- transfers from and to the train station Bischofshofen for free, and free shuttle service from Werfenweng to Bischofshofen;
- night taxi on weekends (10 pm to 4 am);
- free use of the soft-mobile fleet (electric cars, fun-Riders, E-Scooters, Arrows, etc.); hire of bicycles free of charge (depending on availability); vehicles are loaded on the first solar loading station in Austria;
- one guided bicycle trip per week, Nordic Walking guided tour, one guided hiking trip (Alpine meadows);
- free entrance to leisure park Werfenweng with swimming lake, beach volleyball, etc., reduced entrance ski museum Werfenweng, price reduction at the tennis court.

All activities are accompanied by marketing campaigns, e.g. press releases, newsletters, media cooperations, events and a user-friendly website. Since introducing the car-free resort programme (1999–2004), Werfenweg saw a:

- 38% increase in overnight stays in the winter season,
- 101% increase in overnight stays within the special interest group 'holidays from car', 31% increase in overnight stays in the summer season,
- increase in train arrivals from 9 to 25%,
- tripling of passengers for the Werfenweng Dial-a-ride Shuttletaxi.

In Bad Hofgastein, the use of electric vehicles resulted in the reduction of energy use by 5.1%, CO_2 emissions by 6.7%, and NO_x emission by 6.5%. The implementation of all measures in both case studies required financing funds of about €8 million.

Cycle tourism is one option to reduce congestion and pollution, and to disperse tourists and their expenditure into more rural areas (Figure 7.5). There are different forms of cycle tourism: cycling holidays, where cycling is the main purpose; holiday cycling, where cycling forms part of the holiday as one activity among others; and cycling day visits, where the bike is used for short rides. For cycling holidays it is important to offer cycle networks rather than isolated cycle trails. The latter tend to stimulate day visitation, which may even increase transport volumes,

Figure 7.5 Sightseeing of Angkor Watt (Cambodia) by bike

when parts of the journey to the cycle path are undertaken by car. The successful UK National Cycle Network, for example, offers 5000 miles of connected cycling routes on traffic-free trails, traffic-calm roads and minor roads (Lumsdon, 2000).

Besides a connection between cycle paths, the access to a cycle route or network from urban centres is a critical point. The 250-km Danube Cycle Route in Austria from Vienna to Passau (1.5 million cyclists every year) is a good example of the successful integration of a cycle route into public transport. Cycling appears to be a very cost-efficient alternative to motorised road transport: an 8000-mile network would cost the same as three miles of urban motorway (Energy Efficiency and Conservation Authority, 2000). The main barriers perceived by cycle tourists are quality of driving, overall road safety and biking services (Ritchie, 1998). For the development of cycle tourism it is important to address those issues.

Tourism Businesses

Tourism businesses are those that supply products or services to tourists. In addition to businesses serving tourists directly, there is also a wide range of businesses further up the supply chain linked to tourism more indirectly.[5] In the following discussion, the focus is on accommodation and attraction/activity industries, because they are most directly associated with tourism. Tour operators, travel agents and other ancillary services (e.g. insurances) are not discussed. While these industries are an important part of tourism, and have also influencing power through their purchasing behaviour, they are generally small contributors in terms of energy use. The food and beverage industry is another important industry that provides services to leisure participants. Little information is available, however, and this industry is not discussed separately.

Energy use and emissions of tourist accommodation

Typically, tourist destinations offer a mix of accommodation categories, ranging from luxury hotels, all-inclusive resorts, guesthouses, chalets, camping grounds or huts. The diversity of the accommodation sector may explain the limited research undertaken on energy consumption and GHG emissions. The total energy use and amount of CO_2 emissions per year depend largely on the size of the business. Hotels, which are typically large accommodation establishments, have a higher energy demand than smaller businesses. The approximately linear relationship between energy use and size does not apply for campgrounds, where greater capacity does not necessarily lead to a larger annual energy demand (Figure 7.6).

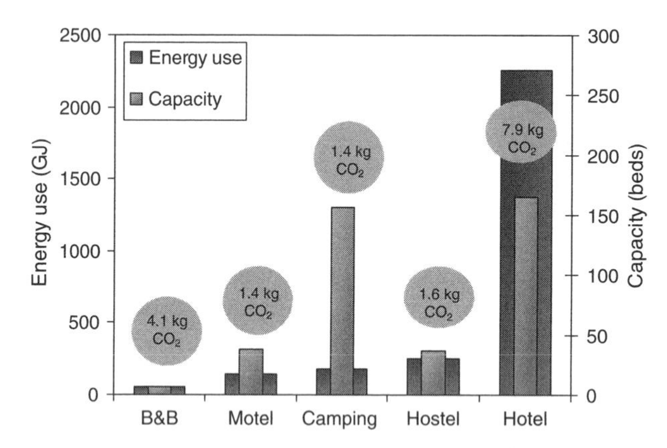

Figure 7.6 Annual energy demand (left axis) and capacity (right axis) for different accommodation categories in New Zealand; and CO_2 emissions per visitor night represented by circles
Source: after Becken *et al.* (2001)

Based on data on annual energy use, fuel mix, maximum capacity and occupancy rates, it is possible to calculate energy use and CO_2 emissions per visitor-night. A number of studies have been undertaken to derive energy intensities for different types of accommodation (Table 7.9). As opposed to manufacturing industries, energy data in tourism, for example in the form of 'energy/unit product' (as used in industrial ecology), are rarely collected by tourism businesses (Schendler, 2003).

Becken *et al.* (2001) described a two-tier distribution of energy efficiency (energy use per visitor-night) in the New Zealand accommodation sector: 'comfort or service-oriented accommodation', including hotels and bed & breakfasts, with a very large energy use per visitor-night, and 'budget or purpose-oriented accommodation', including backpacker hostels, campgrounds and motels, characterised by lower energy use per visitor-night. Warnken *et al.* (2005) also found that the more up-market an accommodation business, the larger the energy use per visitor-night. This is not surprising given that hotels or lodges offer a wide range of energy-intense facilities and services, such as bars, restaurants, saunas and pools, fridges and television in guest rooms, air conditioning and a generous provision of space. Often, hotels also operate their own laundry facilities. It is important to note that some establishments contract those services out. Similarly, visitors staying in the apparently more energy-efficient budget accommodation and motels may satisfy their needs in other locations (e.g. restaurants), thereby increasing the energy demand in other subsectors.

Energy use in the accommodation sector is usually due to heating and cooling, hot water supply, cooling for fridges and freezers and lighting. In

Table 7.9 Examples of energy intensities for different accommodation categories

Accommodation category	*Direct energy use (MJ/ visitor-night)*	*Source*	*Country of reference*
Hotel	256	Gössling (2001)	Zanzibar
Hotel	200	Brunotte (1993)	Germany
Hotel	191	Warnken *et al.* (2005)	QLD, Australia
Hotel	155	Becken *et al.* (2001)	New Zealand
Hotel 4-star	384–602	Trung & Kumar (2004)[1]	Vietnam
Hotel 3-star	160–200	Trung & Kumar (2004)[1]	Vietnam
Hotel 2-star	107–162	Trung & Kumar (2004)[1]	Vietnam
Hotel	51	Simmons & Lewis (2001)	Majorca
Hotel	87	Simmons & Lewis (2001)	Cyprus
Ecoresort	165	Warnken *et al.* (2005)	QLD, Australia
All-inclusive hotel	109	UK CEED[2] (1998)	St Lucia
Non-inclusive hotel	141	UK CEED[2] (1998)	St Lucia
Holiday village	91	Lüthje & Lindstadt (1994)	Germany
B&B	110	Becken *et al.* (2001)	New Zealand
Motel	32	Becken *et al.* (2001)	New Zealand
Backpacker	39	Becken *et al.* (2001)	New Zealand
Campground	25	Becken *et al.* (2001)	New Zealand
Caravan park	32	Warnken *et al.* (2005)	QLD, Australia

[1]Figures were provided for electricity consumption and the average share of electricity was reported to be 91% for 2-star hotels, 90% for 3-star hotels and 76% for 4-star hotels
[2]Figures are for electricity consumption (23 kWh per guest night) and the average share of electricity was reported to be 76% for all-inclusive hotels. The figures for non-inclusive hotels in St Lucia were 34 kWh per guest night of electricity use and an average share of electricity of 87%.

warm holiday destinations, the single largest energy end-use is air conditioning (e.g. Trung & Kumar, 2004; UK CEED, 1998; Table 7.10). In tropical destinations, air conditioning is often switched on even in unoccupied rooms to prevent the growth of indoor moulds and smell. The only alternative is to ensure sufficient ventilation and the use of mould-retardant materials. For existing buildings, this means retrofitting, which is often inhibitively expensive. This highlights the importance of planning for resource use efficiency in the early design phase (Warnken *et al.*, 2005).

Table 7.10 Electricity use in Vietnamese hotels

Electricity consumption	*4-star*	*3-star*	*2-star*	*Resort*
Air conditioning and ventilation (%)	53	47	46	48
Lighting (%)	26	13	17	23
Water heating (%)	17	27	25	12
Other (lifts, pumps, refrigerators, etc.)	4	13	12	17

Source: Trung & Kumar (2004)

The main energy source for accommodation is typically electricity, for example in the order of three-quarters of total energy use. Deng and Burnett (2000), for example, found that 73% of total energy consumed in Hong Kong's hotels is electricity. Herein lays a potential for future sustainable electricity generation based on renewable energy sources. At present, however, much of the electricity consumed by tourist accommodation comes from thermal power plants or diesel generators – as in the case of many island resorts – which are characterised by low efficiencies and high CO_2 emissions (Figure 7.7). Increasingly, accommodation providers use gas, typically LPG for cooking and laundry facilities.

Energy use and emissions of tourist attractions and activities

The recreational component of tourism has rarely been analysed from the perspective of energy use and GHG emissions (Becken & Simmons, 2002; Stettler, 1997). Many tourist activities require little direct energy input; for example going for a walk. An increasing number of activities undertaken by tourists require energy for operating buildings, transporting tourists or for other parts of the business operations.

Recreation facilities used by tourists can constitute a big source of energy demand in the local context. Ski fields in France consume between 2.1 and 2.6 PJ for 4000 ski lifts. Similarly, leisure facilities in urban environments are often main consumers of energy. The opera

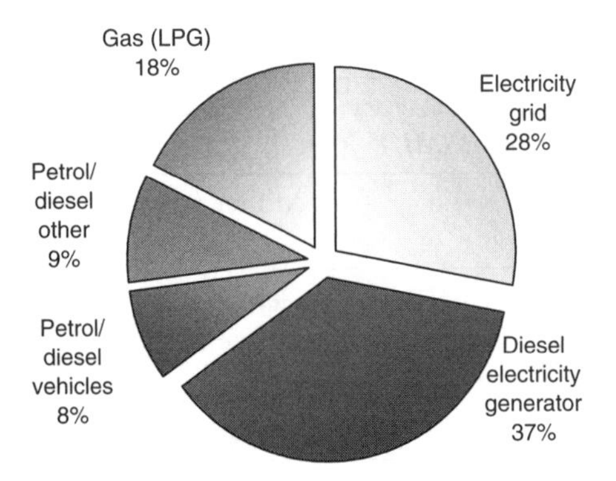

Figure 7.7 Breakdown of energy use into fuel sources for accommodation businesses in Viti Levu, Fiji ($n = 15$)
Source: after Becken (2004a)

theatre in Zurich, for example, consumes about 1190 GJ per year (Becken, 2004c). A study on Japanese leisure activities revealed that playing 'Pachinko' (similar to pinball) consumed a total of 49 PJ in 1995, just behind the energy consumption of eating and drinking places (59 PJ), and well ahead of other leisure activities (e.g. sports facilities, gardens and amusement parks with 9.3 PJ).

The energy use of different recreational attractions or activities is strongly dependent on visitor volumes. The largest amusement park worldwide is Disneyland Tokyo, attracting more than 16 million visitors per year. On the other hand there are many small (tourism) businesses catering for an exclusive market of several hundred visitors per season. Accordingly, when measured on a per visit basis, energy use associated with different activities varies widely. For example, watching TV consumes about 1.8 MJ/day, whereas one day of heliskiing uses well over 1000 MJ (Table 7.11). Generally, mobile activities are more energy intensive than stationary attractions. Activities often build on the use of motorised vehicles, either to get to the attraction or activity or for the activity itself (e.g. snowmobiling) and are therefore reliant on the input of petroleum fuels.

Theoretically, those energy intensities can be converted into CO_2 emissions; however, the energy mix and emission factors need to be known. There is a scarcity of studies providing this kind of information. Aspen Skiing Company (Text Box 1, Chapter 3) reports its energy use and costs, and resulting CO_2 emissions. The average energy cost per skier amounts to US$2.22/day and 21 kg of CO_2 (Aspen Skiing Company, 2004).

A large number of tourist activities are water based and require some form of water transport. Popular activities include short cruises, diving and snorkelling, fishing, whale watching and swimming with dolphins.

Table 7.11 Energy intensities for different leisure activities

Activity	*Energy use per visit (MJ)**	*Source*	*Country of reference*
Watching TV (3.8 h/day)	1.8	Müller (1999)	Switzerland
Gondola ride to Pilatus	9^{+}	Geisel (1997)	Switzerland
Museums, art galleries	10	Becken & Simmons (2002)	New Zealand
Golf	12	Becken & Simmons (2002)	New Zealand
Restaurant meal	18	Müller (1999)	Switzerland
Indoor ice skating	29	Müller (1999)	Switzerland
Experience centres	29	Becken & Simmons (2002)	New Zealand
Rafting	36	Becken & Simmons (2002)	New Zealand
Swimming in public pool	47	Müller (1999)	Switzerland
Skiing	90^{+}	Motiva (1999)	Finland
Swimming in public pool	119^{+}	Motiva (1999)	Finland
Scenic boat cruises	165	Becken & Simmons (2002)	New Zealand
Heliskiing	1300	Becken & Simmons (2002)	New Zealand

*Direct energy use (i.e. secondary energy). Energy use figures with a $^{+}$ refer to primary energy use, i.e. these figures take the embodied energy into account.

For many island destinations and countries with an extended coastline, the marine tour boat subsector is of great importance to the overall tourism product. In Australia, for example, with a coastline of 35,000 km, there are about 1500 tour boat operators with at least one vessel in a tourist operation (Byrnes & Warnken, 2006). Vessels in use include anything from single-hull boats, half-cabin cruisers, purpose-built tourism vessels (e.g. dive boats) and high-speed catamarans. Engine configurations on vessels used by Australian tour boat operators are equally varied. Operators use small two-stroke petrol-engine outboard designs as well as large inbuilt, turbocharged two- and four-stroke diesel engines (outputs of up to 1720 kW). Tour boats also operate generators to provide electricity.

Byrnes and Warnken (2006) estimated that the whole Australian tour operator sector is responsible for the emissions of about 70,000 tons of CO_2-e, or 0.1% of the transport sector in Australia. For every tourist who participates in a boat trip, emissions amount to 61 kg CO_2-e if their travel itineraries included a trip on a boat with a diesel engine, or 27 kg CO_2-e for a trip on a boat with a petrol engine. The authors identified a range of practices that lead to unnecessary waste of energy. For example, some operators were found to use old engines or unfavourable combinations of engine/hull designs. Sometimes, vessels are over-equipped relative to the small number of passengers on board, for example, operating a vessel equipped with two 485 kW engines and a fuel consumption of 300 l/h for an average of 11 passengers per trip results in very high energy intensities. Better communication of technical solutions between operators, naval architects and engineers would help address those inefficiencies. Other options include the introduction of fuel taxes or incentives to improve energy efficiency, for example through government grants.

Emission reduction potentials

Energy management and energy efficiency

There is a wide range of measures for tourism businesses to save energy or improve energy efficiency (Table 7.12); and many practical handbooks exist to help managers implement those. Most information targets accommodation providers. One long-standing institution is the International Hotels Environment Initiative (IHEI, 1993), which was founded in 1992 to support and improve environmental performance by the hotel industry worldwide. IHEI provides benchmarking tools and publishes a quarterly magazine, the *Green Hotelier.* On their website, they provide information on why to benchmark, how to do it and how an individual business compares with other operators. A similarly useful website is the Australian Twinshare: Tourism Accommodation and Environment.

One concrete example of increasing energy efficiency of hotels is provided by Chan and Lam (2003) for the case of hotels in Hong Kong with outdoor swimming pools. The authors suggest that using a heat pump for water heating as opposed to traditional heating systems, such as electric boilers or condensing boilers, could save up to 52% of energy use. For the electricity mix in Hong Kong this equates to savings of almost 12,000 kg of CO_2 and 37 kg of NO_x per heating season (mid-December to late April). Savings double if the whole life-cycle of providing electricity is accounted for. The investment in a heat pump

Table 7.12 Practices for energy saving in accommodation businesses

End-use	*Good practice*	*Inefficient practice*
Room temperature (cooling)	Temperatures in guestrooms are kept between 20 and 25°C; equipment is turned off when not in use. Design allows for natural cooling and ventilation. Air-conditioner thermostats are set depending on outside temperature (e.g. season).	Split-type and window-type air conditioners with low efficiency. Wrong location of air conditioner leads to infiltration of hot air into cooled space. Poor building design and materials.
Lighting	Daylight is made best use of; energy-saving lighting is installed in common areas and guest rooms. Room cards for guest rooms to turn off lights when leaving the room.	Common use of incandescent lamps, only slow uptake of compact fluorescent lamps (CFL). Often, CFL usage is limited to the lobby, restaurants and corridors.
Cooking	Fuel switching and replacement of old equipment with energy-efficient appliances. In kitchens, LPG is used instead of electricity.	Use of inefficient appliances.
Energy management	Conducting an energy audit and assigning staff for monitoring and managing energy use. A room key-card controls energy use in guest-rooms. Purchase of energy-saving equipment.	Few hotels conduct an energy audit. If results of energy-saving measures are not shared with other hotels, hotels do not know how they compare. No staff training.
Water heating	Solar hot water panels; water temperatures set at no more than 60°C; low-flow shower heads.	Individual electric water heaters for water heating in guestrooms lead to high electricity consumption.
Other	Electricity use for drying clothes is reduced by making use of solar energy. Cards to encourage guests to reuse towels. LPG can replace electricity in drying machines.	

Source: after Trung & Kumar (2004)

would be paid back in about 2 years, making heat pumps an interesting alternative both environmentally and financially.

It is possible to provide general guidelines for good or best practice, but it remains difficult to establish benchmarks for energy use or CO_2 emissions against which different accommodation providers can compare their performance. Warnken *et al.*'s (2005) study in Australia showed that the general mode of operation, type of product, existing infrastructure (buildings, access roads, etc.) and local climatic conditions differ widely and at the same time set limits on the achievable level of energy efficiency.

Most of the energy-saving measures outlined for accommodation businesses apply equally for other tourism businesses that operate buildings. Theme parks or large entertainment centres are a good example (see Text Box 15). Most tourist attractions and activity operators operate at least an office building, in which energy conservation and efficiency are highly relevant.

Some tourist activities require very special infrastructure, and energy demand is often high. A good example is the ski industry, which not only relies on energy for lift operations and trail preparation, but also increasingly for snow-making systems and on-mountain entertainment (see also Chapter 3).

Renewable energy sources

A number of renewable energy sources are relevant for tourism. These are wind, PV, solar thermal, geothermal, biomass and waste (for more detail, see Twinshare, 2005; UNEP, 2003). Several studies explored the extent to which renewable energy sources can be used for tourism, in particular in island destinations where energy supply based on fossil fuels is expensive and at risk of supply interruptions (Cavallaro & Ciraolo, 2005; Uemura *et al.*, 2003).

Wind energy is of interest in areas with an annual wind speed of more than 5–5.5 m/s (at a height of 10 m). Evaluation of a site's suitability for wind energy would typically require a 'duration curve', that is the number of hours per year in which a minimum wind speed is exceeded (Cavallaro & Ciraolo, 2005). There are different systems for wind energy, ranging from small scale to medium scale (100–700 kW) and large scale (more than 700 kW output). Tourism businesses, such as hotels, would require electricity supply in the order of 50–100 kW. The capital costs of wind power are generally smaller than those of solar power. A wind turbine at an electrical output of 10 kW, including an 18-m tower and electrical connections, would cost about US$33,000 (Twinshare, 2005). While windmills produce zero-emission electricity and cause no other air pollution, they are sometimes criticised for other environmental impacts, for example noise, visual disruption and impacts on birds.

Text Box 15 Hong Kong Disneyland
Source: Hong Kong Disneyland and Shank *et al.* (2005)

The Hong Kong Disneyland (KHDL) project is a venture between The Walt Disney Company and the Hong Kong SAR Government. Opened in September 2005 it is projected that about 10 million people will visit this Disneyland-style theme park annually. Clearly, energy-related costs would represent a significant cost in the overall-life-cycle cost of the resort. Walt Disney Imagineering Research & Development planned the resort, developing an energy-efficient infrastructure for the resort from the very early design process. An energy simulation model was build to identify efficiency gains and reduce operating costs and capital investment costs by eliminating major pieces of unnecessary or inefficient equipment and infrastructure.

The modelling involved evaluating different scenarios for heating, cooling and powering buildings and then selecting the best combination based on performance and life-cycle cost. The focus was on total resort performance and not on individual buildings. To develop the model it was necessary to:

- Collect energy use data from existing buildings from Walt Disney World; enter current building design information for HKDL into the US Department of Energy's DOE 2.1 programme.
- Model prototypical buildings, and calibrate to operating from existing, comparable buildings; optimise design of prototypical buildings by modelling various energy conservation measures.
- Model central energy plant to supply cooling and heating to optimised buildings; optimise central energy plant by modelling use of thermal energy storage and co-generation systems.
- Document life-cycle-cost elements and complete life-cycle cost analysis.

It is expected that the implemented energy conservation measures and efficient air-conditioning technologies will reduce the energy consumption of HKDL by 20%, compared with the original design. This equates to approximately 5 million kWh, or 6000 tons of avoided CO_2 emissions per year. Implemented technologies include:

- super-efficient windows and insulation to reduce cooling needs,
- dual-path air conditioning (especially effective in Hong Kong's tropical climate), bringing outside air into a building separately from the recirculated air and removing the humidity in the outside air before mixing it with the recirculated air,

- changing the location of buildings and infrastructure to take into account factors such as solar orientation and shading,
- Energy Management System to enable optimum ventilation, temperature and humidity control,
- the Energy Management System also coordinates outdoor lighting installations (taking into account daily dusk and dawn times), and
- constant monitoring of energy consumption.

HKDL does not use renewable energy sources at this stage.

The energy of the sun can be used in three ways: to heat space (passive design for buildings), generate hot water or to produce electricity. Solar thermal systems are probably the most common use of solar energy in tourism. Solar water heaters are useful where water temperatures of less than 100°C are required, for example for showers. Depending on the climate, solar water heaters can meet at least half of the hot water requirements in an establishment. Additional heating (electricity, gas, oil or firewood) will be required on cloudy days or at times when demand for hot water is high. In subtropical destinations this percentage can be much higher. The United Nations Environment Programme (UNEP, 2003) reported that in equatorial regions two hours of bright sunshine on two square metres of collector panel are sufficient to maintain a water temperature of 40–60°C in a 225-l tank, when water is withdrawn at a rate of 8.8 l/min. The effectiveness of solar hot water systems means that investment can be paid back through energy savings in between 2 and 4 years. Water heating in small hotels for laundry, kitchen and guest use can make up between 15 and 30% of electricity costs. Some solar hot water systems can reduce this contribution to 2% (Twinshare, 2005).

Solar water collectors are usually mounted on the roof. Figure 7.8 shows an active system, where water circulates continuously through copper tubes in the collector, exchanging heat in a water cylinder. Solar energy can also be used cost-effectively to heat the water in a swimming pool.

Another way of using solar energy is PV; that is, solar radiation is transformed into electricity by means of a PV cell. PV systems are simple to operate and therefore attractive for a range of tourism applications (see Figure 7.9). PVs have low operating maintenance costs and are reliable in terms of energy production (when equipped with a battery, see following text). PV cells can be used at most locations, but they must be positioned to capture maximum sunlight. An important component of a PV system is a system for energy storage, usually batteries (UNEP, 2003). A back-up diesel-powered generator is often necessary. PV systems are in the range of 1–50 kW; one-kilowatt rooftop cells can be an interesting option for

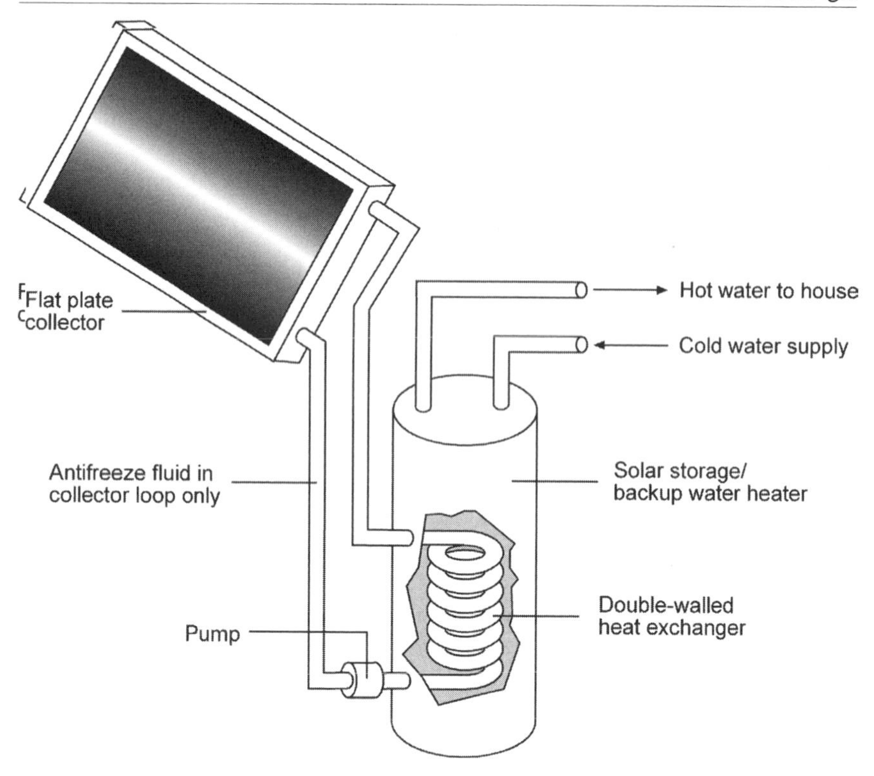

Figure 7.8 Example of a solar water system

tourist bungalows, for example to provide electricity for lighting and smaller appliances (e.g. radios).

The payback time for PV and a back-up diesel generator (i.e. time to recover the higher capital cost through lower operating costs) is 10–15 years. If a PV system is used without a back-up generator the payback period is less than 5 years at the expense of some reliability (Twinshare, 2005). The optimal system depends on the nature of energy demand, in particular the relationship between fixed and variable demand (Text Box 16). Fixed demand is independent of visitation levels (e.g. energy use for running the offices), and variable demand depends on usage, i.e. the number of guest and staff present. Cost of PV cells is decreasing all the time, making them increasingly competitive with diesel generators. This is particularly true in remote areas without connection to electricity grids.

For most tourist operations it would be most practical to exploit a number of renewable energy sources, for example hybrid systems (e.g. wind-solar; Thiakoulis & Kaldellis, 2001). Bode *et al.* (2003) demonstrated for a hypothetical tourist resort how the commodities power, heat, cold

Figure 7.9 Solar panels at the Mauna Lani Bay Resort, Hawaii. The PV system supplies 674 kW of electric power, which makes Mauna Lani the largest solar-powered resort in the world.
Source: Powerlight (2005); Photo: F. Damgaard

air, water and transportation could be supplied with zero CO_2 emissions. The holiday resort in Bode *et al.*'s study generates electricity through wind turbines and solar modules (upper part of Figure 7.10). The demand for water is met by a seawater desalination plant (reverse osmosis). Heat pumps and refrigerating machines provide heat and cold and vehicles are hydrogen powered (see lower part of Figure 7.10). The annual balance of energy demand showed that it is easily possible to supply the hypothetical resort with those renewable energy sources. However, when considering smaller time intervals and peak demand, the above set-up was not able to supply energy reliably at all times, even with large installed power capacities. The solution is to store energy surplus. Bode *et al.* suggest a combination of electrolysers, fuel cells and pressure tanks (see middle part of Figure 7.10). Electrolysers split water into hydrogen and oxygen. Energy is needed for this process. Hydrogen is stored in the pressure tanks and oxygen is released into the atmosphere. In return, fuel cells produce energy by oxidation of hydrogen. The only product is water. Figure 7.10 shows the energy supply of the result with storage facilities for peak times, without the need of a back-up diesel generator.

In their hypothetical case study, Bode *et al.* demonstrated the technical feasibility of a zero-emission tourist resort. They argue that wind energy is competitive and electricity from solar power is also available at

Text Box 16 Tortoise Head Guest House, Australia
Sources: UNEP (2003); Twinshare (2005)

Tortoise Head Guest House is located on French Island, Victoria. The guest house has 11 rooms with a bed capacity of 30–40. The average weekend occupancy is 15–20 and the peak visitation periods are February, March and April.The lodge faces north and receives maximum sun; natural ventilation is maximised and ceiling fans are used both for cooling and to circulate warm air in winter. Walls are now being insulated with R 2.5 glass fibre. French Island has no mains power connection and the guest house is powered by wind turbines and solar energy. The system includes:

- 10-kW wind turbine,
- 840-W PV array,
- two diesel generators (15 and 25 kW),
- battery storage (wired to produce a system voltage of 120 V direct current), and
- 10-kW inverter to convert the direct current into the Australian standard of 240 V alternating current and 50 cycles per second.

About 80% of the electricity generation comes from wind and 10% from PV; the remainder is generated using diesel. In addition, water is preheated using an Aquamax gas storage system (using LPG) and a solar-powered heat pump supplements heating. This pump is only used during the day as noise levels may disturb guests at night. Heating in the cabins is provided by hot water radiators and electric blankets are also provided for guest comfort.

An energy audit identified that the coldroom alone is responsible for about half the electricity consumption (7 kW/day). This could be reduced by replacing it with two domestic refrigerators in the kitchen and a small refrigerator in the dinning room for guest use. This change will reduce refrigerator electricity use by more than 50%. This will have a disproportionately high benefit in reducing diesel fuel use, which is used to supply the shortfall in electricity that cannot be met by solar and wind systems.

reasonable costs. The problem is the price of fuel cells and electrolysers, and because of those installations the expanded system would currently add over €50 per guest-night, which is considered too high. With decreasing prices for fuel cells, the above scheme might become competitive.

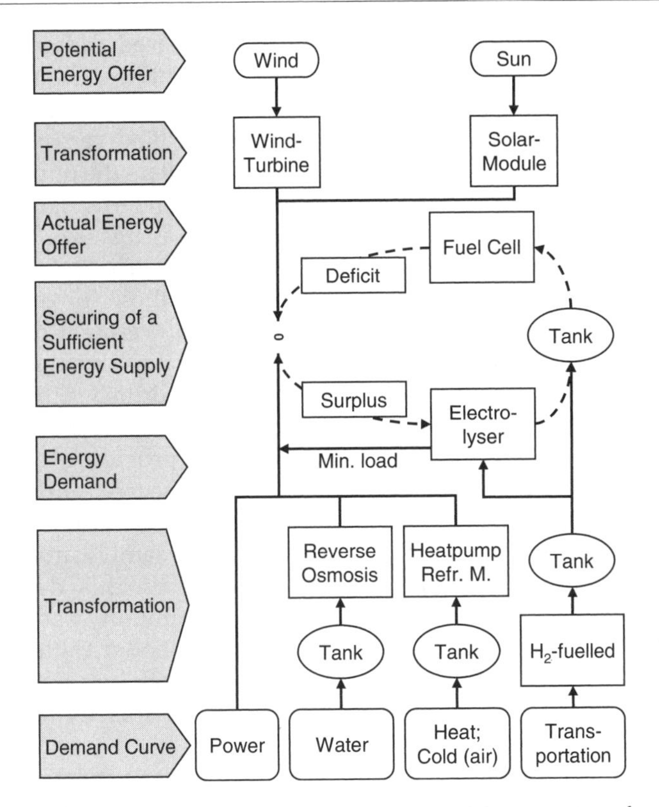

Figure 7.10 Expanded scheme for an autarkic CO_2-emission free energy supply for holiday facilities
Source: Bode *et al.* (2003)

Carbon Compensation Projects for Tourism

From a climate change mitigation point of view, the best strategy is, of course, to reduce emissions at source, but where this is not possible (or too expensive) compensation projects are an acceptable alternative (Sterk & Bunse, 2004). Carbon compensation means that the amount (or a proportion thereof) of GHGs produced by a certain activity or business is reduced elsewhere, where mitigation is cheaper than reducing emissions at the activity in question. Theoretically, compensating achieves an optimal relationship between climate protection and costs. Instead of compensating, the terms 'offsetting' or 'neutralising' emissions are used as well.

Carbon offsetting is now officially promoted by two of the most influential guidebook companies, namely Lonely Planet and Rough Guide. A range of options is available for tourists to compensate their travel. These are available in the form of Web-based carbon offsetting schemes tailored to various markets (e.g. Climate Care, Climate Protection Partnership, Business Enterprises for Sustainable Travel, Future Forests, 500 ppm, Trees for Travellers, Emissions Biodiversity Exchange (EBEX21®)[6]). Most schemes offer individual travellers the opportunity to work out their travel GHG emissions with an online calculator. Tourists can choose to invest either in energy-efficiency measures (e.g. low-energy light bulbs), energy renewal (e.g. hydro-turbines), or carbon sequestration (by projects for restoring forests). Often these projects are in developing countries for the reason of empowering communities through commerce and tackling climate change at the same time. Climate Care also works with tour operators (including British Airways) who include offsetting in their package (Text Box 17). An approach in New Zealand being led by Kaikoura District Council is providing tourists with the opportunity to plant a tree during their visit. The tree is numbered, its exact location recorded and it can be revisited by the tourist.

Compensation projects include carbon sinks (mainly carbon sequestration in the form of biomass), energy efficiency projects, development of renewable energy sources and small-scale community-based projects to reduce GHG emissions. Figure 7.11 illustrates how CO_2 (emitted for example by aeroplanes) is absorbed by trees and soils and thereby taken out of the atmosphere. A potential co-benefit of this form of carbon sequestration is a gain in biodiversity.

Carbon sinks, in particular, are seen as a controversial solution to climate protection, because they potentially divert from the real causes of the problems and therefore avoid structural and technological changes that need to be made to achieve long-term GHG reductions. The usefulness of forests as sinks (as opposed to geological or deep-ocean storage) is debated, mainly because it is difficult to measure carbon uptake, the carbon stored is unstable and not permanent (e.g. forest fires, pests) and the promotion of carbon sinks diverts from the pressing need to reduce the combustion of fossil fuels (Noble & Scholes, 2001). Concerns also relate to the difficulty of measuring carbon uptake, sinks being a short-term solution, insecurity of projects and costs of administration. Moreover, in tree-planting schemes the initial rates of sequestration are low and, therefore, it can be some years before the travel emissions are actually offset. On this issue, both managers of the Lonely Planet and the Rough Guide commented that carbon offsetting schemes were not a perfect solution but they demonstrated people were increasingly aware of the pollution they were causing and were 'better than doing nothing'.

Supporters of carbon sinks put forward that they are an invaluable means to save time on the way towards energy-efficient economies and alternative fuel sources. They are also associated with other benefits, for

Text Box 17 Carbon offsets and air travel
Contributed by Tom Morton, Climate Care, UK

Climate Care, based in Oxford in the UK, is a not-for-profit company in the voluntary offsets market. Climate Care sells emissions reductions and uses the money raised to fund new projects to reduce CO_2 emissions. The majority of offsets are in renewable energy and energy-efficiency projects that reduce emissions at source and the remainder are in a reforestation project. Whilst 'planting trees' is a relatively simple message to convey to potential purchasers, taken alone it will not be the solution to climate change, which is why Climate Care's focus is on 'technology' type projects.These include the installation of biogas digesters in villages in India to provide a renewable fuel for cooking, which reduces pressure on the local forest resources; the installation of 50,000 energy-efficient lamps in houses in South Africa to reduce electricity consumption (based on coal-fired power stations); and replanting of an area of native forest in Uganda that was cut down in the 1970s.

Emissions are offset from a wide range of sources, one of which is air travel. Government and NGOs are increasingly aware of aviation's impact on the climate; however, so far very few tourists know that they can compensate for the damage by buying a carbon offset. Notwithstanding, visits to the Climate Care website are doubling year on year and increasingly people are offsetting the emissions from their flights online.

In the UK a number of independent travel companies recognised the irony of the situation: holidays often depend on the natural environment and, yet, each customer emits on average 2.5 tonnes of CO_2 in getting to the destination. Climate Care currently works with about 20 tour operators and a number include the cost of a CO_2 offset in the price of their holidays (which works out as about £1.00 per person per hour of air travel). This approach ensures 100% take up. However the majority of operators offer Climate Care on an optional basis to their customers and the take-up rates are directly related to how easy it is for customers to take the Climate Care option. Some companies, for example, include the cost of an offset on their invoice and tell customers that they don't have to pay it. In this instance the traveller has to take a conscious decision *not* to take the green option and take up rates have been as high as 70%. Others offer it as an *opt in* option on their invoice and here take up rates tend to be in the region of 15%. Other companies either give customers a leaflet to return or point them to a website and in this case take up rates are very low.

There are two other significant potential markets. The first is to have offsets offered by the large-scale tour operators and travel agents, rather than small niche operators. The industry has resisted this move on the basis of costs and there is also a feeling that the consumer is not ready for such an option. The other market is the corporate sector, where companies tend to offset their aviation emissions for two main reasons. The first is because they are being asked by their clients what they are doing for the environment. For service sector companies with relatively low direct environmental impacts offsets are an attractive way of taking action. Secondly they offset emissions in order to fulfil shareholders' expectations as part of their environmental policy. Both of these have been driven by the rise of the corporate social responsibility agenda.Offsets are gaining recognition as part of the tool box of measures available to individuals, corporations and policy makers to tackle climate change. Emissions trading will be a way to reduce the impact of aviation over the long term and offsets offer an early step in that direction for those who wish to make their air travel climate-neutral.

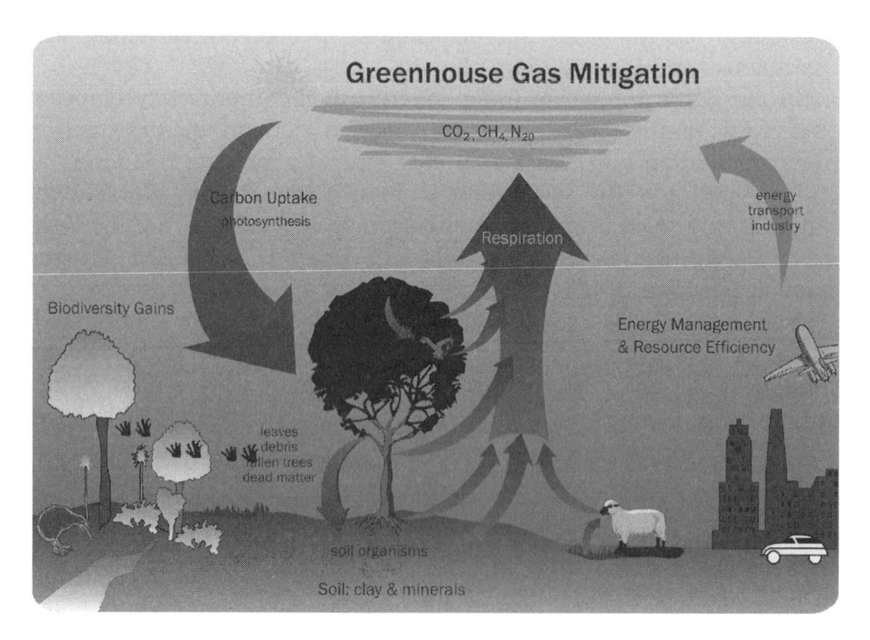

Figure 7.11 Carbon cycle and sequestration of CO_2 through forest sinks

example in relation to biodiversity, hydrology, soil retention and noise reduction. Ultimately, there is also the potential to attract tourists to reforested areas. Carbon sinks absorb and store GHGs in a permanent or semi-permanent form. One form of carbon sink is the sequestration of CO_2 as biomass, usually trees. Carbon sinks have also been suggested as an intermediate solution (Curtis, 2002) to compensate for energy use in the accommodation industry in Queensland, Australia. In the context of air travel, Gössling (2000) calculated a hypothetical land area of 28,800 km^2 that would need to be forested to offset CO_2 emissions resulting just from global tourism air travel in 2000. This area is about the same as the area of Belgium or Taiwan. The enhancement of carbon sinks is recognised in the Kyoto Protocol (Article 3) as a mitigation measure, and it is for this reason that the potential to include tourists in such mechanisms needs to be explored further.

Conclusions

Tourism is comparably energy- and emissions-intensive, especially when a substantial transport component is involved. There are a number of options to improve the energy efficiency of tourism or reduce its carbon footprint. Several energy-efficiency measures have been discussed for transport and other tourism businesses, for example hotels. Also, there is a selection of alternative technologies that build on carbon-neutral or zero-carbon energy sources. The potential for renewable energy sources has not been exploited yet by tourism. Improving tourism's carbon intensity also involves behaviour, in particular that of tourists. Beyond concrete choices relating to their activities (e.g. travelling by train, staying in an environmentally friendly resort), tourists can also opt to offset those emissions that cannot be avoided by using currently available technologies. For example, GHG emissions from air travel can be sequestered by forest sinks.

Future work in the area of energy efficiency and GHG emissions from tourism may benefit from shifting our focus from the use of energy to the demand for specific energy services, such as heating or cooling. In many circumstances, these services can be provided through innovative, low-energy solutions. The following chapter, Chapter 8, looks at the climate change-related risks for tourism and ways for adapting to climate change.

Notes

1. Lift refers to the physical force that supports the weight of the aircraft on the wings; drag means the force that resists motion.
2. The mach number is the ratio of object speed to speed of sound.
3. See: www.fortgoden.com/wjag.html; www.graywizard.net/Personal/Cruise/cruise/html; www.hboi.edu/marineops/pdf/rusj2_specs.pdf; http//buerger.metropolis.de/michael_bruno/e4_umbau.htm.

4. There is also a great range of opportunities for stationary fuel cells.
5. For some activities, energy use associated with upstream activities can be substantial, for example in motor sports and airborne activities. Skiing and snowboarding are other popular activities that require large energy inputs outside the actual activity. Stettler (1997) found that per skier-day, the energy input for infrastructure and equipment amounts to 200 MJ compared with 55 MJ for tennis and 15 MJ for soccer. In these examples, the indirect energy use is more than twice the direct energy use (Becken, 2004).
6. Climate Care (www.climatecare.org.uk); Climate Protection Partnership (www.clipp.org); Business Enterprises for Sustainable Travel (www.sustainabletravel.org); Future Forests (www.futureforests.com); 500 ppm (http://travel.500ppm.com); Trees for Travellers (www.treesfortravellers.co.nz); Emissions Biodiversity Exchange (EBEX21) (www.ebex21.co.nz).

Chapter 8
Climate Change-related Risks and Adaptation

Key Points for Policy and Decision Makers, and Tourism Operators

- Risk considers not only the potential level of harm but also the likelihood that such harm will occur. The familiarity of planners and decision makers with risk management in other areas helps facilitate the mainstreaming of risk-based adaptation to climate change.
- Adaptation can be defined as those actions or activities that people undertake, individually or collectively, to accommodate, cope with, or benefit from, the effects of climate *change*, including changes in climate variability and extremes (beyond the natural variability of the global climate). In the past, tourism has continuously adapted to other risks and changes; adapting to climate change builds on this adaptive capacity.
- Adaptation should be 'mainstreamed' and implemented as an integral part of national and tourism development planning, environmental management and disaster management.
- Climate change has consequences for various components of the tourism system: appeal of a destination, transport infrastructure and operations, the resource base (natural and human), tourist satisfaction and safety, and the viability or sustainability of tourist facilities and destinations.
- Tourism demand is sensitive to changes in climate; this poses a challenge to those destinations that might become less competitive as a consequence of climate change; it also provides an opportunity for those destinations that have low appeal under current climatic conditions.
- 'Climate proofed' attractions and product diversification are being used to reduce disparities in tourist demand between high and low seasons and to increase the attractiveness of destinations that may suffer unfavourable climatic conditions.
- Climate change impacts on the natural resource base of tourism; conservation of biodiversity and maintenance of ecosystem structure and function are therefore important climate change adaptation strategies in relation to tourism's natural resource base. This is

particularly true for coral reefs – a major resource for tourism in many destinations. Reducing pressures on these ecosystems is a key climate change adaptation measure.
- Climate change affects human health directly through increased heat-related mortality and morbidity and climate-mediated changes in the incidence of infectious diseases. Adaptation measures include early warning systems (e.g. for heatwaves), improved hygiene, and international and regional health monitoring systems that provide alerts for disease outbreaks and health risks (e.g. malaria).
- The increasing assimilation of tourism planning into integrated planning initiatives (such as coastal zone management) facilitates the adaptation of the tourism sector to climate change. As it offers advantages over purely sectoral approaches, integrated coastal management has been widely recognised and promoted as the most appropriate process to deal with climate change, sea-level rise and other current and long-term coastal challenges.

Introduction

Chapter 5 described why and how global, regional and local climates are changing, and indicated the likely nature of changes into the future. These future changes are expected despite the effort being made to reduce human-related emissions of GHGs, including initiatives taken by the tourism sector (Chapter 7). The anticipated changes in climate, including those related to climate variability and extreme events, will often increase the level of risk to tourism and most other systems.

Risk considers not only the potential level of harm but also the likelihood that such harm will occur. The level of harm resulting from climate change typically varies from location to location, and within and between sectors. But in general the likelihood component of a climate-related risk is applicable over a larger geographical area, and to many sectors. This is due to the spatial scale and pervasive nature of weather and climate. Thus the likelihood of, say, an extreme climate event or anomaly, is often evaluated for a country, state, small island or similar geographical unit. While the likelihood may well vary within the given unit, there is often insufficient information to assess this spatial variability, or the variations are judged to be of low practical significance.

Observed and other pertinent historic data can be evaluated to determine past and current likelihoods for relevant climate variables such as extreme rainfall, temperatures, wind, sea-level and drought. Climate change scenarios (see Chapter 5) are used to develop projections of how the likelihoods might change in the future and often form part of a climate risk profile (ADB, 2006). For example, Figure 8.1 illustrates that for

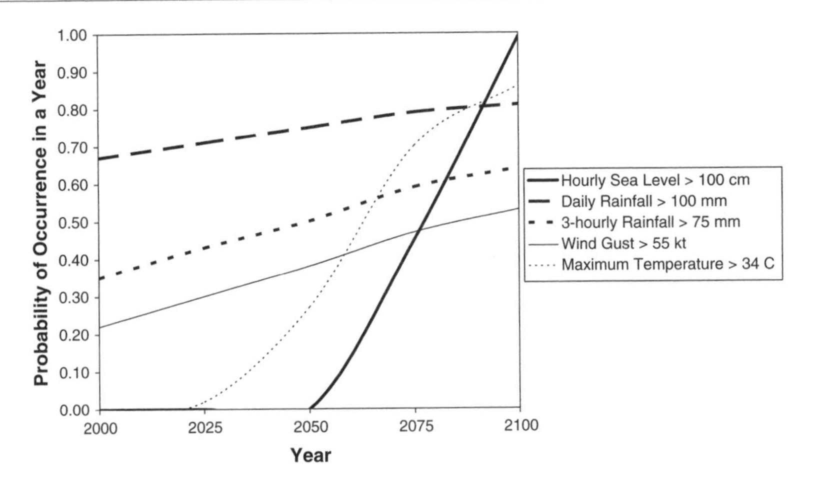

Figure 8.1 The likelihood (0 = no chance; 1 = statistical certainty) of a given extreme event occurring in a year in Maldives

Maldives, a major tourist destination, the chance of a selected extreme event occurring in a given year may increase markedly due to continued strengthening of the greenhouse effect.

Overview of Adaptation

Adaptation can be defined as those actions or activities that people undertake, individually or collectively, to accommodate, cope with or benefit from the effects of climate *change*, including changes in climate variability and extremes. People continually cope with the fluctuations in the current climate. Adaptation to climate change encompasses the additional actions or activities necessary to take the consequences of climate change into account. Natural systems also adapt to climate variability and change – this is termed 'autonomous adaptation of natural systems'.

Adaptation is a dynamic process: the influences people and enterprises adapt to, as well as their needs and wants, are constantly changing. From this perspective, when we refer to adaptation in respect to climate change we are referring to adjustments to existing coping strategies. It thus incorporates a wide range of actions designed to increase the resilience of the tourism system as a whole, as well as its constituent parts, to the possible adverse effects of climate change.

The continuing build up of GHG in the atmosphere, and the serious consequences of climate change for the tourism sector make adaptation unavoidable. Thus it is wise to develop and implement policies and plans that will ensure timely adaptations that reduce or even prevent the adverse effects of climate variability and change. Such an anticipatory approach is prudent, for five principal reasons:

- If adaptation is reactive, as opposed to anticipatory or proactive, the range of response options is likely to be fewer; adaptation may also prove more expensive, socially disruptive and environmentally unsustainable.
- Many adaptation strategies are consistent with sound environmental management, wise resource use, and are appropriate responses to natural hazards and climate variability, including extreme events – such 'no-regrets' adaptation strategies are beneficial and cost effective, even in the absence of climate change.
- Many development plans and projects have a life expectancy that requires future climate conditions and sea-levels to be included in their design.
- Tourism has a heavy dependency on valuable and important ecosystems that are sensitive to climate variability and change, including extreme events – it is easier to enhance the ability of ecosystems to cope with these variations in climate if they are healthy and not already stressed and degraded.
- Adaptation requires enhancement of institutional capacity, developing expertise and building knowledge – all these take time.

Individuals in the tourism sector will, as a result of their own resourcefulness or out of necessity, adapt to climate variability and change. Decisions to act will be based on individual and collective understanding and assessment of the anticipated risks, as well as on the perceived options and benefits for response. This may be considered to be *independent adaptation*. This contrasts with *formally planned adaptation*, which involves deliberate policy decisions, plans and implementation by *external* parties. In many cases independent adaptations will be adequate, effective and satisfactory. However, under some circumstances independent adaptation may not be successful, often for one or more of the following reasons:

- An individual's understanding of climate variability and change effects may be limited or even erroneous.
- Similarly, understanding of the possible adaptation options may be limited or defective.
- Adaptation responses undertaken by one group may impact adversely on another group.
- The needs of future generations may not be taken into account.
- There may be cultural constraints to certain adaptation responses.
- Individual tourism enterprises, or other groups or institutions, may not have adequate resources to implement the most desirable adaptation measures.

- It may be more cost effective, and in other ways more efficient and effective, to implement certain adaptation responses on a more collective basis, rather than at the level of the individual or community.

In *formally planned* adaptation the role of an external entity, such as central or local government or a tourism organisation, can be to facilitate the adaptation process to ensure that the above-mentioned obstacles, barriers and inefficiencies are addressed in an appropriate manner. In the context of tourism this might include:

- Facilitating adaptation through the provision of information about climate variability and change processes, effects and adaptation options.
- Through the provision of financial, technical, legal and other assistance, facilitating the implementation of adaptation options where those affected, such as individual tourism enterprises, do not have the resources to adapt effectively.
- Implementing adaptation options directly where the scale of response is most appropriate at a national or tourism sector level.
- Ensuring that when it is undertaken, adaptation does not have adverse environmental, social, economic or cultural outcomes.
- Ensuring that there is equity in the adaptation process and that individuals are not unfairly affected either by the effects of climate change or as a result of adaptive actions.
- Ensuring coordination, cooperation and equitable partnerships between tourism enterprises, a local authority and central government in formulating and implementing adaptation plans.

A prerequisite to effective adaptation is a continuing dialogue among technical experts, governments, resource managers and key players in the tourism sector. This will facilitate a more thorough understanding of the risks arising from climate variability and change, and identify and explore potential response options, evaluate the success of existing policies and programmes, revise such policies and programmes when necessary and identify new information needs.

A particularly important consideration for *formally planned* adaptation in the tourism sector is to ensure that climate change and sea-level rise considerations are integrated into development activities – they should be mainstreamed and implemented as part of national and tourism development planning, environmental management and disaster management. Throughout the development planning process there are many opportunities for decision-makers to take deliberate action to enhance the resilience of the tourism sector, or reduce its vulnerability, or both. Until recently, failure to grasp the real and pervasive costs of disasters made it difficult to convince most policy and decision makers to divert scarce resources from one part of the national, enterprise or community budget in order to support disaster reduction programmes. Moreover, uncertainties in climate change impact estimates, and even more so in the likely success of adaptive responses, were simply too large for adaptation to be incorporated into national development planning in a meaningful way.

A Risk-based Approach to Adaptation by the Tourism Sector

One consequence of the open nature of the tourism system is a need for multisector and multiagency cooperation when addressing the consequences of climate change. As risk assessment and management procedures have already been embraced by many sectors – e.g. health, financial, transport, agriculture, energy and water resources – a risk-based approach (Figure 8.2) provides a common framework that facilitates coordination and cooperation amongst the various players, including the sharing of information that might otherwise be retained by information 'gate keepers'. Importantly, the existing familiarity of many planners and decision makers with risk management helps facilitate the mainstreaming of risk-based adaptation.

Risk-based methods also facilitate an objective and more quantitative approach, including cost benefit analyses that not only assist in evaluating

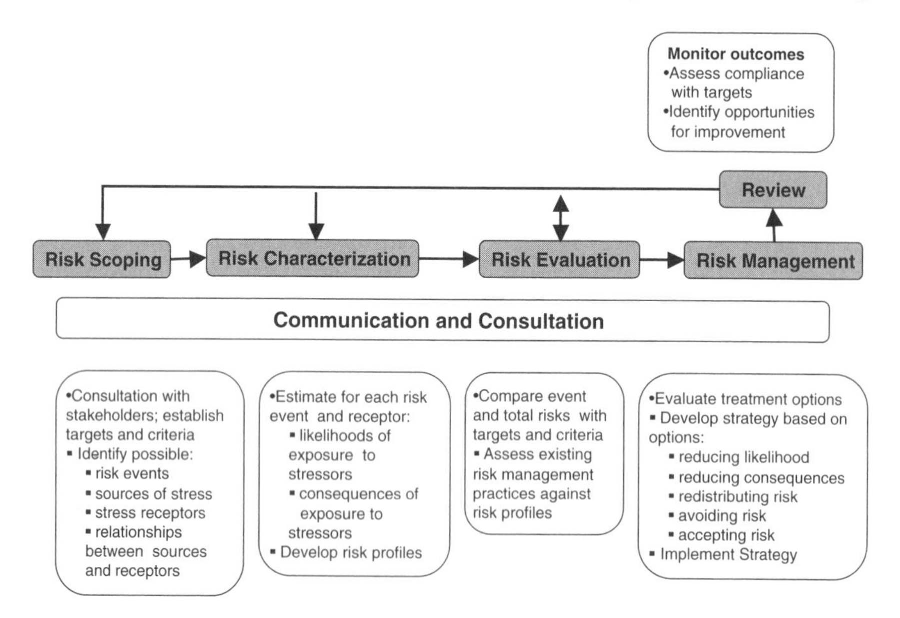

Figure 8.2 Risk-based approaches to identifying and assessing options for managing the adverse consequences of climate change

the incremental costs and benefits of adaptation and but also help in prioritising adaptation options. The approach also links to sustainable development by identifying those risks to future generations that present generations would find unacceptable.

In the case of tourism there is a need to align disaster risk management and adaptation to climate variability and change in a mutually supportive manner. This can best be achieved by having key players recognise that both disasters and climate change are significant impediments to the sustainability of tourism – i.e. they represent real and significant risks to individual enterprises and to people, communities and national economies dependent on tourism. The key is to recognise a continuum of potential events that may all be classed as hazards, ranging from extreme events of short duration (e.g. a tropical cyclone), through events associated with variations in atmospheric and marine conditions (e.g. ENSO-induced drought), to events resulting from long-term changes, such as accelerated coastal erosion as a consequence of sea-level rise (Text Box 18). Risks associated with the full spectrum of hazards should be managed in a holistic manner as an integral part of sustainable tourism development.

The risk scoping and characterisation steps (Figure 8.2) facilitate identification of both the direct and indirect consequences of climate change. Overall, the risk-based approach ensures there is a strong functional link between the assessment of the risk and the identification, prioritisation and eventual selection of the adaptation initiatives (the risk management step) required to reduce risks to acceptable levels. These two attributes of the risk-based approach are illustrated in Figure 3.4 in the context of tourism in small island countries. The illustration is also broadly applicable to coastal tourism in general.

Characterisation and Management of Climate Change-related Risks to Tourism

In this section the future climate-dependent risk factors, and the likely consequences for tourism, are characterised for each of the key components of tourism–climate futures (Figure 8.3). Many of these key components have their equivalent in the chaos model of tourism (Figure 2.7), though here the relationships between climate-related risk, the consequences and management interventions are represented in a more linear and deterministic manner. This is consistent with a more pragmatic, shorter-term approach to managing climate-related risks to tourism. Thus the most appropriate ways to manage the risks, and exploit any opportunities, are also described for each of key components.

Text Box 18 Adaptation for coastal tourism

The importance of coastal tourism worldwide, and the fact that coastal systems are highly at risk from climate change, highlights the need for proactive adaptation in this component of the industry. Five generic approaches for proactive adaptation in coastal areas have been identified (Nicholls & Klein, 2005):

- Increasing robustness of tourism infrastructure and other long-term investments – for example by extending the range of relevant climatic factors that a system can withstand without failure (e.g. extreme wind gusts) and/or changing a system's tolerance of loss or failure (e.g. by increasing financial reserves or insurance).
- Increasing flexibility of vulnerable components of the tourism system – for example by following adaptive management approaches, which explicitly allow for plans and operations to be modified as a result of experience, and/or reducing economic lifetimes (including increasing depreciation).
- Enhancing adaptability of vulnerable natural systems on which tourism depends – for example by reducing non-climatic stresses (e.g. by decreasing discharge of wastes), and/or removing barriers to migration (e.g. no roadways on the landward side of coastal mangrove forests).
- Reversing maladaptive trends – for example, for vulnerable areas prone to repeated flood events ensure the zoning regulations prohibit redevelopment after major damage.
- Improving societal awareness and preparedness – for example, by informing tourism operators about the risks and possible consequences of climate change and/or ensuring all tourism businesses have disaster response plans and early-warning systems.

Implications for destination appeal

As noted in Chapter 2, tourists frequently base their decision to travel to a particular destination on the actual and perceived, current and predicted, climatic conditions at the destination. Climate change will modify the relative attractiveness of many destinations (see for example McEvoy *et al.*, 2006).

Risk factors and anticipated consequences

The relative attractiveness of tourist destinations, both international and domestic, will be modified as a consequence of climate change. As temperature dominates the influences of climate on tourism, the projected increases in temperature worldwide, and especially in the 'climate–tourism hotspots' (Chapter 2), may well bring about substantial changes

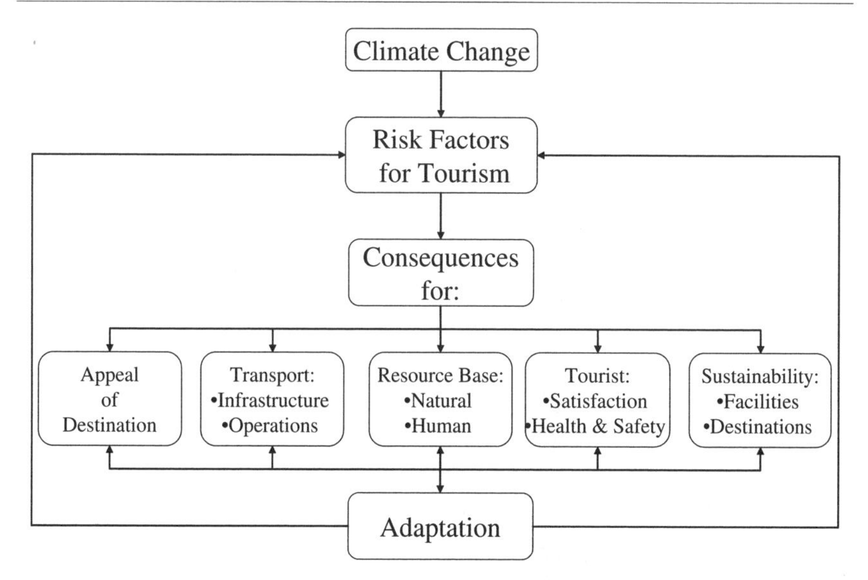

Figure 8.3 The principal components of tourism–climate futures

in global tourism flows. A global perspective on the possible impacts of climate change is provided by the pioneering work of Hamilton *et al.* (2004). They modelled the impact of changes in population, per capita income and climate change on arrivals and departures for 207 countries. Specifically, as a cool country warms it firsts attracts more tourists, but once the mean annual temperature exceeds 14°C, fewer tourists will visit. Similarly, the country will initially generate fewer tourists, but once the temperature reaches 18°C it will generate more tourists as people will seek opportunities to travel to cooler climates. The result is a gradual shift of tourism towards higher latitudes and altitudes. In a warmer world the currently large groups of international tourists from Western Europe stays closer to home, resulting in a relative decline in international tourist numbers (Figure 8.4). But overall, the changes in tourist flows attributable to global warming are smaller than those resulting from population and economic growth.

The economic implications of these spatial variations in tourism demand, as a result of global warming, have been investigated in similarly ground-breaking work by Berritella *et al.* (in press). At a global scale climate change will lead to a welfare loss, unevenly spread across the globe – GDP will change by -0.3% to + 0.5% in 2050. Net losers as a result of destinations losing attractiveness include Western Europe, the Caribbean and other tropical countries, and energy exporting countries. Gains are felt by North America, Australasia, Japan, Eastern Europe and the former Soviet Union.

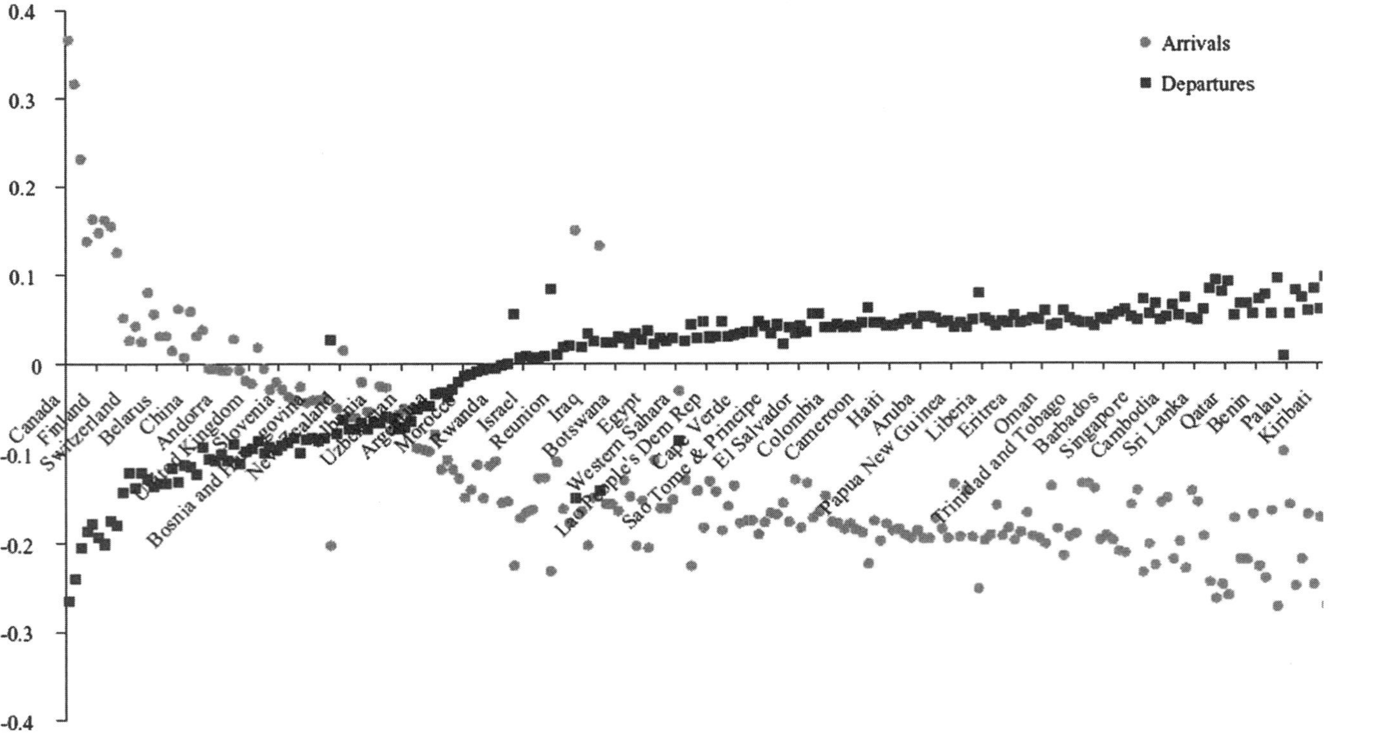

Figure 8.4 Change in international arrivals and departures due to a global mean temperature increase of 1.03°C by 2050, as a percentage of arrivals and departures without climate change; countries are ranked according to their average annual temperature in 1961 – 1990

Source: Berrittella *et al.* (2006)

In a more detailed study the Tourism Climatic Index was used to assess the implications of climate change for tourism in various cities in North America. Scott and McBoyle (2001) found that revenue from tourist accommodation would increase for Canadian cities and that the length and quality of the summer tourism season in Canada's Western mountain parks would both show substantial improvement with climate change, such that in the peak spring and summer tourism seasons the Tourism Climatic Index values would be similar to current ratings for areas 1500 km to the south.

A comprehensive study of consequences for the Mediterranean region and North-western Europe has been conducted by Amelung and Viner (2006). Not only did they assess how projected changes in temperature would modify the values of the Tourism Climatic Index in these regions, but they also incorporated changes in precipitation, wind and sunshine. The results show that the Mediterranean climate will become increasingly unattractive to tourists in summer, but a more pleasant destination in spring and autumn, while Northern Europe will gain appeal in summer. Improvements in the summer climate of both the UK and the Alps are noteworthy. Amelung and Viner stress that the implications for the Mediterranean are not all detrimental – there may be an overall improvement in occupancy rates as a consequence of a lengthened season and more evenly spread tourist demand. However, losing appeal at the time of school holidays in the source countries is a very real concern.

Adaptation

As tourism demand is sensitive to changes in climate, the challenge is to minimise those factors that impact adversely on the competitive advantage of holiday destinations while also maximising those that increase this advantage. Agnew and Palutikof (2001) note that the flexibility to respond to climate variability and change varies between the subsectors of the tourist industry. Suppliers of tourism services and local managers have the least flexibility, but some are already attempting to 'climate proof' the industry through such initiatives as development of indoor tropical holiday parks and artificial snow slopes for skiing and snow-boarding. The tourists themselves have the greatest flexibility, with relative freedom of destination choice and timing of travel. Tourism wholesalers also have some flexibility – if conditions are, or become, unfavourable in one destination region, holidays in more favourable areas can be promoted. Weekend and day tourists, especially, are planning their travel at ever-shorter notice, resulting in the growing influence of actual and forecast weather conditions. Tourism operators and marketers can benefit from taking a similar short-term approach in their advertising campaigns. These same people also play a key role in encouraging tourists to break traditional holiday habits by travelling in the intermediate seasons if regions such as the Mediterranean become unattractive in the summer. Appropriate timing of arts festivals, local fiestas and sporting, food and drink events can also spread demand into the shoulder seasons.

Implications for transport

Elements of the weather and climate can constitute numerous risks for transport infrastructure and operations related to tourism (Figure 8.5). Climate change may modify weather elements such as temperature and precipitation in ways that alter the frequency, duration and severity of risk factors, such as ice and snow cover, extreme heat and drought. Such changes impact adversely on transportation infrastructure, on demand for transportation services and on the services themselves. There is clearly an overlap between the destination preference considerations discussed earlier and transport demand sensitivities such as the mode of transport and the route taken.

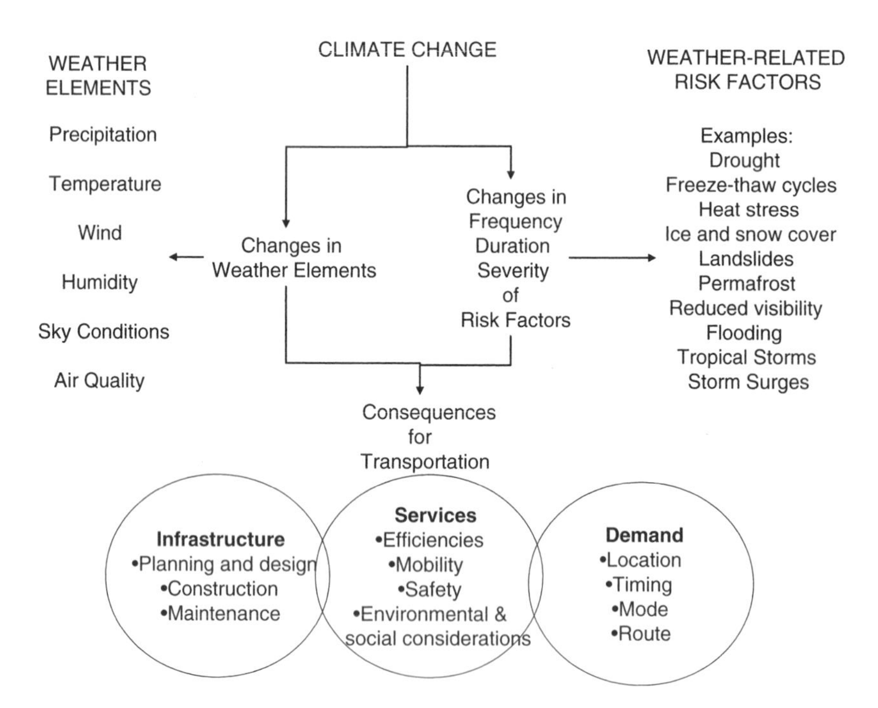

Figure 8.5 Components of the tourism transportation system sensitive to changes in climate

Source: after Mills & Andry (2002)

Risk factors and anticipated consequences

Weather and climate affect the planning, design, construction, maintenance and performance of tourism transport infrastructure, such as roads, railways and airport runways, throughout the service life. While this infrastructure is built to withstand a wide variety of weather and climatic conditions, climate change means that some of the assumptions made about atmospheric conditions during the planned lifetime of the infrastructure may be invalid, possibly leading to premature deterioration or failure of infrastructure. If these changed conditions are not included in the design, expensive reconstruction, retrofit or relocation may be required.

As extreme heat conditions become more severe and frequent, road pavements will suffer increased softening, traffic-related rutting, buckling and flushing or bleeding of asphalt. Railway track is also subject to buckling from extreme heat. On the other hand the anticipated decrease in extreme cold conditions, along with higher minimum temperatures and fewer freeze–thaw cycles, may reduce the incidence of thermal cracking of road pavements and other deterioration in both roads and airport runways. However, in areas of permafrost, warmer temperatures may compromise the stability of paved airport runways, as well as roads and rail beds.

Increases in intense storms, or other weather extremes and their consequences such as flooding, land slides and forest fires, will likely result in increased costs for transport operators and hence the general public. Train, aircraft and ship delays and cancellations, road detours and other interruptions of activities can increase the operating costs for transport companies and state authorities, as well as decreasing customer satisfaction. For example, commercial passenger flight cancellations and diversions are estimated to cost between US$40,000 and US$150,000 per flight (Environmental and Societal Impacts Group, 1997). Adverse weather is a factor in one-third of all aircraft accidents and turbulence is the leading cause of in-flight injuries to passengers and crews. According to Delta Airlines, weather conditions account for up to 75% of all flight delays. Time wasted for this reason is estimated to have cost airlines US$6.5 billion in 2000. The flooding of one section of the rail link between Oxford and London, UK on six days in 2000 incurred time delay penalties for the railway company of at least £1.2 m. This cost does not allow for time lost by train passengers or the cost of repairs to the rail infrastructure (London Climate Change Partnership, 2004). But in a warmer world new transportation routes are likely to open. This includes the Northeast Passage between Norway and Alaska, due to shrinking Arctic sea ice cover.

On the other hand, the reduced cost of snow and ice management is likely to bring major benefits to the transportation industry. Currently annual winter road maintenance costs incurred by government agencies in the USA and Canada are approximately US$2 billion and CN$ 1billion, respectively. A warming of 3–4°C could decrease salt and sand use by between 20 and 70%. Similar savings could be achieved where snow removal, ice breaking and de-icing are at present necessary for railways, airports, harbours and other transport facilities (Mills & Andry, 2002).

In absolute terms, road collisions are by far the most important transport-related safety concern for tourists. Weather conditions are often a contributing factor. For example, each year in Canada weather conditions contribute to approximately 10 train derailments, 10–15 aircraft accidents, over 100 shipping accidents and tens of thousands of road collisions. Precipitation generally increases collision risk by between 50 and 100%, with injury risk increasing by about 45%. Significantly, injury risk is similar for snowfall and rainfall events, relative to normal seasonal driving conditions. As a result, the projected shift from snowfall to more rainfall may have a minimal impact on injury rates. This contrasts with suggestions in earlier studies that the shift would reduce injury rates. However, where precipitation events become more frequent it is likely that injury risk will increase (Mills & Andry, 2002).

Adaptation

Climate change provides additional incentive to ensure that new tourism transport infrastructure is designed and built in such a way that atmospheric conditions anticipated to occur during the planned lifetime of the infrastructure will not lead to premature deterioration or failure. Similarly, existing infrastructure may have to be modified if the performance standards are inconsistent with the changed climatic conditions. In both cases this requires taking into account information such as that contained in a climate risk profile (Figure 8.1). Industry organisations and government (e.g. through regulatory instruments such as building codes) can promote such good practices. Improvements in maintenance practices also provide an opportunity for adaptation to climate change, for instance by the use of more heat-resistant grades of tar and asphalt, and the use of expansion-resistant concrete when resurfacing roads and airport runways.

Tourism transport services also have many opportunities to adapt to climate-related risks. For example, today most ocean-going tourist vessels have access to real time weather information. This is often used in a proactive manner, to plan the safest and most economical route by choosing the optimum track, based on forecasts of weather, sea conditions and a ship's individual characteristics. More advanced systems take into account not only wind and swell conditions but also sea level, currents and ice cover. The optimum track is chosen for maximum safety, passenger and crew comfort, minimum fuel consumption and minimum time underway. Ship weather routing attempts to avoid or reduce the effects of specific adverse weather and sea conditions, by issuing initial route recommendations prior to sailing, recommendations for track changes while underway and weather advisories to alert the commanding officer or master about approaching unfavourable conditions that cannot be reasonably avoided by a diversion.

Avoidance of 'clear air' turbulence and other atmospheric conditions that threaten aircraft safety, reduce passenger comfort and increase operating costs is becoming easier with advances in technology. For example, pilots of Airbus Industry's A380 Superjumbo have the benefit of a state-of-the-art weather radar that automatically scans more than 1.5 million cubic miles of atmosphere every few seconds. The system employs volumetric rather than the conventional sequential slice scanning. The continuous scanning of the aircraft's horizon at ever-changing tilt angles builds up a 3D digital image of weather and terrain ahead. The intuitive displays allow a pilot to quickly diagnose multitiered weather threats by viewing weather images at any altitude. Pilots can detect turbulence and hazardous weather conditions earlier, to provide passengers and crew safer and more comfortable flights. Additionally, the ability to look ahead, and modify the flight path as needed, assists in on-time arrivals and on-time departures, saving time and costs for airlines and operators.

Implications for tourism's resource base

The natural resource base for tourism includes the relatively unmodified components of the environment, such as natural terrestrial, coastal and marine ecosystems and landscapes, surface and ground water, as well as the atmosphere. On the other hand, the human resource base for tourism is generally taken to include individual and collective knowledge, understanding and skills; institutions; financial services; and planning, legislative and regulatory instruments.

Risk factors and anticipated consequences

The natural environment is often very important in determining the demand for tourism, especially for nature-based tourism. Tourists are attracted to parks and similar natural areas because they represent a healthy environment, protected in perpetuity. But the impact of climate change on such natural landscapes may have a negative influence on their amenity value and hence on visitor numbers. Tourism in South Africa, for example, relies heavily on biodiversity. But increases in temperature (2.5°C by 2050) and shifts in precipitation, including aridification of conditions in the Western half of the country and a consequential decline by about half of the current extent of the country's biomes, will cause substantial losses in biodiversity (Simpson, 2003). As a result, the biggest potential economic loss for South Africa due to climate change will follow from the consequences for tourism of a loss of habitats and biodiversity, as well as changes in temperature humidity and malaria risk (in Simpson, 2003).

Scott (2003) has summarised the diversity of consequences of climate change for the natural mountain environments (Chapter 3). Mountain parks are expected to experience both latitudinal and elevational ecotone changes, including loss of high elevation species and emergence of new vegetation communities due to invasion of non-native species. As the tree-line shifts upwards, alpine habitat will diminish and become increasingly fragmented. As many as two-thirds of mammal species on the isolated mountain tops of the Great Basin in the Western USA are threatened by climate change, with the likely extinction of up to 14 species. The brilliant colours of maple trees in the fall in mountains of New England (USA) bring many tourists to the area. However, maple populations are expected to decline in a warmer world and be replaced by less colourful tree species such as oak and hickory. In the same region, coldwater sport fish species, such as salmon, are at their southern limits and are expected to be totally or partially replaced as a consequence of climate change. As a result, the tourists in New England who currently spend an estimated 24 million fishing days, and generate over US$1.3 billion annually, may in future opt to go to Quebec and New Brunswick and other locations in Eastern Canada.

Many plant species are unlikely to be able to migrate fast enough to keep pace with anticipated changes in climate. The problem is exacerbated by the increasing fragmentation of landscapes and habitats, as a consequence of tourism and other development initiatives. As a result, changes in climate have the potential to overwhelm the capacity for adaptation in many plant populations, dramatically altering their genetic composition. Consequences are likely to include unpredictable changes in the presence and abundance of species within communities and a reduction in their ability to resist and recover from further environmental pressures such as pest and disease outbreaks and extreme climatic events. A range-wide increase in extinction risk is likely to result, threatening a major component of tourism's natural resource base (Jump & Peñuelas, 2005).

Coastal tourism is an extremely important component of the tourism sector, with much of the economic values of coral reefs (estimated US$375 billion each year) generated from nature-based and dive tourism (Cesar *et al.*, 2003). Four major categories of socioeconomic links between coral reefs and tourism have been identified (Figure 8.6). In many regions of the world coral reefs provide not only physical protection to significant tourism infrastructure and related assets, but also offer the principal attraction for tourists. For example, through tourism, Florida's reefs contribute US$1.6 billion annually to the state economy. Reefs are valued as fish habitat, for the production of sand for beaches, and for their genetic importance to medicine, science and education. All these benefit tourism (Figure 8.6).

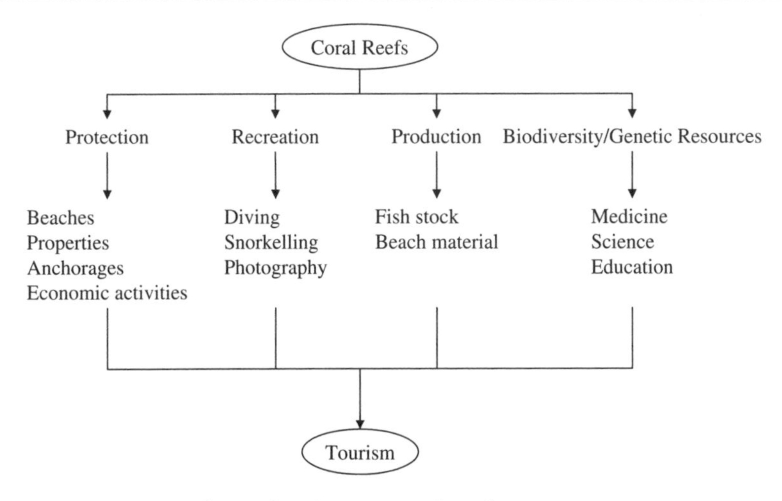

Figure 8.6 Tourism dependencies on coral reefs
Source: after Jackson (2002)

Risks to coral reefs are exacerbated by climate change. Rising ocean temperatures have been implicated in chronic stress and disease epidemics, as well as in the occurrence of mass coral bleaching episodes. Increasing atmospheric CO_2 levels also modify ocean chemistry in ways that inhibit deposition of the calcium carbonate minerals that are the structural building materials of coral reefs. Coral reef communities usually recover from acute physical damage (as a result of tropical cyclones) or coral mortality, but only if chronic environmental stresses (such as reduced water quality) are weak, and if the acute stresses are not strong or overly frequent. Conversely, coral reefs also withstand chronic stresses in the absence of acute stresses. But a combination of acute and chronic stress often results in the replacement of the coral reef community by seaweeds or some other non-reef system. Such ecosystem shifts are well advanced in the Caribbean region. Already two of the major reef-building coral species have been devastated by disease. In the Indo-Pacific region, the repeated and lethal episodes of 'bleaching' associated with unusually high water temperature raise concern that reefs cannot sufficiently recover between such events (Buddemeier *et al.*, 2004).

Projected changes in temperature of the major wine-growing regions is likely to have a mixed consequences for wine tourism, a growing component of the tourism industry (Text Box 19).

Given the combination of increased use of dryland areas for tourism and the projected increase in their aridity as a result of climate change, these areas are likely to be degraded. This will lead to desertification, with surface and subsurface water resources coming under increasing pressure.

Text Box 19 Climate change and wine tourism

Jones *et al.* (2005) studied the nature and trends of climate and wine quality for 27 of the most prominent wine-growing regions in the world. Between 1950 and 1999 the growing season average temperatures in the world's high-quality wine-producing regions increased by 1.26°C. Vintage quality ratings during this same time period increased significantly while year-to-year variation declined.

Improved winemaking knowledge and husbandry practices undoubtedly contributed to the better vintages, but climate also played a significant role in these variations in quality. While the observed warming of the late 20th century appears to have been mostly beneficial for high-quality wine production worldwide, the impacts of future changes in climate will likely be highly variable across varieties and regions. Regions producing high-quality grapes at the margins of their climatic limits may exceed a climatic threshold, such that the ripening of the balanced fruit required for existing varieties and wine styles will become progressively more difficult.

Thus the rule of thumb 'the warmer the better' is not globally applicable. Many wine regions have already reached their optimum temperature, while some have passed the predicted optimum temperature. Projected climate changes could push other regions into more optimal climatic regimes for the production of current varietals. In addition, the warmer conditions could lead to more poleward locations becoming increasingly suited to grape growing and wine production.

The growing scarcity of fresh water supplies in areas experiencing lower summer rainfall or more prolonged drought, and the consequent inability to meet existing, let alone future, tourist demands, may well be a critical constraint on tourism development as well as on current operations. Schröter *et al.* (2005) report that for Europe as a whole, but especially in the Mediterranean and mountain areas, climate change will result in a decreasing supply of ecosystem services of relevance to tourism, including declining water availability. Increased frequency and duration of drought is likely to increase vegetation fires and hence degrade air quality in the affected areas. The fire season in several popular Rocky Mountain parks (USA) is projected to increase by 30–50 days (in Scott, 2003).

Minor increases in sea level are unlikely to have a significant effect on the freshwater lens, and might even increase its volume slightly. However, where rising sea-levels result in a reduction in island area, the volume of the freshwater lens is reduced. These changes, plus flooding of coastal areas due to high waves and storm surges, will typically result in degraded water quality. Competition for water can lead to conflicts between local people and tourist operators.

Good air quality, including unimpaired visibility, is also part of tourism's natural resource base. The interrelations between air pollution and climate change can be categorised as follows: (i) some gaseous pollutants (e.g. ozone in the lower atmosphere) contribute to regional air pollution as well as to climate change; (ii) climate change will impact on regional air quality, and vice versa; (iii) air pollutants and GHGs are often emitted by the same source; and (iv) measures to reduce emissions of GHGs may affect emissions of air pollutants, and vice versa (European Environment Agency, 2004). From a tourism perspective the second category is of greatest immediate relevance, though currently insufficient information is available to assess how climate change will influence the sensitivity of humans to air pollutants.

Climate change may affect the generation, transformation, transport, dispersion and removal of air pollutants, leading in turn to changes in the frequency, duration and intensity of air pollution episodes such as those with photochemical smog and high particulate levels. Such air pollution events can in turn cause human discomfort, health damage and reduced satisfaction. For example, climate change is likely to impact natural emissions of both particles (mineral dust and smoke) and ozone precursor gases such as nitrogen oxides and volatile organic compounds. Collectively this will change the background concentrations of both particles and ozone. Increased temperatures, especially in the summer, will increase ozone concentrations, resulting in more and longer ozone episodes. On the other hand, increased air temperatures result in less extreme winters, with less exposure to fine particulates. Cold weather conditions are usually associated with less wind, and hence with higher particulate concentrations.

Globally, approximately 300 million people suffer from asthma, resulting in some 24.5 million lost work days. By 2025 the number of asthma suffers could increase to 400 million. Asthma rates in the USA have quadrupled since 1980. The fertilisation effect of rising CO_2 levels, plus earlier springs and warmer summers and winters, stimulates ragweed and some flowering trees to produce excessive amounts of pollen. Some soil fungi produce many more spores when grown under elevated CO_2 conditions. Both the pollen and spores are carried deep inside the human lung, a process aided in part by diesel and other fine particles common in urban areas. Enormous dust clouds originate in the deserts of Africa and Asia, which are themselves expanding in part due to increased drought. The particles and microbes from these clouds travel vast distances, but many eventually settle in the lungs of people and animals, causing and aggravating respiratory problems (Epstein & Mills, 2005).

In terms of the human resource base, the pervasive economic, social and environmental consequences of climate change for the tourism sector present serious challenges to institutional structures and working arrangements, not only within the public sector but between it and institutions in the private sector and the wider community. Even at the policy level, addressing the consequences of climate change for tourism will extend beyond the ministry of tourism, or its local equivalent, to government agencies concerned with such matters as natural resource management, transport, energy and border security. At the operational and practical level the implications extend even further, to industry bodies and to community organisations. One major concern is the possibility that the institutional arrangements will preclude a coordinated and effective multisectoral and multiagency response to assessing and managing the climate change-related risks to tourism and that the institutions themselves will prove incapable of meeting the increasingly complex and substantial challenges that these tasks present.

The tourism sector currently places large demands on the financial sector as it strives to limit the adverse consequences of present-day climate variability and extremes, and to exploit any potential benefits. The most obvious financial service is disaster and related insurance (see Chapter 3), but others include loans to finance disaster recovery and to allow investments in technologies that reduce climate-related risks. There has already been some withdrawal of insurance coverage for vulnerable properties such as those on small islands, adding to risk levels. Heavy uninsured losses may also result in loan defaults. As extreme events become more intense, and in many cases more frequent, such financial consequences for the tourism sector will likely become more widespread.

Adaptation

Conservation of biodiversity and maintenance of ecosystem structure and function are important climate change adaptation strategies in relation to tourism's natural resource base (see Figure 8.7). Establishment and enforcement of protected areas is generally considered to be the most appropriate strategy for ensuring that terrestrial, freshwater and marine ecosystems are resilient to the pressures arising from climate change. Habitat fragmentation can be compensated for by establishing biological corridors between protected areas. More generally, it is desirable to develop a mosaic of interconnected terrestrial, freshwater and marine multiple-use protected areas designed to take into account projected changes in climate. Importantly, park and protected area management is often founded on frameworks and analyses that assume climatic and biogeographic stability. At least the former assumption is invalid, given global climate change. Protected area management must now take into account changing climatic conditions. A climate change adaptation

Figure 8.7 Reforestation of mangrove forest at a tourist resort in Fiji

portfolio for protected area agencies is presented in Table 8.1. Many of the suggested interventions are also applicable to ecosystems that are not under a formal system of protection.

Where protected area status is neither possible nor practicable, every reasonable effort should be made to limit both direct and indirect human impacts on the ecosystems. Appropriate specific initiatives include maintaining and restoring natural ecosystems; utilising traditional knowledge and practices; protecting and enhancing ecosystem services; actively preventing and controlling invasive alien species; managing habitats for rare, threatened and endangered species; developing agroforestry systems at transition zones; and monitoring results and, where necessary, changing management practices.

The resilience of coral reefs is of particular concern, as most of the effects of climate change are stressful rather than beneficial. Reef systems that are jointly affected by global climatic and local human stresses will be the most vulnerable. In some instances, climate change has the potential to yield some benefits for certain coral species in specific regions, including expansion of their geographic ranges to higher latitudes. But while the net effects of climate change on coral reefs will be negative, coral reef organisms and communities are not necessarily doomed to total extinction. Survival is made possible by the diversity of coral species comprising existing reefs, the demonstrated adaptation potential of reef organisms, spatial and temporal variations in climate change, and the potential for human management and protection of coral reef ecosystems. Adaptation is further enhanced by having a distributed international system of coral reef refuges and marine protected areas, selected on the basis of biological and environmental diversity, connectivity, potential threats and enforcement feasibility (Buddemeier *et al.*, 2004). At the resort level, specific reef

Table 8.1 Climate change adaptation portfolio for protected area agencies

System planning and policy	Expand the protected areas network where possible and enlarge protected areas where appropriate. Improve natural resource planning and management to focus on preserving and restoring ecosystem functionality and processes across regional landscapes. Selection of redundant reserves. Selection of new protected areas on ecotones. Selection of new protected areas in close proximity to existing reserves. Improve connectivity or protected area systems. Continually assess protected areas legislation and regulation in relation to past, anticipated or observed impacts of climate change.
Management (including active, adaptive ecosystem management)	Include adaptation to climate change in the management objectives and strategies of protected areas Implement adaptive management Enhance the resiliency of protected areas to allow for the management of ecosystems, their processes and services, in addition to 'valued' species. Minimise external stresses to facilitate autonomous adaptation. Eliminate non-climatic *in situ* threats. Create and restore buffer zones around protected areas. Implement *ex situ* conservation and translocation strategies if appropriate. Increased management of the landscape matrix for conservation. Mimic natural disturbance regimes where appropriate. Revise protected area objectives to reflect dynamic biogeography.

Research and monitoring	Make resources available to aid research on the impacts of past (e.g. paleo-ecological change) and future climate change (e.g. projected species composition changes). Utilise parks as long-term integrated monitoring sites for climate change (e.g. monitoring of species, especially those at risk or extinction-prone). Identify specific 'values' at risk to climate change. Regional modelling of biodiversity response to climate change. Incorporate climate change impacts in protected areas 'state-of-the-environment' reporting.
Capacity building and awareness	Strengthen professional training and research capacity of protected area staff with regards to climate change. Capacity building and awareness should proceed with the goal of securing public acceptance for climate change adaptation. Partnerships/collaboration with greater (regional) park ecosystems stakeholders to respond to the need for climate change adaptations. Improved collaboration from local to international scales. Make resources available for investing in active, adaptive management. Develop precautionary approaches (such as disaster preparedness and recovery systems) through forecasting, early warning and rapid response measures, where appropriate.

Source: Scott & Lemieux (2005)

conservation measures include education and awareness programmes, tertiary treatment of sewage and disposal via a long off-shore pipeline; monitoring systems that show the relative efficacy of adaptation measures; enforcement of fishing quotas and designation of exclusion zones and seasons. In addition, it is now a common to ban destructive fishing practices as well as removal of reef resources for the aquarium trade, construction materials, navigation improvements, and medicinal and pharmaceutical applications.

With regard to other components of tourism's natural resource base, alleviating water shortages will require substantial investments in storage facilities and, where appropriate, desalinisation plants. This will be especially the case for resorts some distance, or otherwise isolated from plentiful sources of surface or ground water. Adaptation measures that enhance water security at the resort level involve a mix of infrastructure improvement, sustainability planning, water source management and health and environmental protection. Structural deficiencies in water treatment and distribution systems, such as ageing, breaks and leaks, need to be addressed. Monitoring of water quality for early signs of waterborne diseases should be strengthened; especially to take into account increased risk associated with heavy rainfall, drought and other extreme weather conditions. The resulting information can provide the basis for early warning and prevention systems. Improved protection of water sources and maintenance of storage systems can help prevent contamination and disease outbreaks. Above all, planning, management and infrastructure decisions should take into account the likelihood of significant changes in climate during the lifetime of the facilities.

In terms of air quality, implementing climate policies such as achieving Kyoto Protocol targets can significantly reduce the costs of meeting air quality targets. Conversely, meeting stringent air-quality targets is likely to require measures that go beyond end-of-pipe technological solutions and require broader structural changes that are consistent with climate goals. Thus policies and initiatives targeting activities that release air pollutants and GHGs (e.g. combustion of fossil fuels, intensive agriculture), rather than the emissions of one specific substance, are more likely to achieve synergies. Depending on the nature and choice of the policy instruments, climate change policies developed independently from air pollution policy will either constrain or reinforce air pollution policies. Similar consequences result when air-quality policies operate in isolation from climate change policies. Moreover, from an economic perspective, policies that may not be regarded as cost-effective from a climate changeor an air pollution perspective alone, may be found to be cost-effective if both aspects are considered in tandem. If climate change considerations are taken into account when designing policies and measures to abate local and regional air pollution (and vice versa), the cost-benefit equation and thus the relative priorities of policy options could change considerably; there is as yet no agreed operational framework to evaluate how climate change will influence air quality (European Environment Agency, 2004).

Adaptation via changes in the human resource base includes improved knowledge and understanding of how the climate will change in the future (Text Box 20). This cannot come solely from further increases in the complexity and resolution of global climate models. Such efforts will undoubtedly decrease uncertainties and increase the

Text Box 20 Using climate models to assess the competitiveness of tourist destinations

Contributed by J.P. Ceron, CRIDEAU, Université de Limoges, France

Météo France has built a global model with a concentration of its calculation capacities on France. This model can be used to undertake a scenario analysis for potential impacts of climate change on tourism in France and neighbouring countries. In particular, the climate model could be used to assess the future competitiveness of France as a tourist destination. The model shows that countries to the south of France, which are already hotter than the hottest areas within France, will experience great increases in temperature and therefore become possibly too hot and uncomfortable for tourists. Temperatures in countries north of France (e.g. the UK) will rise by between 1 and 2°C, which is not quite enough to make them competitive, sunny summer destinations. It is concluded that the greatest part of France (possibly except the Mediterranean shores) could gain in relative competitiveness. However, the climate model also indicates growing problems with water resources. These would need to be managed carefully for the viability of tourism.

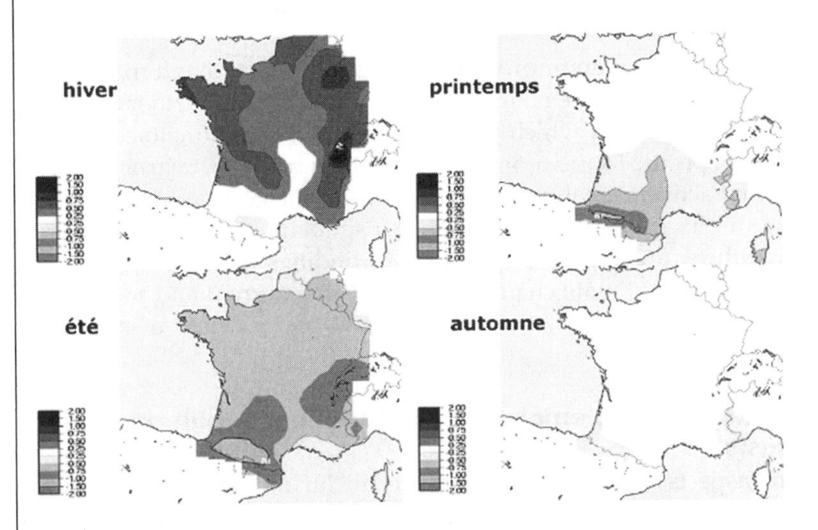

Text Box 20 – Figure 1 difference in precipitation (mm/day) between 2070–2099 and 1960–1999 for all four seasons

Source: Déqué (2004)

These kinds of assessments are acceptable at a macro level but may become fraught at the micro level, for example as in the case of the Côte Basque, the South-western area of France at the border to Spain.

An earlier version of climate scenarios produced by Météo France in 1998 indicated that benefits from higher winter temperatures could be offset by a substantial increase in rainfall. A later analysis, however, shows that rainfall in the Côte Basque would not increase. Hence the conditions for tourism seem to be more favourable compared with the earlier model.

The French example shows that while generally of interest and relevant at a higher level, global models have to be treated with caution when making inferences for tourism development at a local level. Global models need to be augmented with detailed information reflecting local climate conditions and changes over time, possibly monitored by local meteorologists.

potential usefulness of climate projections, perhaps to the stage where they can be called 'climate predictions'. But parallel efforts are required to provide climate change information for specific locations and for an appropriate selection of climate variables. The capability to do this, and make meaningful interpretations of the findings, must be developed in the relevant institutions of the country requiring this more targeted information.

It is also clear that an improved ability to understand and manage the impacts of climate change on the tourism sector requires further empirical studies of the ways in which climate affects destination choice and the viability of specific tourism operations such as resort infrastructure, and passive and active recreational activities. Adequate geographical coverage in these studies is of crucial importance if simulation studies of climate–tourism futures are to have credibility. The findings will help improve the ability to plan for climate change, by prioritising current and future risks and by helping to identify the most effective and efficient management interventions.

Implications for the enjoyment, satisfaction, health and safety of tourists

Many aspects of the comfort, pleasure, health and safety of tourists are climate dependent. Thus climate change is likely to have a significant influence on the health, well-being and pleasure of tourists.

Risk factors and anticipated consequences

As already noted in Chapter 2, thermal comfort is considerably more important to a tourist than are cloud and sunshine conditions, with factors such as wind being of even less importance. The ideal climate-related requirements for water- and land-based recreation activities are summarised in Table 8.2. Another study suggests that ideal atmospheric

Table 8.2 Ideal climate-related requirements for water- and land-based recreation activities

Climate parameters	*Summer*				
	Motor boating	***Fishing***	***Swimming/sunbathing***	***Camping***	***Golf***
Air temperature (°C)	15–35	15–30	15–30	> 10	10–30
Wind (km/h)	< 50	< 15	< 15	< 10	< 20
Precipitation	Nil	Nil	Nil	Nil to light	Nil
Lake depth (m)	1.5–2.5	0.5–1.0	0.5–2.0	–	–
	(b) Winter				
	Nordic skiing	***Alpine skiing***	***Snow shoeing***	***Snowmobiling***	
Snow depth (cm)					
Minimum	20–30	20–30	20–30	30	
Optimum	60	60	60	60	
Snow density (g/cm^3)	< 0.6	< 0.6	0.2–0.6	0.4–0.1	
Air temperature (°C)	– 2 to – 15	5 to – 20	10 to – 40	10 to – 30	
Wind (km/h)	< 20	< 15	< 45	< 45	
Wild chill (Watts/m^2)	700	700	1600	1400	

Source: after Scott *et al*. (2005)

conditions occur when there is slight heat stress ('slightly warm') along with scattered cloud, wind speeds less than 6 m/s and rainfall lasting less than half an hour (de Freitas, 2005).

Warming and precipitation trends due to the human-induced climate changes that have occurred over the past 30 years could already be causing over 150,000 deaths annually, as well as reducing total life expectancies by approximately five million years. Both are a result of increasing incidences of cardiovascular mortality and respiratory illnesses due to heatwaves, the altered transmission of infectious diseases such as diarrhoea and malaria, and malnutrition from crop failures. But the unequivocal attribution of the expansion and resurgence of diseases and illness to climate change is not yet possible. This is due to the lack of long-term, high-quality data sets as well as to the large influence of socioeconomic factors and changes in immunity and drug resistance. Nevertheless, there is strong evidence of two key climatic impacts on human health – direct, heat-related mortality and morbidity and climate-mediated changes in the incidence of infectious diseases. Heat-related mortality is dominated by the difference between temperature extremes and the mean climate – especially in early summer when people have not yet become accustomed to higher temperatures – rather than by gradual increases in mean temperatures.

Projections of future climate suggest such increases in extremes in relation to mean temperatures may occur, particularly in the mid-latitudes. In addition, the effect of heatwaves is exacerbated in large cities owing to the urban heat island effect. Thus, as urban areas and urban populations grow, vulnerability to heat-related mortality will likely increase. Studies of climatic influences on infectious diseases have mainly focused on the influence of ENSO, with relationships found for malaria in South America, rift valley fever in East Africa, dengue fever in Thailand, hantavirus pulmonary syndrome in the South-western USA, childhood diarrhoeal disease in Peru and cholera in Bangladesh. It is currently unclear whether global warming will significantly increase the amplitude of ENSO variability, but if this is the case, the countries surrounding the Pacific and Indian oceans are expected to be most vulnerable to the associated changes in health risks due to large rainfall variability related to ENSO. Regions bordering areas with high endemicity of climate-sensitive diseases, where temperatures at present limit the geographic distribution of disease (such as malaria in the African highlands) could be at risk in a warmer climate (Patz *et al.*, 2005).

Currently many subtropical countries, such as Spain, are considered to be low health risk destinations, not requiring immunisation, prophylactics or other treatments and precautions. However, it is anticipated that by the 2020s suitable conditions for malaria will extend from North Africa into Spain. Increased use of air conditioning is likely as the climate warms and high temperature extremes become more common. The incorrect use of such systems has been associated with legionnaires' disease in travellers (Kovats & Casimiro, 2003).

Tourists travelling from countries with a cooler climate will be less tolerant of high temperatures and more vulnerable to disease at their destination. The increased likelihood of excessively high temperatures, including heatwaves, in the major tourist destinations has already been highlighted. The tourism scenarios described in Chapter 5 identify the increasing dominance of older people in tourist activities. Elderly people are more vulnerable to heat stress and are more likely to have pre-existing health conditions such as cardiovascular disease and reduced immunity. Excessively high temperatures for prolonged periods have also been linked with the development and increase of algal blooms, resulting in beach closures, fish kills, seafood poisoning and a general degradation of beach aesthetics. For example, warming ocean waters are thought to have resulted in Alaskan oysters being contaminated with *Vibrio parahaemolyticus*, the most common cause of seafood-related illness in the USA. When a total of 62 passengers fell ill on four, week-long Alaskan cruises in 2004, the cause was traced to consumption of raw oysters reared on an oyster farm 1000 km north of any previous known source of contaminated oysters. The bacteria is believed to grow in oysters only when water temperatures exceed 15°C. Ocean temperatures in South-east Alaska have been rising gradually since 1976, and have now reached historic highs (McLaughlin *et al*., 2005).

Infectious intestinal disease ('travellers' diarrhoea') is the main ailment experienced by tourists. It is caused by a range of pathogens, with the main transmission routes being contaminated food and water. The incidence of food poisoning is often closely associated with high temperatures. Rapid changes in precipitation and sea surface temperatures also lead to increased nutrient levels in seawater, resulting in contamination of shellfish with marine biotoxins. Higher water temperatures can increase the reproduction and longevity of pathogens found in seawater as a result of sewage discharge and in swimming pools and other recreation waters, increasing health risks for swimmers. Warmer freshwater temperatures favour the growth of cyanobacteria, increasing the risk of skin irritations.

Climate change may also increase the risk of accidents on mountains and at coastal destinations as well as illnesses related to poor urban air quality in summer. A frequent cause of excess mortality in travellers is accidents, mostly due to traffic and swimming mishaps. These are highly correlated with excess alcohol consumption, which in turn is often associated with hotter weather. Sea currents, which may also be modified by changes in the climate, also play a significant role in swimming accidents (Kovats & Casimiro, 2003).

Adaptation

Many adaptation measures are already being employed in order to reduce the ways in which climate jeopardises the enjoyment, satisfaction, health and safety of tourists. One response to the increasing likelihood of excessively high temperatures is illustrated by a decision to address safety concerns by closing a walking track in Uluru-Kata Tjutu National Park, Australia (a major attraction for international visitors as it contains Ayers Rock and the Olgas) on days when maximum temperatures are forecast to exceed 36°C (Skinner & de Dear, 2000). After a legal challenge by tourism operators, the practice was accepted by all parties, based on the findings of detailed heat stress assessments that included calculation of such indices as the outdoor wet bulb globe temperature and the outdoor standard effective temperature (Pickup & de Dear, 2000). Due to a generally strong association between high summer temperatures and increased fire activity, measures such as the closure of forest and parkland areas, to ensure public safety as well as reduce accidental ignition, will likely become increasingly common.

Early warning systems have been shown to be effective in reducing mortality associated with heatwaves. Excessively high night-time temperatures and the duration and early seasonal occurrence of heatwaves have been identified as relevant risk factors that can be incorporated into early warning systems designed to limit the serious health effects of heatwaves for tourists and other individuals. For example, after the 1995 heatwave in the USA, the city of Milwaukee initiated an 'extreme heat conditions plan'. This involves local agencies, communications tests, stepped responses to early forecasts, a 24-hour 'hotline' and other interventions such as opening air-conditioned shopping malls at night-time to those individuals most vulnerable to heat. During a comparable heatwave in 1999, heat-related morbidity and mortality were almost halved, relative to expected levels. Currently, over two dozen cities worldwide have weather-watch warning systems. Such a system successfully forecast most days with excess deaths in Rome during the 2003 heatwave (Patz *et al.*, 2005).

Tourists to destinations that offer both cooler highland areas and beaches, such as Cyprus and Corsica in the Mediterranean, can make day trips to the upland areas when conditions at the beaches and other lowland areas become uncomfortably hot. At the resort level measures that reduce the consequence of excessively high temperatures include allowing night-time access to air-conditioned facilities for individuals vulnerable to heat, roof gardens, vegetation and covered walkways designed to ameliorate temperature conditions, and low-emission vehicles in larger resorts.

National health surveillance systems are generally designed to reduce the vulnerability of the local population to infectious diseases, rather than identify travel-associated outbreaks. However, international and regional health monitoring and reporting networks that provide alerts for disease outbreaks can reduce tourist exposure to those health risks, including those which may be increased by climate change. The Southern African Regional Climate Outlook Forum, comprising 14 countries, provides an informative example. The Forum uses seasonal forecasts to prepare early warnings of health risks, to plan interventions and to allocate resources. With advanced warning, prophylactic treatment can be offered to those most in danger from infectious diseases.

But the regions with the greatest burden of climate-sensitive diseases are also the regions with the lowest capacity to adapt to the new risks. This includes Africa, where an estimated 90% of malaria occurs, and where international tourist arrivals are expected to grow annually by 5.5% to reach 77 million by 2020. The incidence and spread of malaria can be reduced through such measures as preventing deforestation and other forms of land clearance that encourage flooding, creation of mosquito breeding sites and destruction of habitat for mosquito predators. Well maintained urban drainage systems can also reduce the mosquito population and hence the risk of malaria. Resort-based responses include use of pesticide-impregnated bed nets and the eradication of potential mosquito breeding sites, including spraying as a last resort. However, the long-term health, ecological and resistance-generating consequences and financial costs of pesticide use are cause for concern (Epstein & Mills, 2005).

Improved regulation and monitoring, and good practices in food hygiene, including preparation and presentation, will be required if the additional risks to public health due to food contamination in a warmer climate are to be avoided. With respect to wine production, Jones *et al.* (2005) argued that to facilitate planning for and adaptation to climate change, focused research is needed in two main fields: production of finer-resolution climate simulations more appropriate for assessing microclimates critical for grape growing; and improved viticulture modelling incorporating treatment of varietal potential, phenological development and vine management.

Implications for the sustainability of facilities and destinations

In this section emphasis is on the ways in which climate change may threaten the longevity of tourism infrastructure and other facilities, as well as entire tourist destinations. The section also focuses on how such climate change-related risks might best be managed.

Risk factors and anticipated consequences

Until very recently the tourism industry viewed climate change as a longer-term issue of little importance given the relatively short time horizons usually adopted by tourism operators. But this attitude is shifting rapidly, largely as a result of the growing evidence that events consistent with those projected to occur as a consequence of climate change are already having detrimental repercussions for tourism. As already discussed earlier, climate-related risks are particularly evident for tourism in coastal zones and mountain regions.

Climate change-related risks to coastal tourism operations and infrastructure, and the industry as a whole, fall into three main categories: (1) sea-level rise that accelerates beach erosion, and also contributes to inundation, rising water tables, destruction of coastal ecosystems, salinisation of aquifers and, at worst, the total submersion of tourist facilities on low islands and coastal plains; (2) elevated sea temperatures causing coral bleaching, leading to decreased reef protection and amenity loss for divers and snorkellers; and (3) increasing storm intensity and possibly frequency which, especially when combined with rising sea-levels, results in damage to sea defences, protective mangrove swamps and coral reefs, and shoreline buildings, and also further beach erosion (Todd, 2003).

Around the world many beaches are suffering accelerated erosion and extensive coastal wetland losses, not only as a consequence of accelerated sea-level rise but also due to the effects of storm surges and high wave events. In many cases development activities are exacerbating the situation. On open coasts poor management of beach sand and its movement along shorelines contributes to erosion. The average annual erosion rate in the beach communities of Delaware's Atlantic coast is about 1 m per year. It is threatening the sustainability of the area as a major summer tourist attraction (Daniel, 2001). An estimated 5.1 million person-trips are made to the Delaware beaches each year and the consumer surplus (the recreation/aesthetic value of the beach as revealed by beach visitors) for these visitors exceeds US$380 million. Visitors spend more than US$573 million in beach trip-related expenditures each year.

Garcia *et al.* (2000) report that severe storm damage to beaches on Spain's Costa del Sol at the end of the 1980s, together with concern for the tourist industry then suffering stagnation and broader unease over the irreversible transformation of the coast through human activities, all coalesced to prompt the implementation of planning controls and new protective measures. In a prophetic study, Burkett *et al.* (2005) revealed that land subsidence, rising sea-levels and the increased height of storm surges signalled serious losses of life and property in New Orleans and adjacent areas unless flood-control levees and drainage systems were upgraded. As history now shows, Hurricane Katrina proved the predictions to be correct. Climate change will increase coastal erosion. As a result the tourist industry will face the dilemma of maintaining beaches and other coastal assets through the costly process of replacing sand (i.e. beach nourishment) or staging a planned retreat. The availability of sufficient sand and the question of who pays for beach nourishment will be key issues to be assessed (Neuman *et al.*, 2000).

The likely consequences of climate change for the major tourism destination of Phuket, Thailand, have been investigated by Raksakulthai (2003). The South-west monsoon is the largest single factor determining the timing of the high and low seasons for tourism, with the heavy monsoon rains limiting the main activities (diving, snorkelling, sunbathing) that attract tourists to the area. While it is still unclear how climate change will affect the Asian monsoon, the seasonal onset and duration may well shift. Water supply also fluctuates throughout the year, with the supply reservoir sometimes having as little as 10 days of supply. Such shortages may increase in frequency as rainfall amounts in the area are projected to decline. Climate change will increase sea levels, and hence the rate of coastal erosion, and likely increase water temperatures, escalating the risk of damage to reef ecosystems as a consequence of coral bleaching – previous extremes in high water temperatures have caused up to 80% mortality of some reefs.

In mountain regions, it seems very probable that ultimately demand for winter sports will diminish. The season will shorten, opportunities for young people to learn the sports will diminish and demand pressures on high-altitude resorts will increase. This in turn could raise environmental pressures and cause further damage. On the other hand, summer seasons could lengthen, generating increased demand. But this in turn could bring further detrimental environmental consequences. Climate change-related risk factors for mountain and other cold-climate-dependent tourist areas include less snow, receding and downwasting glaciers, melting permafrost, diminished sea ice cover, increased frequency of events such as landslides, rockfalls and avalanches, and modification of alpine fauna and flora, in part through fire and disease. For example, the polar bear population near the Canadian town of Churchill, on the shore of Hudson Bay, is an important tourism resource. Sea ice projections indicate that by the 2030–2040s the polar bear population will have had to relocate away from Churchill, bringing the $CDN300-million annually bear tourism to an end (Vinnikov *et al.*, 1999).

Snow cover modelling using two climate change scenarios for the mountains of North-western USA suggests a 75–125 cm reduction in average winter snow depth and an upward shift in the snowline from 900 to 1250 m above sea-level. For the same scenarios applied to central Ontario (Canada), the ski season would be reduced by between 40 and 100% by the 2050s. Snowmobiling has been found to be even more at risk from climate change than alpine skiing, due to its reliance on natural snowfall. The average reduction in season length for snowmobiling in Southern Ontario is estimated to be as much as 49% by the 2020s (Scott, 2003).

Adaptation

Whatever the specific impacts of climate change on tourism might be, these cannot be seen in isolation as the resulting changes in the pattern of demand will have repercussions for many areas of economic and social policy such as, for example, employment and labour demand and policies related to housing, transport and social infrastructure. Flow-on effects could influence other sectors, such as agriculture and fisheries, handicraft industries, and water supply and sanitation. However, with the probable exception of mountain and other cold-climate-dependent tourism, climate change is unlikely to lead to a net loss in demand for leisure tourism. A loss of demand for a given destination or type of destination may well lead to increases in demand for alternative destinations. Whether a net environmental gain or loss results from such changes will partly depend on the ability of the tourism industry to raise its sustainability.

The resort development path (Butler, 1980) facilitates understanding of when and how adaptation might best be undertaken by the tourism sector. As a tourist area progresses from exploration through to either rejuvenation or decline (Figure 8.8), the options for responding to climate change will vary. For example, managed realignment (i.e. retreat) is

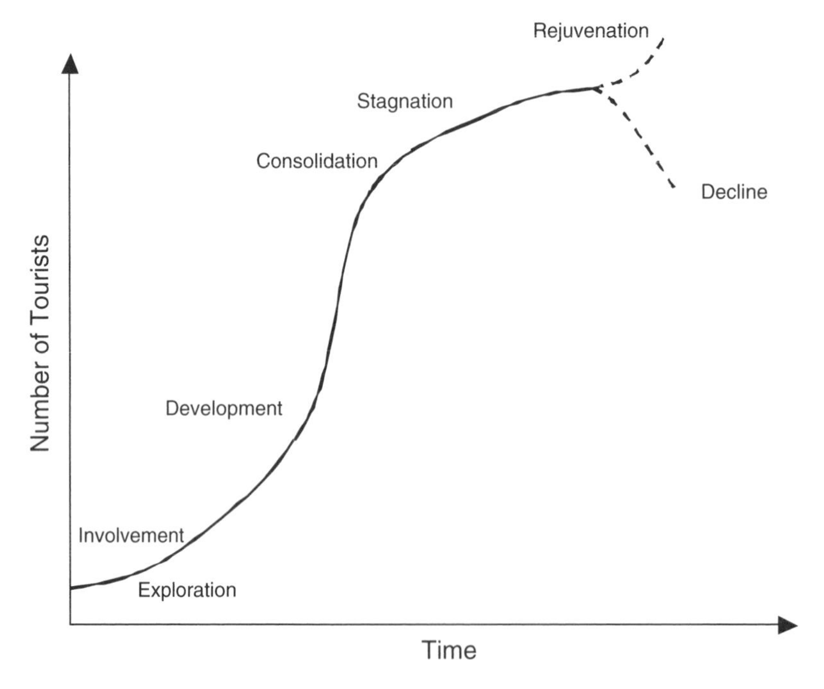

Figure 8.8 The resort development path
Source: after Butler (1980)

probably only viable in the early stages of the cycle, while the value of economic assets is relatively low. After the early development stage is completed, holding the line (i.e. protection and/or accommodation) may be the only economically and politically viable option. However, if the stagnation stage allows for the total redevelopment of an area, realigning the coast may be an option. The rejuvenation stage also provides an opportunity for reducing climate-related risks, by making changes that increase the longevity of tourist areas, infrastructure and operations, in part through improvements to structures but also through integrated coastal planning and management (Jennings, 2004).

Burkett *et al.* (2005) identified several adaptation strategies that would have aided in reducing, but not eliminating, the damage to New Orleans as a consequence of Hurricane Katrina. These included upgrading levees and drainage systems to withstand Category 4 and 5 hurricanes; designing and maintaining flood protection on the basis of historical and projected rates of local subsidence, rainfall and sea-level rise; minimising drain-and-fill activities, shallow subsurface fluid withdrawals and other human developments that enhance subsidence; improving evacuation routes; protecting and restoring coastal defences; encouraging flood proofing of buildings and infrastructure; and developing flood-potential maps that integrate local elevations, subsidence rates and drainage capabilities (for use in the design of ordinances, greenbelts and other flood-damage reduction measures).

Raksakulthai (2003) highlights the fact that the tourism industry in Phuket, Thailand has already developed a number of coping strategies for dealing with climate variability. These are suited to addressing future changes in climate. Moreover, emphasis is on implementing 'no-regrets' actions as these provide benefits even in the absence of climate change. Such initiatives include raising prices in the peak season, or closing down for the monsoon season. The latter reduces operational costs so that it is still profitable to operate with only a six-month income. 'Climate-proofed' attractions and other forms of product diversification are being used to reduce disparities in tourist demand between the high and low seasons. These include medical and cultural tourism, entertainment parks, convention centres, spas, indoor pools and tennis courts, and shopping malls. Reservoir capacity has also been increased to ensure security of water supply. Integrated coastal zone management is improving coordination and cooperation among agencies and between the public and private sectors, with a focus on conserving natural resources, including coral reefs, sandy beaches and clean seas, and water supply.

Larger hotels and tour operators are using daily and weekly weather forecasts to determine what activities they will offer on a given day. Recent improvements in the accuracy of seasonal weather forecasts, including the onset and duration of the monsoon, are allowing more targeted promotion and marketing of weather-dependent activities such as diving and snorkelling. But small and medium tourism enterprises for the most part lack policies and programmes to manage climate risks, reacting to events by coping on a day-by-day basis, rather than anticipating them. Given the dominance of these enterprises in the tourism sector, and involvement of a wide range of people from the resident community, reducing risks related to climate change will bring large local benefits. The risks are also spread among many establishments. Capacity building and public awareness will enhance the ability of people working in local businesses to anticipate and prepare for the variety of risks that can effect the industry and hence their own livelihoods. Financial safety nets such as insurance, and creation of economic opportunities for the low season, when many people are underemployed or idle, will also reduce the adverse consequences of climate change (Raksakulthai, 2003).

Large corporate ski operations may be less exposed to climate-related risks, relative to smaller ski operations, as the former generally have a more broader business mix, (e.g. real estate, warm-weather tourist attractions), are regionally diversified (this reduces risk related to poor snow conditions in one region) and are better capitalised and hence able to make substantial investments in adaptation technologies such as snow-making systems. The serious attention the ski industry is giving to climate change is illustrated in the guidance on how members of the US National Ski Areas Association might respond when asked questions about climate change (Text Box 21). Artificial snow making enhances the

Text Box 21 Climate change and the ski industry

Question and Answer Guidance to Members of the National Ski Areas Association of the USA (selected questions)

1. *Are you concerned about global warming?*

Yes. We view global warming as a long-term problem for winter recreation and our livelihood. The good news is that we know how to solve the problem. But we need to act now to turn things around and put solutions in place.

2. *Do you have specific proof that global warming is affecting your resort?*

No. We don't have specific data on the impacts of climate change on our resort. However, we have general data showing that the 1990s were the warmest decade on record and that global average temperatures have risen one degree over the 20th century. While these trends could continue, we view this as a long-term problem, so our hope is that solutions will be in place before we do see direct impacts.

3. *What is your going forward strategy on global warming?*
The snow is great this season and we want to keep it that way. Our strategy is reduce, educate and advocate:
(a) We are reducing emissions of warming pollutants in our operations through investments in energy efficiency and renewable energy;
(b) We are educating our guests and the public about this problem through an outreach campaign called 'Keep Winter Cool.' (see www.keepwintercool.org)
(c) We are advocating that policy makers take steps to reduce CO_2 emissions, including state initiatives and the mandatory caps provided for in the McCain Lieberman Climate Stewardship Act.
4. *Does the industry have an official policy on global warming?*
Yes. The ski industry has taken a leadership role on this issue and adopted a climate change policy in 2002. The policy acknowledges the problem and adopts the reduce, educate, advocate approach addressed above. See www.nsaa.org for more information on the policy.
5. *Are US resorts well equipped to handle the consequences of global warming?*
Yes. Ski areas in the USA are in a great position to adapt, if need be, to warmer than normal temperatures. We have state-of-the-art snow-making systems and are increasingly offering activities year round that do not depend on snow or winter. But merely adapting to the consequences isn't good enough for us: namely, we want to help solve the problem. In our view, winter is short enough already.
6. *What kind of steps are resorts taking to reduce GHG emissions?*
Resorts are taking advantage of many opportunities to improve energy efficiency and invest in clean energy. We are constructing green buildings, using pollution-free wind and solar energy to run buildings and lifts, retrofitting facilities to save energy (and money, too), replacing inefficient compressors in snow-making operations, exploring alternative fuels, and providing or promoting carpooling or mass transit by guests and employees. Over the long-term, investments in efficiency will actually reduce future energy costs. While we may pay more to be more efficient, over time these investments will reduce our energy bills.

ability of resorts to remain operational in a warmer climate, opening earlier in the autumn and closing later in the spring than would otherwise be possible. For example, by 2050 the average season for a ski area in Canada will be shortened by between 37 and 57% compared to the present, while snow making will reduce the decline substantially, to between 7 and 32%. But snow-making requirements are increased by between 48 and 187%. The resulting cost increases might well threaten the financial viability of a ski area (Scott *et al.*, 2002). As a result, snow-making technology is changing very fast, in order to reduce costs and increase performance (see Alpine Europe case study in Chapter 3).

Conclusions

The tourism sector is exposed to many climate change-related risks. This is especially the case for coastal and cold-climate-dependent tourism. While in the past attention has focused more on longer-term trends in climatic conditions such as temperature and sea level, it is now clear that in the near term, risks related to extreme events and climate variability are likely to be more critical to tourism. Within tourism itself, many options exist for reducing climate-related risks through adaptation. These can be implemented for all three of the major components of the tourism system – the source region for tourists, tourist travel and at the destination – and at all levels, ranging from the individual tourist, operator and tourism-dependent community through to global initiatives such as those already being undertaken by the UNWTO.

But due to strong interdependencies between tourism and local, national and regional economies and societies, both the necessity to adapt and the responsibility for adaptation extend well beyond the tourism sector. Moreover, responses to climate change-related risks should include appropriate combinations of both adaptation and mitigation, the former constituting the risk reduction measures that will be required until attempts to restore atmospheric GHG concentrations to appropriate levels are successful and the global climate system has sufficient time to adjust.

The following chapter will explore the policy and practical dimensions of climate change mitigation and adaptation, at global, national, subnational and tourism enterprise levels.

Chapter 9

Climate Change Policies and Practices for Tourism

Key Points for Policy and Decision Makers, and Tourism Operators

- A number of international organisations deal with either tourism (e.g. UNWTO, WTTC) or climate (e.g. WMO, IPCC); integration between the two strands has been minimal to date.
- The UNFCCC (1992) and the Kyoto Protocol (1997) were major milestones in the international political response to climate change.
- Four main principles are generally considered in international and national climate policies: (a) the precautionary principle, (b) polluter pays, (c) sustainable development and (d) equity. The principles are interrelated.
- There are numerous reasons why tourism businesses should take some interest in, and respond to, climate change; however, several barriers inhibit implementation of climate change initiatives by such businesses. These relate to lack of information, lock-in situations, set behavioural patterns, lack of financial resources and skilled personnel, the large number of small businesses and the international nature of tourism.
- Existing tourism businesses, entrepreneurs, investors and destinations can manage climate-change risks by understanding their company's exposure to risks such as energy use and GHG emissions, new or strengthened climate policies and regulations, vulnerabilities to the direct impacts from climate change, and other vulnerabilities such as changes in the natural resource or customer bases.
- An integrated approach to climate policy formulation and implementation is critical. This involves stakeholders making development more sustainable by emphasising both mitigation and adaptation.
- An iterative policy framework and process for mitigation and adaptation is proposed. The cycle starts with actions designed to support informed decision making with respect to mitigation and adaptation initiatives. A strong enabling environment is critical to the successful implementation of these initiatives. Continuous improvement is facilitated through monitoring and review.
- The enabling environment is strengthened by 'mainstreaming' – that is integrating – climate-related policies, plans and other initiatives into the wider development policies, plans and actions.

- Participatory planning methods are encouraged as these help ensure effective stakeholder involvement in the preparation and implementation of policies and plans that benefit not only tourism itself but also tourism-dependent countries and communities.

Introduction

As a result of its global nature, tourism is inextricably linked to climate change and development issues. Substantial tourist flows from industrialised countries to less developed countries potentially provide an effective mechanism for distributing wealth from rich to poor countries. Restricting tourist flows to, or tourism development in, less wealthy countries raises critical issues of equity in that those countries could be limited in their opportunity to develop both economically and socially. Moreover, less developed countries are typically more vulnerable to climate variability and change. In such situations, tourism development may be hampered by climate change. Also, tourism development can in itself enhance vulnerability and increase stress on environmental, economic and social systems. The relationships between tourism and climate change are thus complex and are also likely to be controversial. Preparing policies for climate change and tourism is therefore a challenging task.

Climate policy has evolved considerably over the last decade. Klein *et al.* (2005) identified three roles for climate policy: (a) to control atmospheric concentrations of GHGs, (b) to reduce the negative consequences of changes in climate and (c) to address development and equity issues. In addition, while issues such as poverty and gender inequality are not the primary concerns of climate change policy, it is increasingly recognised that effective implementation of climate policy requires that these issues also be addressed.

This chapter provides an overview of institutions that address matters related to tourism and climate change as well as of key events. Particular attention is paid to the development of adaptation frameworks. This is followed by a discussion of the underlying principles of climate change policy: the precautionary principle, polluter pays, sustainable development and equity. The barriers to developing and implementing climate change policy relating to tourism are outlined. A framework for incorporating climate change into business decision making and practices illustrates that climate change is already relevant in today's business environment. National policy making in the context of climate change and tourism is discussed. This has a particular focus on support for decision making, enabling environments, and monitoring and review to achieve continuous improvements. The chapter highlights the importance of integrating adaptation and mitigation policies into a wider framework of sustainable development.

Global Context

Institutions and agreements

Tourism-related

There are a number of organisations that deal with travel and tourism at a global or macro-regional level; examples include the UNWTO, World Travel & Tourism Council (WTTC), European Travel Commission (ETC), Pacific Asia Tourism Association (PATA) and International Civil Aviation Organisation (ICAO). Through tourism, the UNWTO aims to stimulate economic growth and job creation, provide incentives for protecting the environment and heritage of destinations, and promote peace and mutual understanding among all nations of the world. Among others, WTTC represents the tourism industry and is therefore a forum for global business leaders in tourism. Its main focus is to raise awareness of the full economic impact of tourism worldwide. ICAO has responsibility for developing operational standards and practices for civil aviation, and for providing guidance on various aspects of civil aviation, including environmental management. Environmental activities by ICAO are undertaken through their CAEP.

To date involvement of the above organisations in international efforts to address climate change has been limited. The Kyoto Protocol gives responsibility to ICAO to address GHG emissions from air travel, including preparation of appropriate guidelines and recommendations. The UNWTO (2003c) made an important contribution to addressing the complex interrelationships between climate change and tourism by convening the First International Conference on Climate Change and Tourism, in April, 2003, in Djerba, Tunisia (Figure 9.1). The conference brought together over 140 delegates from 53 countries, drawn from representatives of the scientific community, various UN agencies (including the UN Environment Programme (UNEP), UNFCCC and the IPCC), the tourism industry, NGOs, national tourism administrators and environmental managers, and local governments. The main outcome of the conference was the Djerba Declaration on Climate Change and Tourism.[1] The Declaration recognises the two-way interaction between climate change and tourism and provides a basic reference and framework for further action by major stakeholder groups. Since the Djerba Conference, the UNWTO has contributed to various events related to climate change and tourism (e.g. research workshops, tourism fairs and the UNFCCC Conference of the Parties).

Climate-related

To date organisations focusing on weather and climate have generally not specifically addressed tourism. However, the WMO recently established an expert team on 'Climate and Tourism'. However, concerns over a

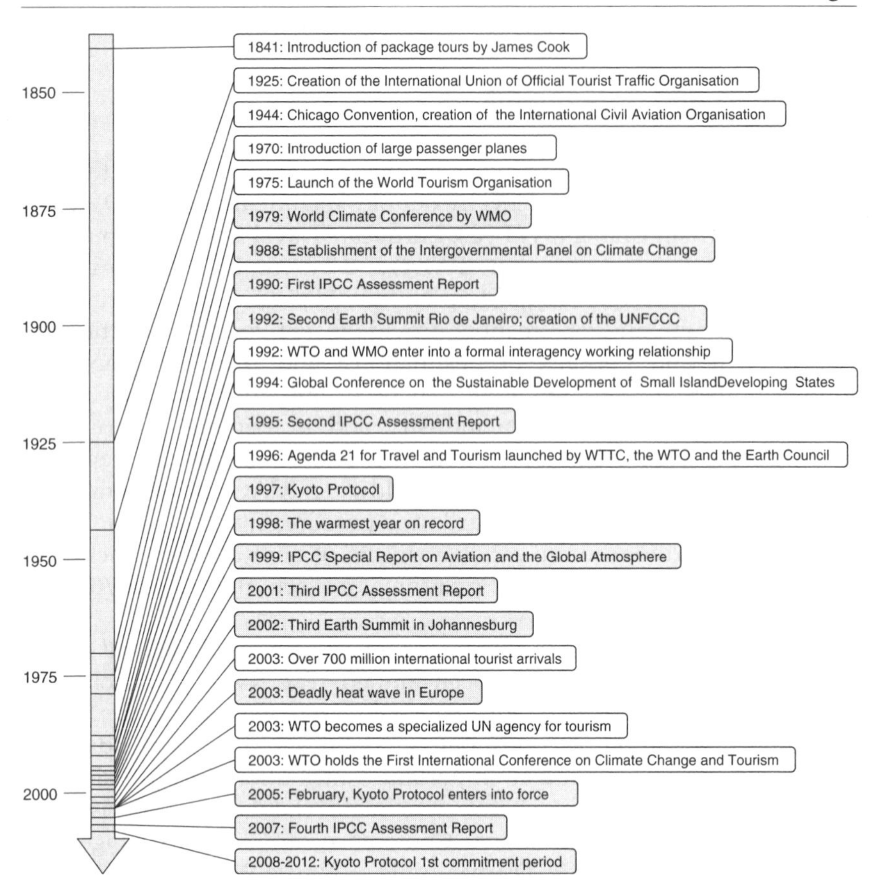

Figure 9.1 Timeline of major events relating to climate change and tourism

changing climate have a relatively long history. They were first expressed at the World Climate Conference in 1979, organised by the WMO (Figure 9.1). A call was made for global cooperation to increase understanding of man-made impacts on the climate and consider potentially negative effects in future planning and development. At the 'Assessment of the role of carbon dioxide and of other GHGs in climate variations and associated impacts' conference in Villach (Austria) in 1985, it was recognised that human activities are likely to lead to a warming of the atmosphere. As a result the Advisory Group on Greenhouse Gases (AGGG) was established and mandated to periodically assess the scientific knowledge of climate change. In 1988 two UN bodies, the WMO and UNEP, established the IPCC. The role of the IPCC is to assess on a comprehensive, objective and transparent basis the scientific, technical and socioeconomic information required for understanding the risk of human-induced climate change. The

IPCC does not carry out research as such. Rather, its assessments are based on the peer-reviewed and published scientific and technical literature. The findings of the IPCC are intended to be 'policy relevant but not policy prescriptive'.

The IPCC has prepared three Assessment Reports (1990, 1995 and 2001), a range of Special Reports (e.g. Aviation and the Global Atmosphere in 1999) and Technical Papers. A fourth Assessment Report will be released in 2007. Preparation of the reports and papers involves numerous scientists from many countries, including many of the world's leading climate scientists. The assessment procedures also involve extensive reviews, including those by government representatives who also participate in preparation of the summaries for policy makers. The latter help ensure the assessments are policy relevant. The fourth IPCC assessment will give more attention to tourism than has previously been the case. This is mainly with respect to Working Group II (adaptation to climate change) and in the chapters on 'coasts and low-lying areas', 'small islands' and 'industry, settlement and society'.

The international political response to climate change took a major step when many countries signed the UNFCCC in 1992. The UNFCCC sets out a framework for action directed at stabilising atmospheric concentrations of GHGs to avoid 'dangerous interference' with the climate system and to achieve the 'stabilization of greenhouse gas concentrations in the atmosphere ... within a time-frame sufficient to allow ecosystems to adapt naturally to climate change'. The UNFCCC came into force in March 1994; it now has 188 parties. The annual meeting of the convention parties, called the Conference of the Parties (COP), is divided between two groups: the Subsidiary Body for Scientific and Technical Advice (SBSTA), a scientific and technical advisory group, and the Subsidiary Body on Implementation, a policy and convention implementation group. There has been limited interaction between these two bodies and tourism organisations; however, lately, the ICAO has provided input to the SBSTA in relation to improving the reporting guidelines for GHG emissions from aviation.

Following intense negotiations culminating at the Third Conference of the Parties (COP-3) in Kyoto, Japan, in December 1997, delegates agreed to a protocol to the UNFCCC that commits developed countries (see EU and other parties in Table 9.1), as well as those with economies in transition to a market economy (e.g. Russia), to achieve quantified emission reduction targets. These countries – 'Annex I Parties' – agreed to reduce their overall emissions of six GHGs by an average of at least 5% below 1990 levels between 2008 and 2012 (the first commitment period). Specific targets vary from country to country. The resulting Kyoto Protocol came into force in February 2005 when 55 parties to the UNFCCC, representing at least 55% of the global CO_2 emissions for 1990, ratified the Protocol.

Table 9.1 GHG emissions and reduction targets for Annex I Parties

EU (15 countries)	*Target (%)*	*GHG in 1990*[a]	*Economies in transition*	*Target (%)*	*GHG in 1990*[a]	*Other parties*	*Target (%)*	*GHG in 1990*[a]
Portugal	27.0	58.4	Russia	0	3050.0	Iceland	10	3.3
Greece	25.0	107.1	Ukraine	0	919.2	*Australia*	8	430.5
Spain	15.0	284.6	*Croatia*	– 5	31.6	Norway	1	52.1
Ireland	13.0	53.4	Poland	– 6	564.4	New Zealand	0	61.6
Sweden	4.0	72.1	Romania	– 8	262.8	Canada	– 6	608.7
Finland	0.0	76.8	Czech Republic	– 8	192.0	Japan	– 6	1187.3
France	0.0	564.2	Bulgaria	– 8	141.8	*USA*	– 7	6129.1
Netherlands	– 6.0	211.4	Hungary	– 6	113.1	Switzerland	– 8	53.1
Italy	– 6.5	509.1	Slovakia	– 8	72.4	Liechtenstein	– 8	0.2
Belgium	– 7.5	146.1	Lithuania	– 8	50.1	*Monaco*	– 8	0.1
UK	– 12.5	742.6	Estonia	– 8	43.5			
Austria	– 13.0	77.7	Latvia	– 8	28.9			

(Continued)

Table 9.1 (*Continued*)

EU (15 countries)	*Target (%)*	*GHG in 1990[a]*	*Economies in transition*	*Target (%)*	*GHG in 1990[a]*	*Other parties*	*Target (%)*	*GHG in 1990[a]*
Denmark	− 21.0	68.8	Slovenia	− 8	20.6			
Germany	− 21.0	1246.8						
Luxemb.	− 28.0	13.4						
EU	− 8.0	4231.4						

[a]GHG emissions in 1990 (unit: million t-CO_2-equivalent)
Countries in italics have not ratified Kyoto yet
EIT Parties, which do not set 1990 as their base-year for the GHG emissions, are Bulgaria (1988), Hungary (1985– 87 average), Poland (1988), Romania (1989) and Slovenia (1986)

The Kyoto Protocol also established three mechanisms to assist Annex I Parties to meet their national mitigation targets in a cost-effective manner: an emissions trading system, Joint Implementation (JI) of emissions-reduction projects and a Clean Development Mechanism (CDM) that encourages projects in developing counties (Text Box 22).

International initiatives for adaptation

Climate change policy has arisen from a concern over the likely or possible consequences from increasing amounts of anthropogenic GHG emissions. For this reason, existing agreements and frameworks concentrate largely on how to reduce these emissions, or more specifically on policies designed to achieve such reductions in an effective manner. Recently, however, it has become clear that regardless of the emissions reduction efforts and success, and in part because of past emissions, there will be a need to adapt to unavoidable changes in climate (see Chapter 8). It is also now recognised that there are enormous differences (and inequalities) in relation to vulnerability to climate change and the capacity to adapt in ways that reduce such vulnerabilities. There is now general agreement, at least in principle, that developed countries will support developing countries in their climate change adaptation efforts. To assist the process and implementation of adaptation – especially in less developed countries – several international organisations have prepared tools or guidelines. For example, the United Nations Development Programme (UNDP) has prepared a users' guidebook, a series of technical papers, case studies, and related tools and resources designed to provide technical guidance to key national players for developing and assessing climate change adaptation policies and measures. The materials are directed by the vision of a society whose response to climate change combines national policy-making with proactive 'bottom-up' or 'grass roots' actions. Thus the goal of the Adaptation Policy Framework is to protect and, when possible, enhance human well-being in the face of climate variability and change. There is also a development objective, namely to facilitate the incorporation of adaptation into a country's national development strategy by promoting sustainable policy processes and reducing climate vulnerability. Four guiding principles underpin the Framework (Figure 9.2):

- Place adaptation in a development context.
- Build on current adaptive experience to cope with future climate variability.
- Recognise that adaptation occurs at different levels – in particular, at the local (bottom-up) level.
- Recognise that adaptation will be an ongoing process.

Text Box 22 Kyoto mechanisms for mitigation

Clean Development Mechanism (CDM). The CDM allows Annex I Parties to invest in GHG emission reduction projects in non-Annex I Party countries that benefit from activities, as CDM projects contribute to achieving sustainable development in the host country. A CDM project activity needs to be *additional*, this means that GHG emissions need to be reduced below those that would have occurred in the absence of the CDM project activity. Afforestation and reforestation projects are also eligible under the CDM. A credit received in a CDM project is called a Certified Emission Reduction (CER) unit. There are simplified procedures that apply for small-scale CDM project activities, for example energy efficiency improvement project activities that reduce energy consumption by up to the equivalent of 15 GWh/year. This option of smaller-scale projects is very important for tourism, as many potential projects (e.g. electricity supply on an island) would be too small to fall into the scope of normal CDM projects. In their principles for integrating tourism in the wider sustainable policy making, UNEP highlights the role of CDM for the transfer of environmentally sound technologies for the tourism sector.

Joint Implementation (JI). Joint Implementation allows Annex I Parties, which have emission caps, to assist other Annex I Parties to implement project activities to reduce GHG emissions or remove those by sinks. Credits will be issued based on net emission reductions achieved by the project activities. The credit is called an Emission Reduction Unit (ERU). As for CDM activities, it is critical that any JI project provides GHG emission reductions or removals by sinks that are *additional* to any that would otherwise occur. Annex I Parties can use ERUs to contribute to compliance of their quantified GHG emissions reduction targets of the Kyoto Protocol. The total amount of emission cap of Annex I Parties will not change, because JI is credits transfer between Parties both of which have emission caps.

Emissions Trading (ET). Annex I countries can engage in emissions trading. The total amount of the emission cap of Annex I Parties will not change, as credits are simply transferred from one party to another. The purpose of emissions trading is to reduce overall costs in achieving emission reduction targets collectively. Emissions trading is already taking place, although there is no single GHG market, but a range of markets. These are, for example, 'International pre-compliance markets' that relate to project-based emission reductions (e.g. through CDM) or national and regional markets such as those in the EU or in the UK. There are also markets for companies or individuals who are concerned about their GHG emissions. It is currently being discussed whether aviation should be included in the European ET market.

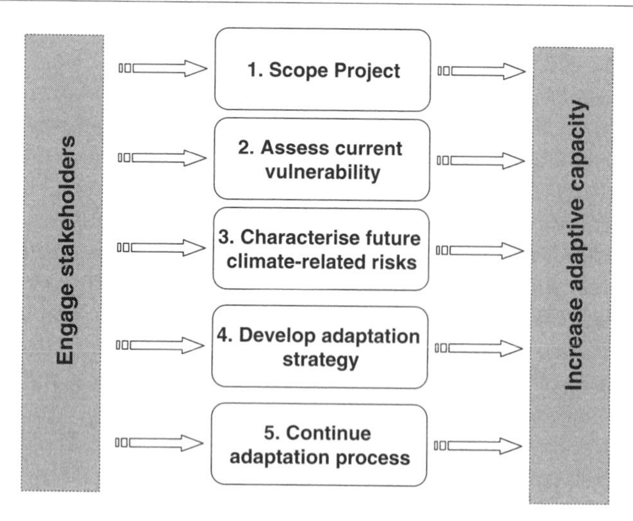

Figure 9.2 Steps of the Adaptation Policy Framework
Source: after UNDP (1998)

The major output from the Framework is strategies that realign national sustainable development goals while realigning poverty reduction programmes. More specifically these strategies include policies and measures. Policies deal with the development of a portfolio of adaptation initiatives and guidelines for implementing specific actions as part of national and local planning. Measures are usually directed at overcoming barriers to adaptation, enhancing adaptive capacity and altering investment plans.

The *Handbook on Methods for Climate Change Impact Assessment and Adaptation Strategies*, developed by the UNEP (1998), provides an introduction to a wide range of methods that can be used to design assessment studies of climate change impacts and related adaptation strategies. The intention is to provide an overview of methods that gives readers sufficient information to select the method most appropriate to their situation. The handbook is organised in two parts: Part I treats generic and cross-cutting issues and includes a 'getting started' chapter, which deals with issues, methods and considerations common to all impact and adaptation studies. The next two chapters discuss how to design scenarios, and where to obtain them. Chapter 5 discusses scenarios of climate change, and Chapter 3 treats scenarios of the socioeconomic context in which climate change impacts and adaptation may occur. Chapter 4 describes the need for integration across sector studies and interaction with stakeholders, and suggests ways to achieve such integration and interaction. Chapter 5 describes adaptation to climate change, the options that exist and how to evaluate them. Part II presents methods for studying impacts and adaptation in the selected sectors of water, coastal resources, agriculture, rangelands, health, energy, forests, biodiversity and fisheries. In Part II a common format is followed. Each of the chapters begins with a brief introduction that defines and describes the scope of the problem. The likely or known climate change impacts in the sectors are briefly described. Against this background an array of methods is presented, in order to illustrate the range of different levels of complexity in the methods. Some of the less demanding methods, in terms of data, modelling requirements and the like, are presented alongside the more complex and demanding methods. The aim is always to be user friendly, and to provide enough information to permit users to make a more informed choice in the design of impact studies, as well as to begin identification and preliminary assessments of adaptation.

To facilitate the mainstreaming of adaptation measures, the Asian Development Bank (ADB) has developed and demonstrated their *Guidelines for Adaptation Mainstreaming* (ADB, 2006). The guidelines are grouped into three categories (Figure 9.3): underlying principles, enhancing the enabling environment and the process of mainstreaming adaptation.

The World Bank Group (2006) is developing a screening and design tool for projects to help raise the awareness of its staff and clients to the risks projects face as a consequence of climate variability and change. The screening tool will indicate the aspects of a project that might be subject to high risk from climate change and a design tool to help identify the elements of best practice in avoiding or adapting to those risks. This work seeks to identify climate-related risks early in project design so that appropriate adaptive options can be incorporated. The project will assist in bringing the rapidly developing knowledge base of community and local responses to current climate variability (including early climate change) into the wider development planning process. This will assist the Bank in its goal of supporting better vulnerability assessments and mainstreaming adaptation into the Bank's activities. The screening/ design tool will be Web- and CD-based and will not require detailed technical knowledge of climate-related issues by the user. The user will describe the nature of the project and its location and the tool will provide an initial assessment of any climate risks and guidance on how to deal with such risks. The tool is open-ended and flexible and will be shared widely with other interested parties.

The UNFCCC has also prepared a compendium of methods and tools to evaluate vulnerability and adaptation to climate change (UNFCCC, 2005).

A. Guidelines Relating to the Principles Underpinning the Mainstreaming of Adaptation
Guideline 1: Manage Climate Risks as an Integral Part of Sustainable Development
Guideline 2: Ensure Intergenerational Equity Related to Climate Risks
Guideline 3: Adopt a Coordinated, Integrated and Long-term Approach to Adaptation
Guideline 4: Achieve the Full Potential of Partnerships
Guideline 5: Adaptation Should Exploit the Potential of Sustainable Technologies
Guideline 6: Base Decisions on Credible, Comparable and Objective Information
Guideline 7: Maximize the Use of Existing Information and Management Systems
Guideline 8: Strengthen and Utilize In-country Expertise
Guideline 9: Strengthen and Maximize Use of Existing Regulations, Codes, Tools
B. Guidelines Relating to Enhancing the Enabling Environment for Adaptation
Guideline 10: "Climate Proof" Relevant Legislation and Regulations
Guideline 11: Strengthen Institutions to Support the "Climate Proofing" of Development
Guideline 12: Ensure Macroeconomic Policies and Conditions Favour "Climate Proofing"
Guideline 13: Ensure Favourable Access to Affordable Financing of "Climate Proofed" Development Initiatives
C. Guidelines Relating to the Process of Mainstreaming Adaptation
Guideline 14: Characterize Climate-Related Risks that Require Sustained Attention
Guideline 15: Replicate the Knowledge, Motivation and Skills that Facilitate Successful Adaptation
Guideline 16: Enhance the Capacity for Continuous Adaptation
Guideline 17: Ensure "Climate Proofing" Activities Complement Other Development Initiatives
Guideline 18: A Process of Continual Improvement in Adaptation Outcomes

Figure 9.3 Guidelines for adaptation mainstreaming (ADB, 2006)

Underlying Policy Principles

Four main principles are generally considered in international and national climate policies: (a) the precautionary principle, (b) polluter pays, (c) sustainable development and (d) equity. The principles are interrelated, and discussed below.

The UNFCCC explicitly recognises the precautionary principle[2] in its Article 3 by suggesting the need to 'take precautionary measures that anticipate, prevent or minimise the causes of climate change and mitigate its adverse effects'. In addition, it is noted in the same document that the 'lack of full scientific certainty should not be used as a reason for postponing such measures...'. Paavola and Adger (2005) noted that setting a maximum limit for CO_2 concentrations in the atmosphere would pay tribute to the precautionary principle. A concentration of 400–500 ppm of CO_2 has been discussed as a threshold to avoid dangerous climate change, even though it is clear that despite such precautionary measures some adverse impacts from increased GHG concentrations will remain.

There are a number of examples of precautionary measures in relation to new tourism developments and climate change. One is designing airport improvements in Pohnpei, Federated States of Micronesia. These take into account projected increases in sea level to 2050. On the other hand, there are examples where climate change appears to be ignored in longer-term designs. The soon-to-be-completed railway between the Tibetan capital, Lhasa, and Golmud in the North-western province of Qinghai, China, is built partly on permafrost soil. There is a risk that a warmer climate may affect the operation of the railway (BBC, 15 October 2005).

The polluter-pays principle is implicit in the Kyoto Protocol in which Annex I countries agreed to reduce their emissions to 5% below 1990 levels. In effect this means that those countries responsible for emissions are asked to take leadership in mitigation and to bear the corresponding costs. As in other areas of environmental policy, the polluter-pays principle acts as an incentive to reduce emissions. Internalising costs resulting from GHG emissions is an important part of operationalising this principle. In addition, the polluter-pays principle contains a dimension of fairness and equity. In international climate policy, developed countries have the duty to assist developing countries in meeting their obligations (e.g. preparing national GHG inventories) and in efforts to adapt to climate change. However, as noted by Paavola and Adger (2005), nothing specific is said about how much assistance should be provided and how it should be distributed.

Tourists and tourism operators pay for their emissions as members of the general public, through levies that are related to energy consumption. A number of European countries already have an energy tax in place (e.g. Sweden), while other countries are currently discussing the possibility of introducing energy- or climate-related taxes on fuel consumption (e.g. Japan). There are only a few examples of the polluter-pays principle specifically related to tourism. A number of destinations charge tourists a compensation for some of the environmental costs associated with tourism (e.g. Balearic ecotax[3]). These charges can take the form of an entry tax (e.g. in Seychelles) or a departure tax (e.g. Cook Islands). Revenue raised from these kinds of environmental taxes is typically used for nature conservation or infrastructure provision rather than for activities specifically related to climate change. There are also numerous examples of user fees or permits collected for the purpose of environmental protection and management. Such is the case for many national parks in Africa – for example there is a US$30/person day fee in the Serengeti national park in Tanzania.

Climate change policies should consider principles already applied in policies for sustainable development (Beg *et al.*, 2002). This means that policies need to strike a balance between economic development, the needs of human communities and the environment. In fossil-fuel-driven economies, economic development will increase emissions and therefore the intensity of climate change. At present, energy is a critical input for economic growth. As a result, GHG mitigation policies will affect sustainable development prospects for less developed countries, for example as a consequence of increased costs for low-carbon energy sources or technologies. In turn, the detrimental effects of climate change, for example extreme events, water scarcity and diseases, are likely to hamper sustainable development. Policies that assist adaptation to climate change are therefore critical for the sustainable development of these economies.

For climate change policy to be successful it is critical to achieve integration with development policies, as well as attain a desirable mix of climate and development policies. Munasinghe (2001) pointed out that climate change policy and research needs to take into consideration concepts that have been applied in sustainability research, for example, carrying capacities, non-linearity, complexity and irreversibility (see Chapter 2).

The New Zealand Tourism Strategy 2010 is one example of a strategy for sustainable tourism development that explicitly highlights resource efficiency and carbon neutrality as one of the key priorities. More often, however, tourism development that does consider dimensions of sustainability largely neglects global climate impacts resulting from the burning of fossil fuels. The focus on local rather than global issues in the context of sustainable tourism development is also reflected in the academic literature, where ample studies investigate environmental impacts at a site (e.g. ecosystem disturbance) or local level (e.g. waste or water management), but very few studies deal with GHG emissions as a sustainability indicator.

Equity in climate change policy refers to national or social, international and intergenerational equity (Metz, 2000). Intergenerational equity is also implicit in the principle of sustainable development through its reference to maintaining a resource base for future generations. International equity acknowledges that past emissions are largely due to industrialisation in developed countries, and that for this reason the main responsibility lies with those developed countries (see also, polluter-pays principle). The task of allocating emissions (or the right to emit) is difficult in practice. Several options are being discussed, for example allocating emission rights according to past emission levels. Another option is to allocate a carbon budget to every citizen of the earth. A sustainable budget could be in the order of 1 tonne of carbon per person per year. This would require substantial reductions for citizens of industrialised countries (e.g. from about 5.5. t per capita in the USA) and allow an increase of emissions for developing countries (e.g. currently 0.3 t per capita in India) (Houghton, 2002). Carbon trading might be used to transfer funds from developed to developing countries. While this allocation mechanism might appear just, it fails to take into account past emissions, geographical coincidence and present levels of development (Paavola & Adger, 2005).

To be equitable, climate change policy also needs to take into account the fact that countries in the developing world will be more affected by climate change than industrialised countries. This relates partly to higher vulnerabilities (e.g. small island states), but also to lack of capital and capacity to adapt to climate-related impacts. To assist vulnerable, developing countries in climate change adaptation, three international funds have been created: Adaptation Fund for Assisting Adaptation Projects, Special Climate Change Fund and the Least Developed Countries Fund. The principle of protecting the most vulnerable first applies. It remains unclear at this stage how this assistance should be distributed between recipient countries and adaptation measures. Munasinghe (2001) also pointed out that to be equitable, climate change policies should be participatory and decisions should be made collectively. This means that not only do policies consider 'distributional justice', but 'procedural justice' (Paavola & Adger, 2005). The latter refers to the way in which parties are involved in making decisions and how power is distributed among those parties. Metz (2000) noted that a large degree of the equity considerations in the climate change debate is based on value judgements with a plurality of legitimate perspectives.

The taxation of global air travel (see Chapter 3) provides a useful example for examining equity issues relating to tourism. An increasing number of developing countries rely on tourism, with tourism being the most important foreign exchange earner. Increased airfares as a result of a charge on fuel or emissions would result in a decline in global tourist flows, in particular to long-haul destinations. The impact on a particular country depends on the source market and, among others, on price elasticities for specific market segments. Even within Europe some destinations are potentially more affected by a tax on air travel. These are the destinations that are located at the periphery and where air travel is a major mode of arrival. For the case of Greece, where 82% of arrivals in Greece are by air, Wit *et al.* (2002) calculated that a charge of €50 per tonne of CO_2 in European airspace would result in a decrease of total tourist receipts by about 0.15% per annum.

Equity is also an issue in relation to the differentiation between those who are privileged to travel and those who are affected by impacts resulting from this travel. Only 6% of the global population participate in international air travel. As a result, the study of tourism has often been referred to as 'the study of the rich'.

There are a number of broader principles that are also applied in climate change policies. These refer, for example, to taking measures that are proactive rather than reactive (Table 9.2) and to favouring no-regrets measures. 'No-regrets' measures for mitigation and adaptation are consistent with sound environmental management and wise resource use, and are thus appropriate responses to natural hazards and climate variability, including extreme events; they are therefore beneficial and cost

Table 9.2 Adaptive responses in the case of tourism

Response	*Proactive*	*Reactive*	*Repercussions of inaction*
International	Education of tourists and operators; awareness raising; increase insurance options; funding incremental costs of adaptation by tourism sector in developed countries; adaptation policy frameworks and tools for tourism	Assist stranded passengers; recovery assistance for tourism-dependent countries; information on how tourism is responding to changes in current climate; validate predictive climate–tourism models using historical analogues	Intergovernmental and international tourism organisations become marginalised; escalation of future adaptation costs; unanticipated and detrimental changes in tourist flows; maladaptation of tourism sector
National	Strengthen enabling environment for adaptation of tourism sector; diversify tourism products, meet information requirements; support research and development for new technologies; 'climate proof' tourism-related policies and regulations; include disaster and adaptation responses in tourism training curricula	Extra-budgetary expenditures on disaster relief and recovery of tourism operations; media campaigns to counter adverse publicity; unplanned amendments to tourism-related policies, plans and regulations; increase travel insurance premiums to cover added costs	Loss of competitiveness as a tourist destination and fewer arrivals; increased likelihood of government subsidies and intervention in tourism operations; increased liability for potential damages; loss of voter and investor confidence; increased expenditure on tourism support systems
Local	Locate new tourist facilities in low-risk areas; reduce risks for existing tourism assets and operations (e.g. through coastal protection)	Relocation of tourism facilities; rescue missions and assistance to tourism-dependent communities; increase charges to recover increased cost of services	Contingency plans to deal with disenchanted tourists; seek other drivers of community development; suffer degradation of tourism facilities and attractions

(*Continued*)

Table 9.2 (*Continued*)

Response	*Proactive*	*Reactive*	*Repercussions of inaction*
Individual	Reduce dependency on tourism for livelihood; invest in 'no regrets' adaptation measures; support 'climate proofed' tourism businesses	Improve travel insurance cover, to address added health and other risks; seek employment outside tourism; implement higher-risk and more costly adaptation measures	Develop alternative sources of income; learn to tolerate poor travel experiences; become prepared to pay higher costs, as an operator and as a tourist

effective, even in the absence of climate change. Proactive responses and no-regret options are in line with the precautionary principle.

Business Practices and Managing Climate Change

Existing tourism businesses, entrepreneurs, investors and destinations can improve their management of climate-change risks by approaching the problem in a number of ways, and at a number of levels (Figure 9.4). These include energy use and GHG emissions, policies and regulations associated with emissions, vulnerabilities to direct impacts from climate change and other vulnerabilities such as changes in the resource or customer bases.

The interaction between a business and climate change involves revenue and cost, assets and liabilities, as well as other opportunities and risks. But it also encompasses the wider value or supply chain, as well as the exposure of potential competitors to climate change, for example tourist destinations that might serve as alternatives. In order to identify their exposure to energy-related risks and GHG reduction policies a tourism business or destination needs to determine their energy consumption and GHG emissions, and define baselines. Monitoring is then required to detect the opportunities and mechanisms for possible improvements or

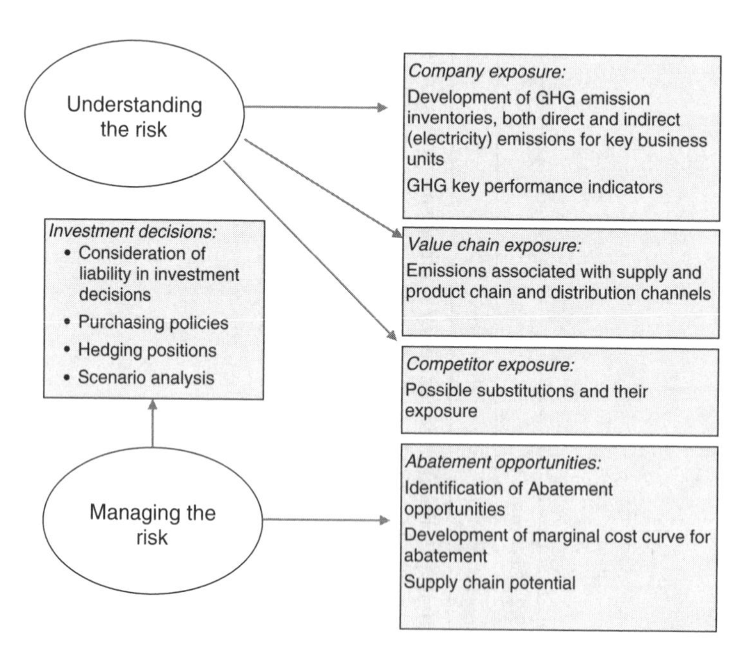

Figure 9.4 Framework for assessing company climate change exposure
Source: after AMP Capital (2005: 26)

deterioration. Useful indicators for monitoring are CO_2 emissions per visitor night or per passenger kilometre. Larger companies may also consider equity exposure, past, present and future, through investments.

Management of risk follows understanding of risk and exposures (Figure 9.4). In terms of GHG mitigation there are three options: internal emission reductions; purchasing emission credits; and generating credits by investing in external qualifying projects. Tourism businesses and destinations might invest in any one of the three options, or develop a portfolio of several options. For example, tourist destinations could consider restoration of native forest to sequester carbon while also capitalising on improving biodiversity for ecotourism operations. It is important to understand the criteria under which emission reduction projects are eligible for carbon credits under the Kyoto Protocol (AMP Capital, 2005).

As also indicated in Figure 9.4, climate change will influence investment decisions. For example, hotel developers will need to consider investment in energy-efficient building design or new technologies, such as wind turbines, and their suitability for tourist resorts. Decision making related to a new destination for a large tour operator or wholesaler could benefit from scenario analysis in relation to climate change policies for aviation. There is a higher risk associated with more distant destinations compared to closer destinations that are accessible by a wider range of transport modes, for example trains.

Climate change has direct implications for financial analyses and balance sheets. In this respect, the British Carbon Trust has developed a summary of drivers related to climate change that could be relevant in financial models. For example, operating costs benefit from reduced consumption and therefore costs for energy. If a company decides to engage in emissions trading or other emission reduction projects, the costs of doing so would need to be incorporated into accounts. The same applies to revenue generated from selling excess carbon allowances (Table 9.3). The impact of climate change on asset values also needs to be addressed in financial models, especially when there is the potential for substantial damage resulting from extreme events such as flooding or storms.

Jones (2003) advocates the systematic integration of climate change into decision making at the various levels of the wider process of planning the development of tourism infrastructure and other assets. Tourism developers need to consider climate change and other environmental issues during their planning process, for example as outlined in Figure 9.5 in the case of accommodation and resort construction. Finances and budgeting, for example, need to include costs associated with potential climate change impacts, including insurance premiums to protect against such costs. The exact location of the development, as well as the design (e.g. building

Table 9.3 Including climate-change risk into financial models for tourism businesses

Model element	*Potential drivers*	
Operating costs	Emissions trading and reporting	Purchase cost of any additional carbon allowances required Exposure to electricity costs Other costs associated with trading, reporting, compliance
	Emissions reduction	Additional operating costs as a result of mitigation measures Reduction in cost as a result of improved energy efficiency
	Other	Building compliance costs (e.g. cyclone proofing) Transport costs (e.g. aviation) Potential costs associated with supply chain risk due to weather exposure
Revenue	Emissions trading	Sale of excess carbon allowances, allocated but not required
	Market dynamics	Climate change-related change in demand (e.g. in response to temperature increase) New markets (e.g. renewable energy sources, domestic bicycle tourism)
	Pricing	To what extent can additional costs be passed through to tourists? Does this affect volumes?
Capital expenditure	Emissions reduction	Capex requirements (e.g. energy-efficiency investment)
Balance sheet	Weather risk and other	How exposed is the company's asset base to weather risk? Any potential litigation risk?
Other	Intangibles	How exposed is a company's or destination's reputation to its perceived response to climate change?

Source: after AMP Capital (2005)

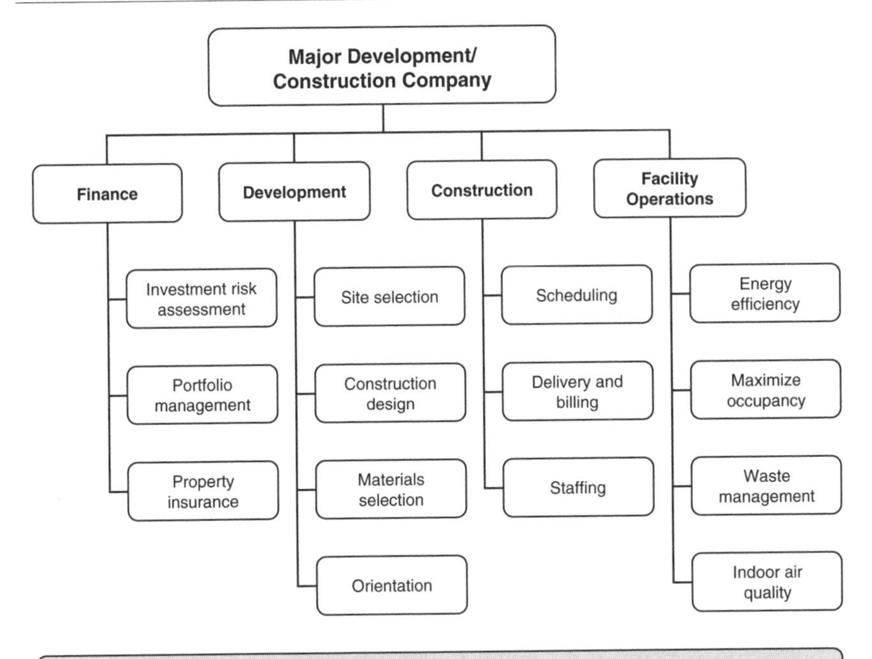

Figure 9.5 Decisions made during the construction process and climate change-related factors that need to be considered
Source: Jones (2003)

materials, orientation, structures) and landscaping, can help reduce risks related to climate change. Other factors to consider during the planning are the need for future maintenance, regulatory compliance, public relations and marketing, for example when environmentally friendly building designs are used as a competitive advantage.

Given the important consequences of climate change for tourism, one might expect tourism businesses to be active in endeavours to reduce the rate at which the climate is changing. Other reasons in favour of such mitigation initiatives have also been identified, including: (a) reducing costs; (b) gaining a marketing advantage; (c) conserving nature (on which many tourism businesses depend); (d) meeting customer expectations; and (e) complying with laws (Carlsen *et al.*, 2001: 282). However, a number of barriers exist.

The Third Assessment Report of the IPCC (2001) provides a comprehensive overview of such barriers, with the barriers relating to transport and buildings being particularly applicable to tourism. Lack of information is often discussed as a key barrier to both climate change mitigation and adaptation. However, as noted above, access to relevant information does not necessarily alter people's perceptions and behaviours. This includes attitude to car usage or the attractiveness of alternatives such as public transport. Car usage is also often offered as an example of 'lock-in' situations, because the developments over the last century – both in terms of technology and infrastructure – lead to a dominant position of the car and its combustion engine, making it almost impossible for alternatives to enter the market. Present settlement structures, cities, lifestyles and recreational patterns have evolved around the car. This makes it very difficult to change established patterns. Increasing demand for individualised and flexible holidays is likely to underpin the dominance of the car (often in combination with air travel) as the preferred transport mode for tourists.

A lack of financial resources is also commonly mentioned as a major barrier to undertaking climate change adaptation or mitigation. This is not only true for the tourism industry but also the public sector, which often lacks funding for undertaking studies (e.g. in the area of renewable energy sources) or measures such as developing hazard maps or a risk management plan. Moreover, environmental management is sometimes perceived as being too expensive for small tourism businesses, especially when it comes to new technologies that often have to be imported at great cost, as is the case in many developing countries. The construction of a sewage treatment plant to service a tourism operation on a small island is a good example of how difficult it is for small tourism operators to raise the required capital costs associated with advanced technology. In some instances, however, there are less costly, creative solutions that are workable at a small scale (e.g. sand filtration, Becken, 2004a).

Lack of capacity and skilled personnel is another barrier identified as being relevant to tourism. Allocating staff to initiatives such as energy management is particularly difficult for small or family-owned businesses that are already working on very tight schedules. Both high staff turnover and the large number of unskilled and low-paid workers in the tourism industry aggravate this barrier. Moreover, the existence of 'numerous centres of decision-making' within small tourism businesses often leads to conflict or suboptimal coordination (Carlsen *et al.*, 2001: 283).

The very nature of tourism as an industry consisting of many small businesses may result in tourism being overlooked in national policies. Also, the small size of many tourism operations often gives rise to 'free riding' – Carlsen *et al.* (2001) suggested that many rural tourism businesses take advantage of common resources, while at the same time failing to take responsibility for damage caused by their use. Some small businesses may perceive their own impact as insignificant, believing that small operations are automatically sustainable.

Other tourism-specific barriers to implementing climate change policies relate to the international nature of tourism. This makes it difficult to regulate tourism at a national level. The introduction of a carbon tax in one country, for example, raises issues of competitiveness when alternative destinations (potential substitutes) do not introduce such a tax and can thence offer more affordable packages. The regulation of international air travel (as discussed in Chapter 3) is another example of how difficult it is to negotiate a global, or at least supraregional, agreement on climate change mitigation. The key issue here is that so far no mechanism has been found for allocating emissions resulting from international air travel to individual countries.

Finally, another possible barrier to climate change action is the tendency for tourists to feel less responsible for their actions while on holiday compared with everyday life at home. This may make it difficult to not only raise general awareness of climate change but also to achieve acceptance of specific mitigation or adaptation measures that might affect the holiday experience. Similarly, investors in tourism often do not have close ties with the destination or with the tourism-dependent community. As a result they may feel less inclined to engage in local adaptation or mitigation initiatives that do not contribute directly and immediately to the financial bottom line.

National Policy Making

A tourist destination is characterised by numerous players and resources. All have roles that are influenced, to a greater or lesser extent, by climatic conditions. For example, capital and labour are likely to be attracted preferentially to destinations with a more favourable climate, as are the tourists themselves. Tourism infrastructure is best designed, and the associated services such as air conditioning are best operated in ways that reflect climatic conditions, including the likelihood of extreme events such as potentially damaging winds and heatwaves. Natural tourism assets, such as reef ecosystems, rain forests, mountain glaciers and beaches, are also subject to climatic influences. It is desirable that policy makers and planners give consideration to these interactions in national and local development policies.

An integrated approach to climate policy formulation and implementation is beneficial. Often such integrated approaches pursue a portfolio and phased approach, are participatory in nature by involving governments and stakeholders groups, and emphasise the importance of both mitigation and adaptation as part of sustainable development (see Text Box 23). Simeonova and Diaz-Bone (2005) established numerous linkages between climate-change issues and other policies (Figure 9.6).

Text Box 23 Integrating adaptation and mitigation

There are two basic approaches to climate change. The first aims to protect the climate by reducing net GHG emissions, i.e. mitigating humans' effect on the climate, while the second strategy consists of anticipating changes and adapting to them through technology or socioeconomic changes.

The wider climate change debate and the work by the IPCC has until recently mainly focused on mitigation for several reasons: (a) lack of data on possible or likely climate change-related impacts and vulnerabilities, (b) a recognised need for global action on mitigation compared with more local initiatives on adaptation and (c) the need to reduce the problem at its source rather than pursuing an end-of-the pipe approach. In the tourism context, research specifically dealing with climate change has largely concentrated on tourism's vulnerability and adaptation to climate change. Climate change mitigation and adaptation differ fundamentally for the following reasons. First, the spatial and temporal scales involved are very different (e.g. mitigation is a long-term, global effort, whereas adaptation can be short-term and local in its effect). Second, costs of mitigation and adaptation measures are difficult to compare, especially given that there is no common unit for adaptation measures as is the case for mitigation measures (i.e. $ per unit of CO_2 saved). Third, actors and types of policies involved are likely to be very different for both approaches (Klein *et al*., 2005).

Notwithstanding those differences and potential conflicts, it is now advocated that adaptation and mitigation should be integrated into a wider framework for sustainable development. This is particularly critical for developing countries that have to allocate scarce resources strategically across both adaptation and mitigation measures. Developing countries are typically not bound to agreed international emission targets through the Kyoto Protocol, and for this reason mitigation seems to be less pressing than adaptation. This order of government and industry priorities has to be recognised when trying to implement any climate change-related measures.

Wilbanks (2003) suggested integration of climate change into sustainable development by first identifying the key local problems and then linking those to climate change. Measures that have the potential to address these key issues in addition to climate change offer no-regret or win–win solutions and are therefore more likely to be funded (e.g. by donor agencies) and taken up by local agencies, stakeholders and industry members. A good example of combining local environmental problems with both climate change adaptation and mitigation is reforestation. Trees reduce vulnerability to cyclones, improve microclimates and enhance landscapes used for tourist activities. Moreover, trees function as carbon sinks, although Dang *et al.* (2003) note that those species preferred for adaptive measures (e.g. erosion control or watershed management) are not necessarily the ones most suited for carbon sequestration. There is potential to include forest sinks in carbon trading schemes, whereby carbon emitters (e.g. tourist resorts) purchase carbon credits from land-owners who restore forest on (marginal) land. Some activities proposed under the Kyoto Protocol to remove GHG emissions by sinks may have an adverse impact on biodiversity. Thus, problems of compatibility with other international agreements related to forests, such as the Convention on Biological Diversity (CBD), may arise. It is important to exploit potential synergies between adaptation, mitigation and sustainable development, and at the same time identify and address possible conflicts between different sustain-ability goals.

Energy

- Promoting economically efficient energy supply and energy use;
- Enhancing energy security and diversification of energy sources;
- Promoting energy-sector reform to increase economic efficiency and competition in supply and distribution;
- Promoting efficient use of energy resources through "green tax" reform.

Transport

- Air-quality management;
- Congestion management; improvement of security and road safety;
- Regional development and improvement of accessibility.

Industrial processes

- Reducing gases emitted as by-products in industrial processes;
- Improving efficiency of industrial processes;
- Improving health and safety conditions.

Agriculture

- Improving environmental performance of agriculture; promoting sustainability through, e.g. for improved food quality, organic farming, and land-use planning;
- Land-use change and forestry management; enhancing forest sink capacity through afforestation and reforestation;
- Protection and sustainable management of forests;
- Conservation of biodiversity, wildlife, soil, and water.

Waste

- Reduction of environmental impacts of waste management, such as impacts on air, soil, and underground waters;
- Waste minimisation and recycling.

Figure 9.6 National policies that have an important impact on climate-change risks

Source: after Simeonova & Diaz-Bone (2005)

There are many potential synergies when mitigation and adaptation are undertaken in an integrated and transparent manner, especially if this is done in the context of both national sustainable development and implementation of the many multilateral environmental agreements rather than just the UNFCCC. But while there is increasing emphasis on the integration of climate change mitigation and adaptation (Becken, 2005), there is also a risk associated with creating synergies that are suboptimal. Klein *et al.* (2005) argue that as a result of inherent dissimilarities between mitigation and adaptation policies a synergetic approach could result in greater institutional complexity. This might be because of the larger number of stakeholders involved at different levels, resulting in less successful implementation of such policies. Moreover, the net effect of investing in synergetic measures may be lower than investing smaller amounts in independent but more effective mitigation and adaptation projects.

Policy framework and process

In countries where tourism is an important part of the national economy there is a need for a policy framework and policy development process to address, through adaptation, the additional risks to tourism imposed by climate change and, through mitigation, the contributions tourism makes to climate change. Figure 9.7 illustrates the many dimensions to such a policy framework and process. Firstly, the framework acknowledges the iterative nature of the policy development process. The first iteration of the

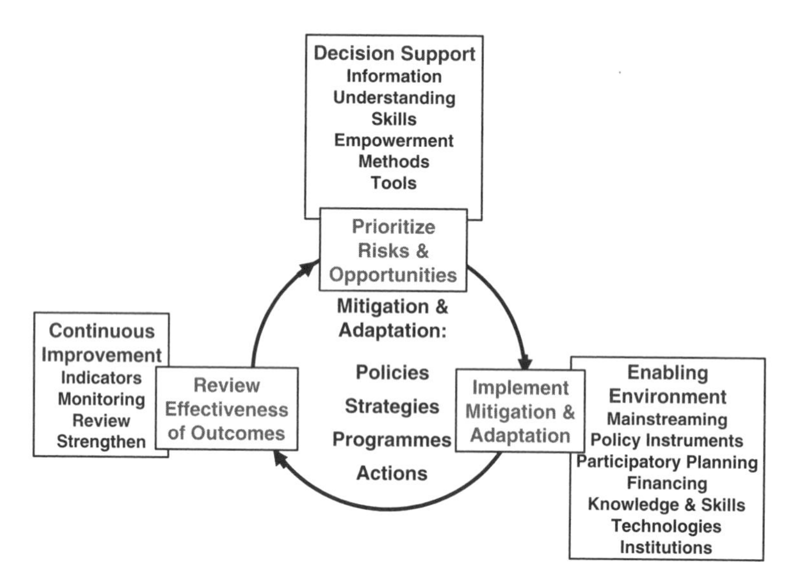

Figure 9.7 Policy framework for climate change mitigation and adaptation

cycle will normally start with activities designed to support informed decision making related to the mitigation and adaptation initiatives to be undertaken. A strong enabling environment is critical to the successful implementation of those initiatives; while continuous improvement is ensured by reviewing the effectiveness of the implemented measures. The information gained in the monitoring process will be used as input into the decision making in the next iteration.

Decision support

Informed decision making requires the capacity to undertake assessments such as that described in Figure 9.2. The outcomes of this phase of the cycle are preferred adaptation and mitigation strategies. These are selected on the basis of the costs and benefits in reducing both unacceptable climate-related risks and GHG emissions and on the extent to which they contribute to sustainable development.

Enabling environment

Importantly, the framework also recognises the importance of the 'enabling environment' for adaptation and mitigation – i.e. the high-level and robust systems and capabilities that foster adaptation and mitigation processes. This includes the mainstreaming of climate change, policy instruments (e.g. legislation and regulations), participatory planning, enhancement and use of human knowledge and skills, technology transfer (including revitalisation and application of traditional knowledge and practices), institutional capacities and financing. While most tourism-related mitigation and adaptation initiatives are undertaken at the enterprise and community level, their success will often depend on the supportive nature of the enabling environment. This is usually strengthened through longer-term actions by government, often with support from the international community, especially in the case of developing countries.

Mainstreaming

A key component of the enabling environment in the context of addressing climate change and related issues is the ability to 'mainstream' – that is infuse or integrate – policies and measures to address climate change into ongoing and new development policies, plans and actions. Mainstreaming aims to enhance the effectiveness, efficiency and longevity of initiatives directed at reducing climate-related risks through mitigation (in the longer term) and through adaptation (in the interim), while at the same time contributing to sustainable development and improved quality of life. Mainstreaming thus endeavours to address the complex tensions between development policies aimed at immediate issues and the aspects of climate policy aimed at both current and longer-term concerns. The tensions often become most apparent when choices have to be made on the disbursement of limited government funds – for example, choices between supporting education and health programmes on the one hand, and funding climate change adaptation and mitigation initiatives on the other. Indeed, mainstreaming is largely about reducing tensions and conflicts, and avoiding the need to make choices, by identifying synergistic win–win situations. Thus mainstreaming focuses on 'no-regrets' measures for adaptation and mitigation.

Policy instruments

In general terms, policy instruments can be regulatory, market-based or based on voluntary action (soft tools) (Table 9.4). There are numerous options within each category. Emission standards and building codes are examples of regulatory policies, while vehicle taxation based on engine size, general fuel taxes or subsidies towards renewable energy sources are examples of market-based instruments. Voluntary initiatives in the tourism industry include ecolabelling.

The motivation behind market-based instruments is to achieve environmental goals at lower costs than those associated with regulatory measures. Governments increasingly recognise that 'optimal taxes' in an economic sense are difficult to implement. For this reason taxes are now often of a political nature (e.g. awareness building) rather than an economic one (Pearce & Pearce, 2000).

Environmental taxes are often designed to modify behaviour, by increasing prices and therefore reducing demand. If consumers are price sensitive, suppliers will be forced to factor the cost of any new tax (e.g. aircraft emission taxes) into their cost structure. When taxes are used as policy instruments, they need to be equitable and simple. This means

Table 9.4 Policy instruments for environmental improvements

Regulations	*Market approaches*	*Soft tools*
Planning	Ownership	Tourism ecolabelling
Environmental Impact Assessment	Taxes, subsidies and grants	Certification/award schemes
Laws and regulations	Tradable rights and permits	Guidelines, treaties and agreements
Special status designation	Deposit-refund schemes	Citizenship, education and advertising
	Product and service charge	

Source: Tribe (2005)

that the political, social and economic impacts of the tax need to be distributed fairly, while the total transaction or collection costs should not exceed the revenue generated through the tax. Sweden, for example, raises US$52 million per annum from taxing energy, petroleum and emissions, including NO_x and CO_2 emissions from commercial airlines. These monies are used to fund research on new energy-related technologies.

There are only a few examples of environmental taxes collected from tourists. Seychelles, for example, impose a US$100 tax on every visitor, with the revenue used for conservation (Tribe, 2005). Often climate change policies focus on large and energy-intensive companies; one exception is Sweden (Krarup & Ramesohl, 2002). As the tourism industry is mainly composed of SME, there are few specific policies that assist tourism businesses to reduce their emissions or adapt to climate change. To address this issue, the New Zealand Climate Change Office introduced an energy-intensive business policy that is specifically targeted at energy-intensive small businesses. Tourism transport is identified as one of nine sectors eligible to receive government support as a result of this policy.

Planning, legislative and regulatory instruments, even those specifically related to tourism activities and facilities in coastal and other at-risk areas, have until recently ignored the need to plan for, and in other ways reflect, climate-related risks. But the situation is changing rapidly, largely as a result of the growing awareness of the substantial costs climate-related events are imposing on individuals, communities, businesses and economies and societies as a whole (Kay & Alder, 2005). However, there are a number of practical constraints on implementing timely and effective planning, legislative and legal instruments. Such constraints add to risk levels. For example, a common planning and regulatory response to increased risks related to storm surges, extreme waves and sea-level rise is to require that new or renovated buildings be set back an appropriate distance from the high-water mark. Compliance with such a requirement is frequently impeded, if not prevented, by existing land uses and ownership on the landward side of the setback line. This gives rise to 'coastal squeeze', a term coined to describe the difficulties and consequences of retreat-based responses. In this regard, Jennings (2004) argues that mature tourism resorts entering a rejuvenation stage of their development have limited options for coastal defence strategies, due to the high value of their built assets, unless redevelopment allows for coastal realignment (for example, the seawall of a tourist resort in Figure 9.8). On the other hand, earlier in their development cycle resorts have greater flexibility to implement more sustainable strategies that increase their harmony with the environment.

Figure 9.8 Seawall to protect the (artificial) beach and tourist accommodation at a tourist resort

Participatory planning

Participatory planning actively engages relevant stakeholders in the planning process, including objective setting, collective enquiry and decision making. Participatory techniques (or approaches) generate constructive collaboration among stakeholders who may not be used to working together, often come from different backgrounds, and may have different values and interests. Planning practitioners use a wide variety of methods, tailored to different tasks and situations, to support participatory planning. These include workshop-based and community-based methods for collaborative decision making, methods for stakeholder consultation, and methods for incorporating participation and social analysis into plan development (Text Box 24).

Text Box 24 Methods to support participatory planning
Source: The World Bank Participation Sourcebook
http://www.worldbank.org/wbi/sourcebook/sbhome.htm

Workshop-based methods

Collaborative decision making often takes place in the context of stakeholder workshops. Sometimes called 'action-planning workshops', they are used to bring stakeholders together to develop plans. The purpose of such workshops is to begin and sustain stakeholder collaboration and foster a 'learning by doing' atmosphere. A trained facilitator guides stakeholders, who have diverse knowledge and interests, through a series of activities to build consensus. Appreciation Influence Control (AIC), Objectives-Oriented Project Planning (ZOPP) and Team Up are three such methods.

Community-based methods

In many planning situations task managers and other staff leave their offices and undertake participatory work with local communities. Task managers work with trained facilitators to draw on local knowledge and begin collaborative decision making. In such settings, local people are the experts, whereas outsiders are facilitators of the techniques and are there to learn. The techniques energise people, tap local knowledge and lead to clear priorities or action plans. Two such techniques – participatory rural appraisal and SARAR – an acronym based on five attributes the approach seeks to build: self-esteem, associative strength, resourcefulness, action planning and responsibility – use local materials and visual tools to bridge literacy, status and cultural gaps.

Methods for stakeholder consultation

Beneficiary Assessment (BA) and Systematic Client Consultation (SCC) are techniques that focus on listening and consultation among a range of stakeholder groups. BA can be used to engage with the spectrum of stakeholders or with a focus on specific marginalised groups, such as through participatory poverty assessments (PPAs). SCC is a set of related techniques intended to obtain client feedback and to make planning interventions more responsive to demand. Both methods intend to serve stakeholders better by making donors and service providers aware of stakeholder priorities, preferences and feedback.

Methods for social analysis

Social factors and social impacts, including gender issues, should be a central part of all planning and action, rather than 'add-ons' that fit awkwardly with the universe of data to be considered. Social Assessment (SA) and Gender Analysis (GA) are methods that incorporate participation and social analysis into the planning process. These methods are also carried out in country economic and sector work to establish a broad framework for participation and identify priority areas for social analysis. Such methods evolved to meet the need to pay systematic attention to certain issues that traditionally had been overlooked by planners.

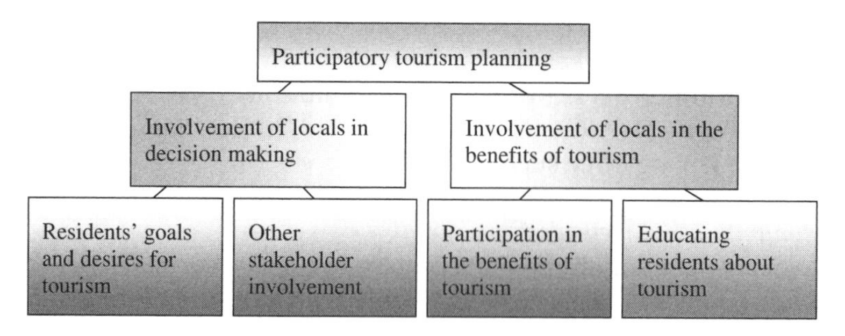

Figure 9.9 A normative model of participatory tourism planning
Source: Timothy (1999)

Participatory planning has been advocated in the tourism development literature, mainly to ensure positive impacts from tourism occur at the community level. Research in this field maintains that approaches to tourism, particularly in rural areas, must be inclusive and ensure meaningful public participation. There is still a need to identify and assess methodologies that can be used to improve public participation in tourism planning activities. Ways also need to be found to share such progress with other researchers and planners, including those from the climate change community. Timothy (1999) notes that the normative (from a Western perspective) approach to participation in tourism development, as shown in Figure 9.9, contains the risk that important inhibiting factors are overlooked. These include cultural and political traditions, poor economic conditions and lack of understanding by locals.

Even though there has been little integration between the areas of tourism development and climate change in relation to participatory planning approaches, the essential steps are very similar:

- assessing current situation;
- reviewing effectiveness of past practices;
- reviewing national and trans-national policies, legislation and regulations;
- identifying all stakeholders and their values and goals;
- recognising the available resources (financial, human, technological, institutional) and shortfalls in capacity;
- reaching a consensus on intentions (vision and goals) and principles for planning and management; and
- identifying strategies to maximise the collective benefits to all stakeholders, including marginalised groups and future generations.

The agreed intentions should be:

- challenging;
- clear and understandable;

- reasonable, achievable and financially viable;
- expressed in quantitative terms, through use of measurable performance targets, where practicable;
- prioritised and consistent; and specific with respect to methods and to timing of implementation and completion.

Financing

At present banks and other lending institutions are very reluctant to finance adaptation. The same is the case for mitigation, except for large projects where carbon credits and other factors ensure the viability of the financial arrangements. Investment conditions are not yet conducive for win–win partnerships. This is due to limited awareness, low technical and institutional capacity and financial constraints, including the small scale of most CDM projects in small island countries.

Financing difficulties are often best addressed through the promotion of institutions and mechanisms that provide innovative financing, including micro-financing, green finance, secured loans, leasing arrangements and public–private partnerships. Under such arrangements adaptation and mitigation can thrive, even without direct government intervention. But often a mix of government, private sector and donor funding, including public–private partnerships, is required simply because no one funding source has the capacity, will or mandate to be the sole source of financing. The GEF, the financial mechanism for the UNFCCC and other international environmental agreements, has a relatively long history of funding mitigation projects in developing countries. Lately the GEF has begun funding adaptation assessments and, even more recently, substantive, on-the-ground adaptation interventions. For both mitigation and adaptation projects GEF funding is usually restricted to meeting the additional project costs that can be attributable to addressing climate change as well as costs of generating specific environmental benefits. As a result, co-financing is required for such projects. This often poses a challenge given that small businesses usually comprise the majority of the tourism industry.

Timely and adequate financing of adaptation and, to a lesser extent, mitigation projects in developing countries is often undermined by one or more perverse incentives. These are situations that favour a far less satisfactory outcome than should occur (Bettencourt *et al.*, 2005). Many developing countries consider it a rational decision not to reduce climate-related risks by undertaking adaptation and mitigation so long as donors show a willingness to respond generously to disasters, irrespective of whether or not preventive efforts have been taken. For donors, such 'Good Samaritan' behaviour is highly visible and warmly accepted, creating a moral hazard around risk prevention – donors face intense media and public pressure to respond rapidly and generously to disasters. Furthermore, the benefits of prevention may not become apparent for years, and hence it is common for climate-risk-reduction initiatives to compete unfavourably with other short-term domestic priorities, such as health and education. Uncertainties related to the rate and consequences of climate change also work against allocating resources in ways that allow adaptation and mitigation to be undertaken in a timely and effective manner. Finally, policy makers in some developed countries have an aversion to self-funding of even a small proportion of the costs of adaptation and mitigation measures, in the belief that, as 'innocent victims', the onus is on developed countries to provide full compensation for climate change. But all three global adaptation funds overseen by the GEF rely on voluntary contributions, and donors want to see clear signs of a matching commitment from the recipient country.

Information, knowledge, skills and motivation

Figure 9.10, which is based on the Argincourt/Kolb learning cycle, highlights why knowledge, skills and motivation are so critical to the successful response of the tourism sector to climate change. As noted in Chapter 2, there are numerous players in the tourism sector who act, formally or informally, as information providers and skill developers.

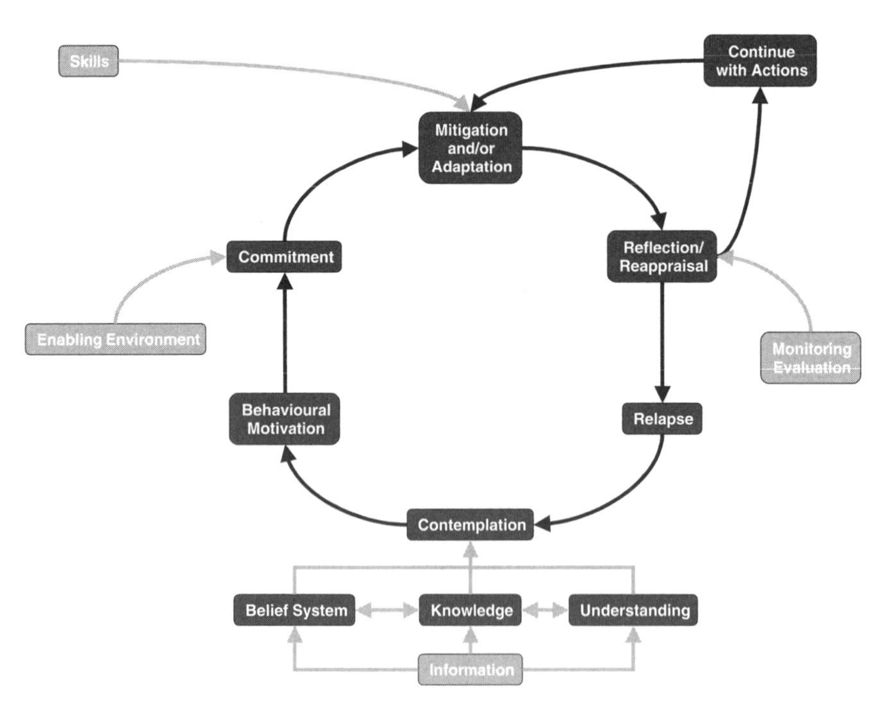

Figure 9.10 The roles of knowledge, skills and motivation in responding to climate change

Information, knowledge and understanding support the development of behaviours that, in combination with practical skills and a strong enabling environment, result in actions that follows on from the commitment to address climate-related risks through adaptation and/or mitigation. At the individual level, and also at higher community, business and organisational levels, the process of monitoring and evaluation signals the appropriateness of either continuing with the mitigation and/or adaptation activities or using additional knowledge and skills to undertake actions that are more appropriate.

Technology transfer and uptake

Transfer of technology is more than just the moving of high-tech equipment from the developed to the developing world, or within the developing world. Moreover, it encompasses more than equipment and other so-called 'hard' technologies. It also includes total systems and their component parts, such as know-how, goods and services, equipment, and organisational and managerial procedures. Thus technology transfer is the suite of processes encompassing all dimensions of the origins, flows and uptake of know-how, experience and equipment amongst, across and within countries, stakeholder organisations and institutions (Figure 9.11).

An important challenge is to ensure that decisions regarding mitigation and adaptation technologies for use within the tourism sector are made and implemented by people who are not only well informed but who also operate within an enabling environment that enhances the possibility their technology investment will be simultaneously environmentally sound, socially acceptable and economically viable.

If the transfer of inadequate, unsustainable or unsafe technologies is to be avoided, technology recipients should be able to identify and select technologies that are appropriate to their actual needs, circumstances and capacities. Therefore, a key element of this wider view of technology transfer is choice. There is no single strategy for successful transfer that is appropriate to all situations. Moreover, a technology that is assessed to be environmentally sound in a given locale, culture, economic setting or stage in its life cycle, may not be in another. Its performance may be influenced markedly by the availability of supporting infrastructure and by access to the expertise necessary for its successful installation, operation, maintenance and monitoring.

Furthermore, a technology that qualifies as being environmentally sound at one point in time, may not be so at another – the performance criteria against which it is assessed may change as a consequence of new information or changing values or attitudes; a technical breakthrough may give rise to more desirable alternatives. It is therefore vital that recipients and users of a given technology are able to choose an option that meets their specific needs and capacities, while also being

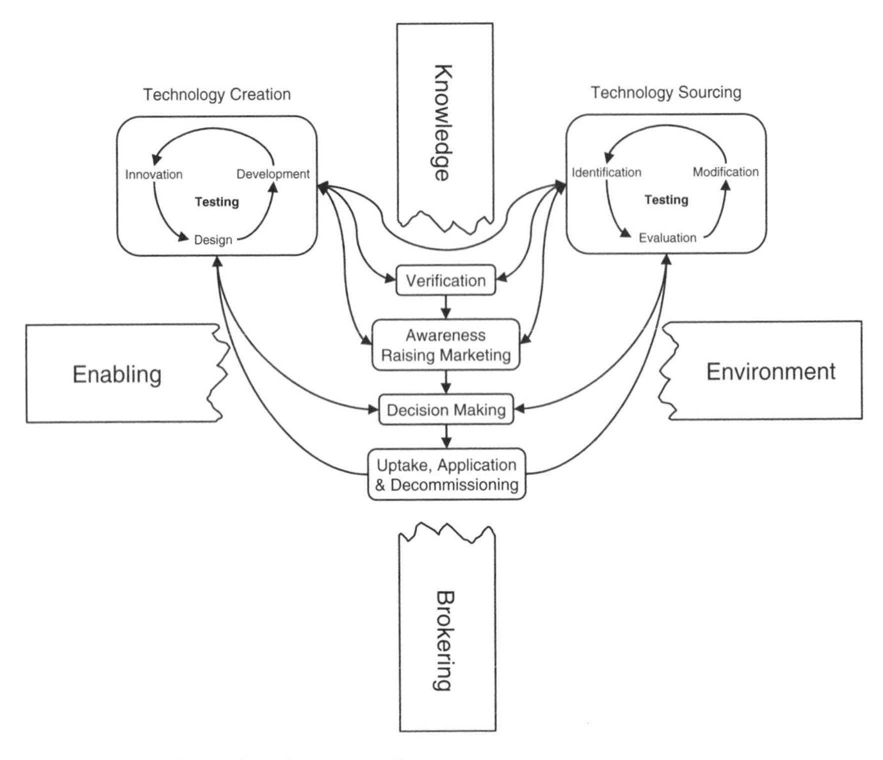

Figure 9.11 The technology transfer process

environmentally sound in its operating locale and over its operational life cycle. It is, of course, highly desirable that the technology is also found to be economically viable and socially acceptable, and hence sustainable.

There are many barriers to successful technology transfer. All along the transfer path, from the supply side of technology transfer (the innovators and developers) to the demand side (the recipients and users), impediments occur at every node. Due to restrictions on the movement of information and materials, barriers also occur at every linkage in the technology transfer chain. Examples of such barriers include shortfalls in technology creation and innovation, underperformance in technology sourcing, suboptimal enabling environments, and insufficient and unverified information. SME, common in the case of tourism, are disproportionately impacted by these challenges.

Institutional strengthening

Effective and efficient institutions, especially governmental institutions, are an important part of the enabling environment for adaptation and mitigation by the tourism sector. Institutional capabilities and cooperation are perhaps key in this respect. Climate change, including identification and implementation of appropriate technology-based mitigation and adaptation responses, is often a new and intimidating area of responsibility for officials based in tourism ministries. For this and other reasons there is often a tendency to defer to colleagues in environment and related ministries who have more experience, at least at a general level. The resulting scattered and uncoordinated approach to such a pervasive issue as climate change and tourism is far from optimal. Even in the absence of climate change, the cross-cutting nature of tourism highlights the need for mechanisms that facilitate cooperation and collaboration between all ministries that have either a direct or indirect interest in the efficient and sustainable operation of the tourism sector. Climate change simply adds another level of complexity to addressing this need. Thus, while the conventional overview of tourism by a single lead government agency is appropriate, mechanisms must be found to facilitate cooperation and collaboration between institutions, and between individuals working in them.

No one institutional arrangement for oversight of policy development and implementation related to climate change and tourism will be appropriate to all countries. However, international experience would favour a two-tiered structure comprising an interministerial committee, with non-governmental representation, serviced by an interagency task force made up of senior officials and other individuals with relevant technical and policy-making experience. Further, there is evidence to support the fact that these bodies should be directed and coordinated at the highest possible level of government (Bettencourt *et al.*, 2005). Lesser players are all too often ineffective in dealing with powerful ministries such as Public Works, Finance, Health and Natural Resources. The additional risks climate change inflicts on tourism are in many cases indirect. The direct impacts are more often imposed on the water resources sector, agriculture, human health, infrastructure, and vulnerable terrestrial and marine ecosystems. All these have significant consequences for the national economy and for society at large. Thus the leaders and senior officials of these key and influential ministries have abundant reason to engage with colleagues to ensure that appropriate action is taken to ensure that tourism is not impacted severely by climate change and that mitigation and adaptation initiatives are implemented, consistent with sustainable development priorities.

Monitoring and review

As shown in Figure 9.7, monitoring and review are also integral parts of any policy response to climate change. If the policy responses are proving effective, the results should be used to encourage greater efforts with respect to both existing and new initiatives. If the responses are partially or totally ineffective, assessment of existing policy initiatives should be undertaken in order to reinforce successful measures and revise or abandon those that are not bringing benefits to the tourism sector, or its constituent parts. Adaptation and mitigation strategies should themselves adapt – to such developments as new information and understanding, to successes and failures in past efforts, and to the availability of new technologies, or re-adoption of traditional technologies. In this way a cycle of continuing increases in the effectiveness of mitigation and adaptation interventions will be established, for the betterment not only of the tourism sector but also for tourism-dependent communities and the national economy at large.

National planning

Tourism developers have a decided preference for investing in seaside or beachfront properties, reflecting in the main the general desire of tourists themselves for beach, sun and sea vacations. As a result of the large economic benefits flowing from tourism, many governments have responded to these preferences with policies and plans that concentrate on tourism facilities in coastal areas. But the growing risks to coastal tourism infrastructure as a consequence of climate change are requiring revisions to planning, legislative and regulatory measures in order to reduce these risks to acceptable levels. There is growing use of the five generic approaches for proactive adaptation of relevance to coastal zones (Text Box 16). These are a major advance on the previous overly simplistic response options of protection, accommodation or retreat, and are thus much more aligned with preferred coastal zone management and planning approaches (Kay & Alder, 2005).

Furthermore, adaptation to climate change by the tourism sector is benefiting from the increasing assimilation of tourism planning into integrated planning initiatives. This reflects in part the fact that not every tourism facility is best located on the coast. Often such needs of the tourist industry as access to coastal land and efficient transport links are best addressed through integrated regional planning. Integrated planning seeks to bring together the various competing demands for coastal resources, including tourism demands, into an overall planning framework. As it offers advantages over purely sectoral approaches, integrated coastal management has been widely recognised and promoted as the most appropriate process to deal with climate change, sea-level rise and other current and long-term coastal challenges. For example, transformation of London into a 'World City', partly as a result of it being a major tourist destination, has resulted in progressive encroachment into the tidal Thames River, exacerbating the flood risk to new and adjacent land, despite continual upgrading of flood defences. The pressure for further encroachment is now combining with the new threat of human-induced climate change, including more intense precipitation events, increased peak river flow in the winter, accelerated sea-level rise, and possibly more severe storms and resulting storm surges. Costly upgrading of defences is inevitable, but recently managed realignment (i.e. retreat) is also being considered as a flood management tool. It offers a range of potential benefits, including increasing storage for flood water, improving ecological functioning, improving riverside access for people and reversing the long-term trend of encroachment into the river. Several projects within urban London have shown that realignment can achieve these benefits even though implementation of urban managed realignment is hindered by fragmentation between flood defence and spatial development policies, lack of political support and riverside encroachment. Successful application requires long-term (i.e. decades) planning that fully exploits all opportunities for landward realignment offered by redevelopment.

The strengthening of environmental impact assessment and similar regulatory and voluntary instruments can also facilitate adaptation if requirements to minimise climate-related risks are included. For example, the Town and Country Planning Office in Barbados uses such coastal standards as a building setback from the shore line of 30 m in beach areas and 10 m in cliffed areas. But Jackson (2002) urges flexibility in imposing such measures as they need to be fair and affordable to all parties. Moreover, strict application of such standards can adversely affect the viability of an investment project. He also suggests minimum finished floor levels for habitable rooms, and for facilities used to house critical equipment or store hazardous substances.

Conclusions

This chapter provides an overview of the evolution of climate change policy and important principles considered in the policy regime. While there is limited explicit reference to tourism, many of the existing policy frameworks and initiatives are very relevant for tourism. It is a challenge to ensure effective development, communication and implementation of climate change policy given that the tourism sector is composed of a wide range of actors (among others the public and private sectors and tourists), has a predominance of small businesses and has important international dimensions. As for general climate change policies, tourism-specific ones should maximise win–win situations and minimise trade-offs between mitigation and adaptation. One key component of this is to 'mainstream' policies and measures to address climate change into existing and new tourism development policies, plans and actions.

Notes

1. The archives of the Djerba Conference can be found at http://www.world-tourism.org/sustainable/climate/brochure.htm.
2. For a discussion of the precautionary principle and uncertainty in relation to climate change, see Dupuy and Grinbaum (2005).
3. The tax was introduced in 2002 in an effort to offset some of the damage caused by mass tourism. Tourists to Majorca, Minorca, Ibiza and Formentera paid an average of €1/day to raise revenue for environmental management such as upgrading footpaths and cycle routes, and the planting of native species of olive and almond trees on land previously reserved for tourist development. The tax was abolished in 2003, because of its impact on visitor numbers and fairness issues: as it was collected by hotels, it ignored the 25% of tourists who stay in unlicensed accommodation or with family and friends.

Chapter 10

Conclusion

The relationships between tourism and climate are many and complex. Tourism interacts with the climate in a wide variety of ways: tourism uses climate as a natural resource; it depends on the nature and predictability of the weather in the short term and of the climate in the longer term; and tourists and tourism operations can be harmed by anomalous weather conditions, including extreme events. They face the risk of substantial losses. Tourism also contributes to GHG emissions that alter the global climate. As a result, tourism is being increasingly influenced by international policies to slow the rate of climate change. Less directly, tourism depends on the natural environment and healthy ecosystems; climate is often a critical determinant of the well-being of these ecosystems. But climate change can also benefit tourism, though any gains are likely to be comparatively small. For example, reduced costs of snow and ice management are likely to bring major benefits to the transportation industry. The Stern Review Report (Stern, 2006) notes that in higher latitude regions, such as Canada, Russia and Scandinavia, climate change may provide a possible boost to tourism. But the report also points out that these same regions will experience the most rapid rates of warming, placing infrastructure, human health, local livelihoods and biodiversity at extreme risk.

The multiple interactions between tourism and other global or local systems mean that interventions designed to reduce climate-related risks to tourism, or take advantage of opportunities, are likely to have consequences beyond the tourism–climate system. It is therefore crucial to understand these linkages and their effects, if only to help ensure that climate change adaptation or mitigation initiatives undertaken by one part of the tourism system do not have undesirable effects on other parts of the system, or on related systems. For example, regulating air travel to reduce GHG emissions can result in serious equity issues. These can be between socioeconomic groups within one society, or between wealthy tourist-generating countries and those countries that depend on continuing tourist flows to aid their economic development. The relationships between tourism, climate change and biodiversity are another important example of wider interactions. Many climate change mitigation and adaptation measures offer real opportunities to help meet biodiversity goals, including restoration of native forest for carbon sequestration or replanting of mangrove forest for coastal protection. But there is also the possibility that measures to reduce the climate-related risks will have detrimental effects on biodiversity (Choudhury *et al.*, 2004). An example is the use of non-native species or chemical fertilisers to facilitate revegetation.

The tourism system exhibits not only the characteristics of complexity, such as non-linearities, but also thresholds. This means that professionals working in the tourism sector cannot control tourism's evolution. Nor can they predict with absolute accuracy what might happen as a result of their planning and management activities. Similarly, the complexities and non-linearities in the tourism–climate system mean that again the consequences of change are difficult to predict and to manage. The tourism–climate system may also undergo abrupt and pervasive changes. These are particularly challenging for those who seek an orderly evolution of tourism, at national or business enterprise levels. Apparently appropriate policies and management procedures, be they tourism or climate related, will not necessarily reduce climate-related risks or maximise opportunities created by climate change. However, the tourism system as a whole will likely respond in a more orderly and hence predictable manner, governed by a number of underlying principles and relationships that determine how the overall system will respond, within broad parameters. Moreover, for short-term planning it is common to assume some sort of linearity and hence predictability.

For these and other reasons provided in the foregoing chapters, tourism should be an important consideration in climate change research and policy making. But it is usually difficult, and often undesirable, to single tourism out from other activities. Firstly, in efforts to implement climate change mitigation or adaptation, tourism faces problems and barriers similar to those experienced by other sectors. Secondly, tourism is one component of a much bigger system that needs to be considered when trying to understand tourism's societal drivers, its policy implications, its technological needs and potentials, and its diverse roles in different parts of the world. Such a holistic approach requires new research methods that are able to account for the complexity and multitude of interrelationships, as well as the up- and down-stream effects of interventions. This research also has to go beyond scientific curiosity and be policy relevant in order to inform decision making. Part of the challenge is to integrate controversial and non-consensual research results and achieve a way forward under a plurality of world views and priorities.

This book has emphasised the need for strengthened and more comprehensive risk management in the tourism sector. One long-term risk that tourism faces is its dependency on fossil fuels. Fuel is often delivered through a long supply chain with various associated risks, for example the risk of interruptions to supplies as a result of a tropical storm. Substantial risks are associated with the ongoing availability of (cheap) oil and the prospect of increased oil prices as a result of internalising the environmental costs of GHG emissions. Tourism's greatest 'Achilles heel' is aviation. To date no viable substitute has been found for the current aviation fuels. At the same time the demand for air travel is increasing. Overall, climate change provides tourism with incentives to redesign products towards carbon efficiency or perhaps even carbon neutrality. The latter can be achieved by investing in renewable energy sources and sequestering carbon (semi-permanently) in the form of forest biomass.

Recently Richard Branson, the founder of the Virgin Group of aviation and other companies, announced his intention to invest in the research and development of biofuels. His aim is to not only lessen the aviation sector's current dependency on fossil fuels, but to also substantially reduce its GHG emissions. The main candidate is bioethanol. This can be produced from corn and agricultural wastes. Branson concedes that use of such fuel substitutes by airlines could be a decade away. Many industry commentators and experts are far less optimistic, For example, the IPCC suggests that any such substitution is decades away. Others emphasise that the low flash point and high cloud point (the temperature at which the fuel begins to congeal) of bioethanol present major safety problems. Moreover, at the low power settings used during taxiing, the exhaust gases can be toxic, creating the potential for localised health problems around airports. Another of Branson's suggestions, to use tow vehicles to move aircraft to and from holding areas close to the runway, may well have more merit – at least for now.

Increasing attention is given to voluntary schemes for offsetting emissions from airline travel, but note that it is still unclear whether such schemes can make a significant contribution to enhancing the sustainability of tourism. However, current voluntary emissions reductions need to be increased dramatically to become relevant. Importantly, such schemes do nothing to reduce aviation emissions directly. In fact they may encourage people to believe that there is no need to change their behaviour, thus creating irreversibility in currently unsustainable consumption and production patterns. The schemes also give limited incentive to airlines to increase their fuel efficiency, in contrast to incentives such as emissions taxes or regulated, tightly capped carbon trading systems.

Risks are also associated with the anticipated and unexpected consequences of a changing climate. Locally present climate-related risks, including their management, are often more relevant to tourism businesses than are global considerations such as reducing GHG emissions. The need to address the latter often appears to be less immediate and concrete. Risks that dominate locally include devastation of tourist facilities by tropical storms, the loss of critical natural tourism resources through coral bleaching and other phenomena, damage to tourism infrastructure as a result of flooding and direct harm to tourists as a consequence of climate-related disasters. There are opportunities to reduce such risks by adapting to changing conditions. This involves collaboration between a wide range of stakeholders and communities, often involving tourists themselves, as is the case for water conservation.

The growing risk levels for tourism as a consequence of climate change are driving the need for greater understanding of the ways in which the climate will change, not only in general terms but specifically with respect to location, climatic variables and time frames, and for average as well as extreme conditions. Chapter 5 showed how far the scientific community is away from meeting such requirements and identifies the barriers to making substantive improvements. As a result, while uncertainties in climate projections will undoubtedly be reduced as forecasting skills increase, in the coming years and decades it will always be necessary to acknowledge and work within the residual uncertainties.

Actions by the tourism sector to reduce climate-related risks will on occasions be inadequate. This is especially the case when there is an absence of detailed information on the consequences of the anticipated changes in climate for specific components of the tourism sector. For example, estimating how climate change will modify future global international tourist flows reveals major gaps in our knowledge. These start with uncertainties in characterising the climatic changes, and include the shortage and limited geographical coverage of studies into the effects of climate change on tourist destination choice as well as tourism demand. As Hamilton *et al.* (2004) note, these shortcomings weaken the empirical basis of the models developed to make such estimates. To date, the models have also neglected changes in preferences, age structure, working hours and lifestyles. Studies that focus only on the tourism sector miss important economic consequences when the implications of variations in tourism demand induced by climate change are considered. Current modelling capabilities extend to quantifying only the direct economic effects of climate change on tourism, and then only in terms of changes in tourism expenditure in the destination country. Thus the economic consequences of changes in tourism travel itself are ignored, as are indirect effects such as a rise in sea level and the implications for beach erosion, requirements for beach nourishment and for the possible inundation of popular atolls and other islands, changes in the water cycle and the spread of diseases. In addition, no allowance is made for the many feedbacks that characterise the tourism system. With an increasing number of players globally, a diversification of the tourism product and specification of demand, the tourism system becomes more and more complex and studies must be increasingly sophisticated.

Addressing the multiple issues of climate change and tourism requires effective communication and collaboration between researchers, planners, policy makers, tourism operators and the wider public, including hosts and guests. Tourism traverses all scales – from local to global. While this poses a challenge, it is not unique to tourism. But tourism is unique in that it is a truly global issue that encompasses more than an industry. As a business, tourism requires inputs from the public sector, communities and the natural environment, making tourism in many ways a social phenomenon rather than an economic activity. Yet tourism's economic contributions and implications are significant.

The numerous interactions between climate and tourism have significant policy implications as well as immediate, practical consequences for tourism businesses, decision makers and tourism operators. Climate change will modify these consequences. In many, but certainly not all cases, this will be to the substantial detriment of the tourism industry. Thus climate change presents a significant risk to the tourism sector, a risk that can only be reduced in the short term through adaptation. Mitigation, the reduction of GHG emissions, represents a strategy for risk reduction that has a longer timescale. It also requires a globally meaningful commitment to reduce such emissions, not only by the tourism sector, but through the efforts of all emitters, both large and small. One benefit of mitigation initiatives undertaken by tourism businesses is that they can enhance financial and environmental performances, and hence the sustainability of the business.

The tourism sector is already experiencing disruptive changes consistent with many of the anticipated consequences of climate change – heatwaves, receding glaciers, extreme and prolonged water shortages, super typhoons and hurricanes, to name only a few. Thus efforts to reduce climate-related risks are not simply something to contemplate for the future. Moreover, the scientific evidence of more serious consequences for tourism in the future is now overwhelming. This demands urgent responses, individually through to globally. Importantly, an increasing number of studies are highlighting an important message – the benefits of the tourism sector taking meaningful early action far outweigh the costs of avoiding such severe consequences in the future.

References

Allen, M.R. and Ingram, W.J. (2002) Constraints on future changes in climate and the hydrologic cycle. *Nature* 419, 224–232.

Agnew, M.D. and Palutikof, J.P. (2001) Impacts of climate on the demand for tourism. In *Proceedings of the First International Workshop on Climate, Tourism and Recreation*. In A. Matzarakis and C.R. de Freitas (eds) International Society of Biometeorology, Commission on Climate Tourism and Recreation, Porto Carras, Halkidiki, Greece, December 2001, WP4, 1–10.

Air Transport Action Group (2002) *Industry as a Partner for Sustainable Development. Aviation*. On WWW at www.atag.org. Accessed 12.06.05.

Alley, R.B., Marotzke, J., Nordhaus, W.D., Overpeck, J.T., Peteet, D.M., Pielke Jr., R.A.R., Pierrehumbert, T., Rhines, P.B., Stocker, T.F., Talley, L.D. and Wallace, J.M. (2003) Abrupt climate change. *Science* 299 (5615), 2005–2010.

Altalo, M., Hale, M., Anastasia, O. and Alverson, H. (2002) *Requirements of the US Recreation and Tourism Industry for Climate, Weather and Ocean*. On WWW at http://ioc.unesco.org/goos/docs/tourismreport.pdf. Accessed 20.11.05.

Amelung, B. and Viner, D. (2006) The sustainability of tourism in the Mediterranean. Exploring the future with the Tourism Climatic Index. *Special Issue Journal of Sustainable Tourism: Climate Change and Tourism* (edited by D. Viner) 14 (4), 349–366.

Ammann, W. (1999) Schnee und Lawinen – Bestimmende Wirtschaftsfaktoren im Alpenraum. In: Université de Genève et Institut Universtaire Kurt Bösch (ed.) *Die Rolle des Wassers in der sozio-ökonomischen Entwicklung der Alpen* (pp. 139–162). Sion: Institut Universtaire Kurt Bösch.

AMP Capital (2005) *Climate Change and Company Value. A Guide for Company Analysts*. November 2005. On WWW at www.igcc.org.au/ProdImages/Climate_Risk_and_Company_Value.pdf. Accessed 02.12.05.

Asia Pacific Foundation of Canada (2002) *Canada Asia Commentary*. Number 23, March 2002. On WWW at www.asiapacific.ca. Accessed 01.06.03.

Asian Development Bank (2006) *Guidelines for Adaptation Mainstreaming*. Manila: Asian Development Bank.

Aspen Skiing Company (2004) *2003–2004 Sustainability Report. Aspen*. On WWW at http://www.aspensnowmass.com/environment/programs/2004_ASC_Sustainability_Report.pdf. Accessed 20.05.05.

Association of British Insurers (2004) *A Changing Climate for Insurance. A Summary Report for Chief Executives and Policymakers*. On WWW at www.abi.org.uk/climatechange. Accessed 20.11.05.

Australian Institute of Energy (no date) *Energy Value and Greenhouse Emission Factor of Selected Fuels*. On WWW at http://www.aie.org.au/melb/material/resource/fuels.htm. Accessed 12.07.05.

Baines, J.T. (ed.) (1993) *New Zealand Energy Information Handbook.* Christchurch: Taylor Baines & Associates.

Ball, C. (2004) Will technology threaten meetings industry associations? *International Congress and Convention Association – Intelligence* 6. On WWW at http://icca.webportalasp.com/newspub/story.cfm?id=704. Accessed 18.01.05.

BBC (15 October 2005) *China completes railway to Tibet*. On WWW at http://news.bbc.co.uk/2/hi/asia-pacific/4345494.stm. Accessed 16.10.05.

Becken, S. (2002) Analysing international tourist flows to estimate energy use associated with air travel. *Journal of Sustainable Tourism* 10 (2), 114–131.

Becken, S. (2004a) *Climate Change and Tourism in Fiji – Integrating Adaptation and Mitigation. Final Report*. Suva: University of the South Pacific.

Becken, S. (2004b) How tourists and tourism experts perceive climate change and forest carbon sinks. *Journal of Sustainable Tourism* 12 (4), 332–345.

Becken, S. (2004c) Leisure, energy costs of. *Encyclopaedia of Energy* 3, 623–634. Academia Press.

Becken, S. (2005) Harmonising climate change adaptation and mitigation. The case of tourist resorts in Fiji. *Global Environmental Change – Part A* 15 (4), 381–393.

Becken, S. and Cavanagh, J. (2003) *Energy Efficiency Trend Analysis of the Tourism Sector*. Research Contract Report: LC02/03/293. Prepared for the Energy Efficiency and Conservation Authority.

Becken, S. and Frame, B. (2005) *Scenario Planning for Climate Change and Tourism.* Tourism Futures Workshop, 19 May 2005, Wellington.

Becken, S., Frampton, C. and Simmons, D. (2001) Energy consumption patterns in the accommodation sector – the New Zealand case. *Ecological Economics* 39, 371–386.

Becken, S. and Patterson, M. (2006) Measuring national greenhouse gas emissions from tourism as an important component towards sustainable tourism development. *Special Issue Journal of Sustainable Tourism: Climate Change and Tourism* (ed. D. Viner), 14 (4), 323–338.

Becken, S. and Simmons, D. (2002) Understanding energy consumption patterns of tourist attractions and activities in New Zealand. *Tourism Management* 23 (4), 343–354.

Becken, S., Simmons, D. and Frampton, C. (2003) Segmenting tourists by their travel pattern for insights into achieving energy efficiency. *Journal of Travel Research* 42 (10), 48–56.

Beg, N., Morlot, J.C., Davidson, O., Afrane-Okesse, Y., Tyani, L., Denton, F., Sokona, Y., Thomas, J.P., Lebre La Rovere, E., Parikh, J.K., Parikh, K. and Rahman, A.A. (2002) Linkages between climate change and sustainable development. *Climate Policy* 2, 129–144.

Belle, N. and Bramwell, B. (2005) Climate change and small island tourism: Policy maker and industry perspectives in Barbados. *Journal of Travel Research* 44, 32–41.

Benestad, R. (2006) Can we expect more extreme precipitation? *Journal of Climatology* 19 (4), 630–637.

Bengtsson, L. (2001) Enhanced hurricane threats. *Science* 293, 440–441.

Beniston, M. (1997) Variations of snow depth and duration in the Swiss Alps over the last 50 years: Links to changes in large-scale forcings. *Climatic Change* 36, 281–300.

Beniston, M. (2003) Climatic change in mountain regions: A review of possible impacts. *Climatic Change* 59, 5–31.

Beniston, M. (2004) The 2003 heat wave in Europe. A shape of things to come? *Geophysical Research Letters* 31, L02022.

Beniston, M. (2005) Warm winter spells in the Swiss Alps: Strong heat waves in a cold season? *Geophysical Research Letters* 32, L01812.

Beniston, M. (in press) Future extreme events in European climate; an exploration of RCM projections. *Climatic Change* in press.

Beniston, M. and Diaz, H.F. (2004) The 2003 heat wave as an example of summers in a greenhouse climate? Observations and climate model simulations for Basel, Switzerland. *Global and Planetary Change* 44, 73–81.

Berrittella, M, Bigano, A., Roson, R. and Tol, R. (2006) A general equilibrium analysis of climate change impacts on tourism. *Tourism Management* 27 (5), 913–924.

Berz, G.A. (1999) Catastrophes and climate change: Concerns and possible countermeasures of the insurance industry. *Mitigation and Adaptation Strategies for Global Change* 4, 283–293.

Bettencourt, S., Croad, R., Freeman, P., Hay, J., Jones, R., King, P., Lal, P., Mearns, A., Miler, G., Pswarayi-Riddihough, I., Simpson, A., Teuatabo, N., Trotz, U. and Van Aalst, M. (2005) *Not if But When: Adapting to Natural Hazards in the Pacific Islands Region. A Policy Note*. Pacific Islands Country Management Unit, East Asia and Pacific Region, World Bank.

Biesiot, W. and Noorman, K.J. (1999) Energy requirements of household consumption: A case study of The Netherlands. *Ecological Economics* 28, 367–383.

Bigano, A., Hamilton, J.M., Lau, M., Tol, R.S.J. and Zhou, Y. (2004) A global database of domestic and international tourist numbers at national and subnational level. *Working Paper FNU–54*.

Bode, S., Hapke, J. and Zisler, S. (2003) Need and options for a regenerative energy supply in holiday facilities. *Tourism Management* 24, 257–266.

Bohdanowicz, P. and Martinec, I.M. (2001) *Thermal Comfort and Energy Savings in the Hotel Industry*. Proceedings of the 15th Conference on Biometeorology and Aerobiology, Kansas City, Missouri, 28 Oct.–1 Nov. 2002. Boston, MA: American Meteorological Society.

Boodhoo, S. (2003) The value of weather, climate information and predictions to the tourism industry in small island states and low lying areas. *First International Conference on Climate Change and Tourism*, Djerba, Tunisia, April, 2003.

Bradley, R.S., Keimig, F.T. and Diaz, H.F. (2004) Projected temperature changes along the American cordillera and the planned GCOS network. *Geophysical Research Letters* 31, L16210.

Brauner, C. (2002) *Opportunities and Risks of Climate Change*. Zurich: Swiss Reinsurance Company.

Brewer G.D. (1991) *Hydrogen Aircraft Technology*. London: CRC Press.

Briedenhann, J. and Wickens, E. (2004) Tourism routes as a tool for the economic development of rural areas – vibrant hope or impossible dream? *Tourism Management* 25 (1), 71–79.

Brons, M., Pels, E., Nijkamp, P. and Rietveld, P. (2002) Price elasticities of demand for passenger air travel: A meta-analysis. *Journal of Air Transport Management* 8, 165–175.

Brunotte, M. (1993) *Energiekennzahlen für den Kleinverbrauch*. Freiburg: Studie im Auftrag des Öko-Instituts.

Buckley, R.C. and Aranjo, G.F. (1997) Environmental management performance in tourism accommodation. *Annals of Tourism Research* 24 (2), 465–468.

Buddemeier, R.W., Kleypas, J.A. and Aronson, R.B. (2004) *Coral Reefs. Potential Contributions of Climate Change to Stresses on Coral Reef Ecosystems*. Arlington, VA, USA: Pew Center on Global Climate Change.

Buerki, R., Elasser, H. and Abegg, B. (2003) Climate change – impacts on the tourism industry in mountain areas. Proceedings of the First International Conference on Climate Change and Tourism. 9–11 April, Tunisia. Madrid, Spain: World Tourism Organization.

Bundesamt für Statistik (1999) *Schweizer Tourismus in Zahlen*. Bern, Switzerland: STV.

Bureau of Transportation Statistics (2004) Homepage. On WWW at http://www.bts.gov/. Accessed 20.02.06.

Burkett, V.R., Zilkoski, D.B. and Hart, D.A. (2005) Sea-level rise and subsidence: Implication for flooding in New Orleans. In Prince, K.R. and D.L. Galloway (eds) *US Geological Survey Subsidence Interest Group Conference, Proceeding of the Technical Meeting*, Galveston, Texas, 27–29 Nov 2001. US Geological Survey, Water Resources Division Open File Report 03-308.

Butler, R. (1980) The concept of a tourist area cycle of evolution: Implications for management of resources. *Canadian Geographer* 24, 5–12.

Byrnes, T.A. and Warnken, J. (2006) Greenhouse gas emissions from marine tours: A case study of Australian tourboat operators. *Journal of Sustainable Tourism* 14 (3), 255–270.

CALSTART (2002) *Passenger Ferries, Air Quality, and Greenhouse Gases: Can Systems Expansion Result in Fewer Emission in the San Francisco Bay Area?* On WWW at www.maradat.dot.gov. Accessed 14.03.05.

Canada CCME (2003) *Climate, Nature, People: Indicators of Canada's Changing Climate*. Climate Change Indicators Task Group of the Canadian Council of Ministers of the Environment.

Carlsen, J. (1999) A systems approach to island tourism destination management. *Systems Research and Behavioral Science* 16 (4), 321–327.

Carlsen, J., Getz, D. and Ali-Knight, J. (2001) The environmental attitudes and practices of family businesses in the rural tourism and hospitality sectors. *Journal of Sustainable Tourism* 9 (4), 281–297.

Carlsson-Kanyama, A. and Linden, A.L. (1999) Travel patterns and environmental effects now and in the future: implications of differences in energy consumption among socio-economic groups. *Ecological Economic* 30, 405–417.

Carter, R.W., Baxter, G.S. and Hockings, M. (2001) Resource management in tourism research: A new direction? *Journal of Sustainable Tourism* 9 (4), 265–280.

Cavallaro, F. and Ciraolo, L. (2005) A multicriteria approach to evaluate wind energy plants on an Italian island. *Energy Policy* 33, 235–244.

Ceballos-Lascurain, H. (1996) *Tourism, Ecotourism and Protected Areas. The State of Nature-Based Tourism Around the World and Guidelines for its Development*. Gland: IUCN Protected Areas Programme.

Cesar, H., Lauretta, B. and Pet-Soede, L. (2003) *The Economics of Worldwide Coral Reef Degradation*. Prepared for the World Wildlife Fund for Nature, Netherlands. On WWW at www.panda.org/coral. Accessed 20.04.04.

Chan, W.W. and Lam, J.C. (2003) Energy-saving supporting tourism: A case study of hotel swimming pool heat pump. *Journal of Sustainable Tourism* 11 (1), 74–83.

Chen, C. (1998) Rising overseas travel market and potential for the United States. *Advances in Hospitality and Tourism Research* (Vol. III). On WWW at www.

hotelonline.com/Trends/AdvancesInHospitalityResearch/ChineseTravelMarket998.html. Accessed 27.06.03.

Choudhury, K., Dziedzioch, C., Haeusler, A. and Ploetz, C. (2004) *Integration of Biodiversity Concerns in Climate Change Mitigation Activities. A Toolkit.* Commissioned by the German Federal Environmental Agency. On WWW at www.umweltbundesamt.de. Accessed 20.11.04.

Christensen, J.H., Carter, T.R. and Giorgi, F. (2002) PRUDENCE employs new methods to assess European climate change. *EOS* 83, 147.

Collier, A. (1999) *Principles of Tourism. A New Zealand Perspective.* Auckland: Addison Wesley Longman New Zealand Limited.

Corbett, J. and Fischbeck, P. (1997) Emissions from ships. *Science* 278, 823–824.

Corbett, J.B., Durfee, J.L., Gunn, R.D., Krakowjak, K.M. and Nellermoe, J.T. (2002) Testing public (un)-certainty of science: Media representations of global warming. *7th International Conference on Public Communications of Science and Technology*, 5 Dec. 2002, Capetown, South Africa.

Corsi, T.M. and Milton, E.H. (1979) Changes in vacation travel in response to motor fuel shortages and higher prices. *Journal of Travel Research* 17 (4), 7–11.

Cox, P.M., Betts, R.A., Collins, M., Harris, P.P., Huntingford, C. and Jones, C.D. (2004) Amazonian forest dieback under climate-carbon cycle projections for the 21st century. *Theoretical and Applied Climatology* 78, 137–156.

Crouch, G.I. (1994) Demand elasticities for short-haul versus long-haul tourism. *Journal of Travel Research* 2, 2–7.

Curtis, I.A. (2002) Environmentally sustainable tourism: A case for carbon trading at Northern Queensland hotels and resorts. *Australian Journal of Environmental Management* 9, 27–36.

Dang, H.H., Michaelowa, A. and Tuan, D.D. (2003) Synergy of adaptation and mitigation strategies in the context of sustainable development: The case of Vietnam. *Climate Policy* 3S1, 81–96.

Daniel, H. (2001) Replenishment versus retreat: The cost of maintaining Delaware's beaches. *Ocean and Coastal Management* 44, 87–104.

de Freitas, C. (2001) Theory, concepts and methods in tourism climate research. In A. Matzarakis and C. de Freitas (eds) *International Society of Biometeorology Proceedings of the First International Workshop on Climate, Tourism and Recreation.* On WWW at www.mif.uni-freiburg.de. Accessed 20.11.04.

de Freitas, C. (2005) The climate–tourism relationship and its relevance to climate change impact assessment. In C.M. Hall and J. Higham (eds) *Tourism, Recreation and Climate Change* (pp. 29–43). Clevedon: Channel View Publications.

Deng, S. and Burnett, J. (2000) A study of energy performance of hotel buildings in Hong Kong. *Energy and Buildings* 31, 7–12.

Department for Transport (UK) (2004) *The Future of Transport*. White Paper CM 6234. On WWW at www.dft.gov.uk. Accessed 14.03.05.

Déqué, M. (2004) *Impact des activités humaines sur le climat*. On WWW at medias.dsi.cnrs.fr/imfrex/web/documents/downloads/md_marseille.pdf. Accessed 10.01.06.

Deutsche Bahn AG (2004) *Environment Mobility Check*. On WWW at www.reiseauskunft.bahn.de. Accessed 12.12.04.

Dickinson, J. and Dickinson, J. (2006) Local transport and social representations: Challenging the assumptions for sustainable tourism. *Special Issue Tourism and Transport: The Sustainability Dilemma* (ed. S. Becken & B. Lane) 14 (2), 192–208.

Dincer, I. (1999) Environmental impacts of energy. *Energy Policy* 27, 845–854.

Dupuy, J. and Grinbaum, A. (2005) Living with uncertainty: From the precautionary principle to the methodology of ongoing normative assessment. *Geoscience* 337, 457–474.

Dwyer, L. and Forsyth, P. (1996) The economic impacts of cruise tourism in Australia. *Journal of Tourism Studies* 7 (2), 36–45.

Dwyer, L., Forsyth, P. and Spurr, R. (2004) Evaluating tourism's economic effects: New and old approaches. *Tourism Management* 25, 307–317.

East Japan Railway Company (2005) *Sustainability Report 2004.* On WWW at http://www.jreast.co.jp/e/environment/pdf_2004/report2004_allpages.pdf. Accessed 13.06.05.

easyJet (2003) *Annual Report and Accounts 2003*. On WWW at www.easyjet.com. Accessed 19.01.05.

Eaton, B. and Holding, D. (1996) The evaluation of public transport alternatives to the car in British National Parks. *Journal of Transport Geography* 4 (1), 55–65.

Elliott, E. and Kiel, L.D. (2004) A complex systems approach for developing public policy toward terrorism: An agent-based approach. *Chaos, Solitons & Fractals* 20 (1), 63–68.

Elsasser, H. and Buerki, R. (2002) Climate change as a threat to tourism in the Alps. *Climate Research* 20, 253–257.

Emanuel, K. (2005) Increasing destructiveness of tropical cyclones over the past 30 years. *Nature* 436, 686–688.

Energy Efficiency and Conservation Authority (2000) *Energy-wise News*. Issue 67, March 2000. Wellington.

Energy Efficiency and Conservation Authority (2004) *Energy-wise News*. Case-study 04/01, January 2004. Wellington.

Energy Information Administration (EIA) (2002) *Average Electricity Emission Factors by State and Region*. On WWW at http://www.eia.doe.gov/oiaf/1605/e-factor.html. Accessed 20.12.04.

Enquête-Kommission 'Schutz der Erdatmosphäre' des Deutschen Bundestages (ed.) (1994) *Verkehr Band 4. Studienprogramm Teilband II.* Bonn: Economica Verlag.

Environmental and Societal Impacts Group (1997) Workshop on the Social and Economic Impacts of Weather. April 2–4, Boulder, Colorado, USA. On WWW at http://www.earthscape.org/r1/ucar01/index.html. Accessed 01.03.06.

Epstein, P. and Mills, E. (2005) *Climate Change Futures. Health, Ecological and Economic Dimensions*. A Project of the Centre for Health and the Global Environment, Harvard Medical School.

European Environment Agency (2004) *Air Pollution and Climate Change Policies in Europe: Exploring Linkages and the Added Value of an Integrated Approach*. Luxembourg: Office for Official Publications of the European Communities.

Farrell, A., Corbett, J. and Winebrake, J. (2002) Controlling air pollution from passenger ferries: Cost-effectiveness of seven technological options. Technical Paper. *Journal of Air and Waste Management Association* 52, 1399–1410.

Farrell, B. and Twining-Ward, L. (2005) Seven steps towards sustainability: Tourism in the context of new knowledge. *Journal of Sustainable Tourism* 13 (2), 109–122.

Faulkner, B. and Russell, R. (1997) Chaos and complexity in tourism: In search of a new perspective. *Pacific Tourism Review* 1, 93–102.

Faulkner, B., Moscardo, G. and Laws, E. (2001) *Tourism in the 21st Century. Lessons from Experience.* London, New York: Continuum.

Fink, A., Brücher, T., Krüger, G, Leckebusch C., Pinto, J.G. and Ulbrich, U. (2004) The 2003 European summer heatwaves and drought – synoptic diagnosis and impacts. *Weather* 59 (8), 209–216.

Fodness, D. and Milner, L. (1992) A perceptual mapping approach to theme park visitor segmentation. *Tourism Management* 13 (2), 95–101.

Foran, B., Lenzen, M., Dey, C. and Bilek, M. (2005) Integrating sustainable chain management with triple bottom line accounting. *Ecological Economics* 52, 143–157

Forsyth Research (2000) *New Zealand Domestic Travel Study 1999*. Auckland: Takapuna.

Franke, M. (2004) Competition between network carriers and low-cost carriers – retreat battle or breakthrough to a new level of efficiency? *Journal of Air Transport Management* 10, 15–21.

Funtowicz, S.O. and Ravetz, J.R. (1993) Science for the post-normal age. *Futures* 25 (7), 739–755.

Gao, X.J., Zhao, Z.C. and Giorgi, F. (2002) Changes in extreme events in regional climate simulations over East Asia. *Advances in Atmospheric Sciences* 19, 927–942.

Gao, X., Zhao, Z. and Giorgi, F. (2004) Application of a regional climate model in climate change studies in China. *Climate Change Newsletters*, China Meteorological Administration, 38–40.

Garcia, G.M., Pollard, J. and Rodriguez, R.D. (2000) Origins, management, and measurement of stress on the coast of Southern Spain. *Coastal Management* 28, 215–234.

Geisel, J. (1997) *Ökologische Aspekte zum Reisebustourismus in Luzern und Ausflugstourismus zum Pilatus*. Diplomarbeit im Studiengang Geooekologie. Luzern. On WWW at http://cobra.hta-bi.bfh.ch/Home/gsj/downloads/uni_ka/da_goek.pdf. Accessed 19.04.01.

German Advisory Council on Global Change (2002) *Charging the Use of Global Commons*. Berlin. On WWW at http://www.wbgu.de/wbgu_sn2002_engl.html. Accessed 20.11.04.

Giles, A. and Perry, A. (1998) The use of a temporal analogue to investigate the possible impact of projected global warming on the UK tourist industry. *Tourism Management* 19, 75–80.

Gillen, D. and Lall, A. (2004) Competitive advantage of low-cost carriers: Some implications for airports. *Journal of Air Transport Management* 10, 41–50.

Giorgi, F., Bi, X. and Pal, J. (2004) Mean, interannual and trends in a regional climate change experiment over Europe. II: Climate Change scenarios (2071–2100). *Climate Dynamics* 22 (6–7), 733–756.

Glantz, M. (2000) *Reducing the Impact of Environmental Emergencies through Early Warning and Preparedness: The Case of the 1997–98 El Niño*. Jointly published by UNEP, NCAR, UNU, WMO.

Golan, A., Judge, G. and Miller, D. (1996) *Maximum Entropy Econometrics: Robust Estimation with Limited Data*. Chichester, UK: John Wiley & Sons Ltd.

Goldenberg, S.B., Landsea, C.W., Mestas-Nuñez, A.M. and Gray, W.M. (2001) The recent increase in Atlantic hurricane activity: Causes and implications. *Science* 293, 474–479.

Gössling, S. (2000) Sustainable tourism development in developing countries: Some aspects of energy use. *Journal of Sustainable Tourism* 8 (5), 410–425.

Gössling, S. (2001) The consequences of tourism for sustainable water use on a tropical island: Zanzibar, Tanzania. *Journal of Environmental Management* 61 (2), 179–191.

Gössling, S. (2002a) Global environmental consequences of tourism. *Global Environmental Change* 12, 283–302.

Gössling, S. (2002b) Human–environmental relations with tourism. *Annals of Tourism Research* 29 (4), 539–556.

Gössling, S., Borgström Hansson, C., Hörstmeier, O. and Saggel, S. (2002) Ecological footprint analysis as a tool to assess tourism sustainability. *Ecological Economics* 43 (2–3), 199–211.

Gössling, S., Bredberg, M., Randow, A., Sandström, E. and Svensson, P. (2005a) Tourist perceptions of climate change: A study of international tourists in Zanzibar. *Current Issues in Tourism* 9 (4&5), 419–435.

Gössling, S. and Hall, M. (2006) Uncertainties in predicting tourist flows under scenarios of climate change. Editorial Essay. *Climatic Change* 79 (3–4), 163–173.

Gössling, S., Lindén, O., Helmersson, J., Liljenberg, J. and Quarm, S. (2007) Diving and global environmental change: A Mauritius case study. In B. Garrod and S. Gössling (eds) *New Frontiers in Marine Tourism: Diving Experiences, Management and Sustainability.* Amsterdam: Elsevier.

Gössling, S., Peeters, P., Ceron, J.P., Dubois, G., Patterson, T. and Richardson, R. (2005c) The eco-efficiency of tourism. *Ecological Economics* 54 (4), 417–434.

Government of China (2004) *The People's Republic of China Initial National Communication on Climate Change, Executive Summary*. Submitted to the United Nations Framework Convention on Climate Change. Beijing, China.

Government of Maldives (2003) First National Communication of the Republic of Maldives to the United Nations Framework Convention on Climate Change. Malé, Maldives.

Gregory, J.M., Stott, P.A., Cresswell, D.J., Rayner, N.A., Gordon, C. and Sexton, D.M.H. (2002) Recent and future changes in Arctic sea ice simulated by the HadCM3 AOGCM. *Geophysical Research Letters* 29, 281–284.

Gregory, J.M, Huybrechts, P. and Raper, C.B. (2004) Threatened loss of the Greenland ice-sheet. *Nature* 428 (8), 616.

Haeberli, W. and Burn, C.R. (2002) Natural hazards in forests: Glacier and permafrost effects as related to climate change. In R.C. Slide (ed.) *Environmental Change and Geomorphic Hazards in Forest*. IUFRO Research Series 9. Wallingford/New York: CABI Publishing.

Hagg, W. and Braun, L. (2005) The influence of glacier retreat on water yield from high mountain areas: Comparison of Alps and Central Asia. In C. De Jong, R. Ranzi and D. Collins (eds) *Climate and Hydrology in Mountain Areas* (pp. 263–275). Chichester: Wiley & Sons.

Hamilton, J. and Lau, M. (2004) The role of climate information in tourist destination choice making. *Working Paper FNU-56*. Centre for Marine and Climate Research, University of Hamburg. On WWW at http://www.uni-hamburg.de/Wiss/FB/15/Sustainability/climinfo.pdf. Accessed 12.12.05.

Hamilton, J., Maddison, D. and Tol, R. (2004) The effects of climate change on international tourism. *Working Paper FNU-36.* Centre for Marine and Climate Research, Hamburg (submitted).

Hanlon, P. (2007) *Global Airlines. Competition in a Transnational Industry.* Third edition. Oxford: Butterworth-Heinemann.

Hantel, M., Ehrendorfer, M. and Haslinger, A. (2000) Climate sensitivity of snow cover duration in Austria. *International Journal of Climatology* 20 (6), 615–640.

Hay, J.E. and Sem, G. (1999) *Evaluation and Regional Synthesis of National Greenhouse Gas Inventories.* Apia, Western Samoa: South Pacific Regional Environment Programme.

Hay, J., Mimura, N., Campbell, J., Fifita, S., Koshy, K., McLean, R., Nakalevu, T., Nunn, P. and de Wet, N. (2003) *Climate Variability and Change and Sea-level Rise in the Pacific Islands Region. A Resource Book for Policy and Decision Makers, Educators and other Stakeholders.* South Pacific Regional Environment Programme. Ministry of the Environment, Japan.

Hegerl, G., Crowley, T.J., Hyde, W.T. and Frame, D.J. (2006) Climate sensitivity constrained by temperature reconstructions over the past seven centuries. *Nature* 440, 1029–1032.

Heinze, G.W. (ed.) (2000) Germany. European Conference of Ministers of Transport (ECMT) Transport and Leisure. Report of the hundred and eleventh round table on transport economics, Paris, 15–16th October 1998 (pp. 5–1). Paris: ECMT.

Hetherington, R. (1996) An input–output analysis of carbon dioxide emissions for the UK. *Energy Conversion and Management* 37, 979–984.

Hofstetter, P. (1992) *Berechnungsgrundlagen für den Energieverbrauch von Reisen*. Zürich: Büro für Analyse & Ökologie.

Hite, J.D. and Laurent, E.A. (1971) Empirical study of economic-ecologic linkages in a coastal area. *Water Resources Research*, 7 (5), 1070–1078.

Houghton, J. (2002) The challenges of energy – response to Moody-Stuart. *Ethics in Science and Environmental Politics* September 26, 47–51.

Høyer, K.G. (2000) Sustainable tourism or sustainable mobility? The Norwegian case. *Journal of Sustainable Tourism* 8 (2), 147–160.

Hsieh, S., O'Leary, J.T. and Morrison, A.M. (1992) Segmenting the international travel market by activity. *Tourism Management* 13 (2), 209–223.

Hunter, C. (2002) Sustainable tourism and the touristic ecological footprint. *Environment, Development and Sustainability* 4, 7–20.

Intergovernmental Panel on Climate Change (1995) *IPCC Second Assessment. Climate Change 1995*. A Report of the Intergovernmental Panel on Climate Change. Cambridge: Cambridge University Press.

Intergovernmental Panel on Climate Change (1996) *Revised 1996 IPCC guidelines for National Greenhouse Gas Inventories*. Reporting Instructions. On WWW at http://www.ipcc-nggip.iges.or.jp/public/gl/invs4.htm. Accessed 03.0.2.01.

Intergovernmental Panel on Climate Change (2000) *IPCC Good Practice Guidance and Uncertainty Management in national Greenhouse Gas Inventories*. Hayama, Japan: Institute for Global Environmental Strategies.

Intergovernmental Panel on Climate Change (2001) *IPCC Third Assessment. Climate Change 2001*. A Report of the Intergovernmental Panel on Climate Change. Cambridge: Cambridge University Press.

International Civil Aviation Organisation (2004) Assembly–35th Session Executive Committee. Draft text for the report on Agenda Item 15. A35-WF/314, EX/116.

International Hotels Environment Initiative (IHEI) (1993) *Environmental Management for Hotels.* The Industry Guide to Best Practice. Oxford: Butterworth-Heinemann.

Jackson, I. (2002) Potential impact of climate change on tourism. Issues Paper, prepared for the OAS – Mainstreaming Adaptation to Climate Change (MACC) Project. 40pp.

Jackson, T. (2005) *Motivating Sustainable Consumption. A Review of Evidence on Consumer Behaviour and Behavioural Change*. University of Surrey, UK: Sustainable Development Research Network.

Janić, M. (1999) Aviation and externalities: The accomplishments and problems. *Transportation Research Part D* 4D (3), 159–180.

Jennings, S. (2004) Coastal tourism and shoreline management. *Annals of Tourism Research* 31 (4), 899–922.
Johnson, D. (2002) Environmentally sustainable cruise tourism: A reality check. *Marine Policy* 26, 261–270.
Jones, G.V., White, M.A., Cooper, O.R. and Storchmann, K. (2005) Climate change and global wine quality. *Climatic Change* 73, 319–343.
Jones, P. (1983) The practical application of activity-based approaches in transport planning: An assessment. In S.M. Carpenter and P. Jones (eds) *Recent Advances in Travel Demand Analysis* (pp. 56–77). Aldershot, Hants., England: Gower.
Jones, T. (2003) Impacts of climate change in coastal and marine areas. In *Climate Change and Tourism*, World Tourism Organization, Proceedings of the 1st International Conference on Climate Change and Tourism, Djerba, Tunisia, 9–11 April.
Jorgensen, M.W. and Sorenson, S.C. (1997) *Estimating Emissions from Railway Traffic. Report for the Project MEET: Methodologies for Estimating Air Pollutant Emissions from Transport*. Department of Energy Engineering, Technical University of Denmark. On WWW at www.inrets.fr/nojs/infos/cost319/MEET Deliverable17.pdf. Accessed 10.03.05.
Jump, A. and Peñuelas, J. (2005) Running to stand still: Adaptation and the response of plants to rapid climate change. *Ecology Letters* 8, 1010–1020.
Jurowski, C. and Reich, A.Z. (2000) An explanation and illustration of cluster analysis for identifying hospitality market segments. *Journal of Hospitality & Tourism Research* 24 (1), 67–91.
Kamp, B.D., Crompton, J.L. and Hensarling, D.M. (1979) Reactions of travellers to gas rationing and increases in gasoline prices. *Journal of Travel Research* 18 (1), 37–41.
Kasperson, R., Bohn, M.T. and Goble, R. (no date) Assessing the risks of a future rapid large sea level rise: a review. *Climatic Change.* On WWW at http://www.uni-hamburg.de/Wiss/FB/15/Sustainability/annex4.pdf. Accessed 20.01.06.
Kay, R. and Alder, J. (2005) *Coastal Planning and Management* (2nd edn). London, UK: E & FN Spon.
Kester, J.G.C. (2002) Preliminary results for international tourism in 2002, air transportation after 11 September. *Tourism Economics* 9 (1), 95–110.
Kjellström, E. (2004) Recent and future signatures of climate change in Europe. *Ambio* 23, 193–198.
Klein, R.J.T., Schipper, E.L.F. and Dessai, S. (2005) Integrating mitigation and adaptation into climate and development policy: Three research questions. *Environmental Science & Policy* 8, 579–588.
Knikkink, J.A.P.M. (1982) Dutch tourism in the light of the energy problem. *Revue de Tourisme – The Tourist Review – Zeitschrift für Fremdenverkehr* 3, 23–27.
Knisch, H. and Reichmuth, M. (1996) *Verkehrsleistung und Luftschadstoffemissionen des Personenflugverkehrs in Deutschland von 1980 bis 2010 unter besonderer Berücksichtigung des tourismusbedingten Luftverkehrs*. Im Auftrag des Umweltbundesamtes. Reihe TEXTE 16/96. Berlin. (in German)
Knoflacher, H. (ed.) (2000) Austria. European Conference of Ministers of Transport (ECMT) Transport and Leisure. Report of the hundred and eleventh round table on transport economics (pp. 53–88), Paris, 15–16th October 1998. Paris: ECMT.
Knutson, T.R. and Tuleya, R.E. (2004) Impact of CO_2-induced warming on simulated hurricane intensity and precipitation: Sensitivity to the choice of

climate model and convective parameterization. *Journal of Climatology* 17, 3477–3495.

König, U. (1998) Climate change and the Australian ski industry. In K. Green (ed.) *Snow: A Natural History. An Uncertain Future* (pp. 207–223). Canberra: Australian Alps Liaison Committee.

Kovacs, P. (2006) Hope for the best and prepare for the worst: How Canada's insurers stay a step ahead of climate change. *Options Politiques* December 2005/Janvier 2006, 52–56.

Kovats, S. and Casimiro, E. (2003) *Tourism, Climate Change and Human Health, The ESF LESC Exploratory Workshop on Climate Change, The Environment and Tourism: The Interactions*, Milan, 4–6 June 2003.

Krarup, S. and Ramesohl, S. (2002) Voluntary agreements on energy efficiency in industry – not a golden key, but another contribution to improve climate policy mixes. *Journal of Cleaner Production* 10, 109–120.

Kreisel, W. (2003) Trends in der Entwicklung von Freizeit und Tourismus. In C. Becker, H. Hopfinger and A. Steinecke (eds) *Geographie der Freizeit und des Tourismus: Bilanz und Ausblick*. Muenchen, Wien: Oldenburg Verlag.

Kwon, T.H. (2005) A scenario analysis of CO_2 emission trends from car travel: Great Britain 2000–2030. *Transport Policy* 12 (2), 175–184.

Lambert, S.J. and Fyfe, J.C. (2005) Changes in winter cyclone frequencies and strengths simulated in enhanced greenhouse warming experiments: Results from the models participating in the IPCC diagnostic exercise. *Climate Dynamics* 26 (7–8), 713–728.

Leckebusch, G.C. and Ulbrich, U. (2004) On the relationship between cyclones and extreme windstorm events over Europe under climate change. *Global and Planetary Science* 44, 181–193.

Lee, D.S., Lim, L.L. and Raper, S.C. (2005) The role of aviation emissions in climate stabilization scenarios. Poster at *Avoiding Dangerous Climate Change Conference*, 1–3 February, Exeter, UK.

Lehner, B., Döll, P., Alcamo, J., Henrichs, H. and Kaspar, F. (2006) Estimating the impact of global change on flood and drought risks in Europe: A continental, integrated analysis. *Climatic Change* 75 (3), 273–299.

Leiper, N. (1995) *Tourism Management.* Melbourne: RMIT Press.

Lemmen, D.S. and Warren, F.J. (eds) (2004) *Climate Change Impacts and Adaptation: A Canadian Perspective.* Climate Change Impacts and Adaptation Directorate, Natural Resources Canada.

Lenzen, M. (1998) Primary energy and greenhouse gases embodied in Australian final consumption: An input–output analysis. *Energy Policy* 26, 495–506.

Lenzen, M. (1999) Total requirements of energy and greenhouse gases for Australian transport. *Transportation Research Part D* 4D (4), 265–290.

Lenzen, M. and Dey, C.J. (2002) Economic, energy and greenhouse emissions impacts of some consumer choice, technology and government outlay options. *Energy Economics* 24, 377–403.

Leontief, W. (1947) Introduction to a theory of internal structure of functional relationships. *Econometrica* 15, 361–373.

Leontief, W. (1951) *The Structure of the American Economy, 1919–1939: An Empirical Application of Equilibrium Analysis*. New York, NY: Oxford University Press.

Leontief, W. (1970) Environmental repercussions and the economic structure: An input–output approach. *The Review of Economics and Statistics* 52, 262–271.

Leontief, W. (1972) Air pollution and the economic structure: Empirical results of input–output computations. In A. Brody and A.P. Carter (eds) *Input–Output Techniques*. Amsterdam: North-Holland.

Leontief, W., Carter, A.P. and Petri, P. (1977) *Future of the World Economy*. New York, NY: Oxford University Press.

Li, C.Y. and Xian, P. (2003) Inter-decadal variation of SST in the North Pacific and the anomalies of atmospheric circulation and climate. *Climatic and Environmental Research* 8 (3), 258–273.

Lise, W. and Tol, R. (2002) Impact of climate on tourism demand. *Climatic Change* 55 (4), 429–449.

Löfstedt, R.E. (1991) Climate change perceptions and energy-use decisions in Northern Sweden. *Global Environmental Change* September, 321–324.

Lohmann, M. and Kaim, E. (1999) Weather and holiday destination preferences, image attitude and experience. *The Tourist Review* 2, 54–64.

London Climate Change Partnership (2004) *London's Warming, A Climate Change Impacts in London Evaluation Study*. London: London Climate Change Partnership.

Lumsdon, L. (2000) Transport and tourism: Cycle tourism – a model for sustainable development? *Journal of Sustainable Tourism* 8 (5), 361–376.

Lumsdon, L., Downward, P. and Rhoden, S. (2006) Transport for tourism: Can public transport encourage a modal shift in the day visitor market? *Special Issue Tourism and Transport: The Sustainability Dilemma* (ed. S. Becken and B. Lane).

Lüthje, K. and Lindstadt, B. (1994) *Freizeit- und Ferienzentren. Umfang und regionale Verteilung. Materialien zur Raumentwicklung*, Heft 66. Bonn: Bundesforschungsanstalt für Landeskunde und Raumordnung.

Maddison, D. (2001) In search of warmer climates? The impact of climate change on flows of British tourists. *Climatic Change* 49 (1–2), 193–208.

Maibach, M., Peter, D. and Seiler, B. (1995) *Ökoinventar Transporte. Grundlagen für den ökologischen Vergleich von Transportsystemen und für den Einbezug von Transportsystemen in Ökobilanzen*. SPP Umwelt, Modul 5. Schweiz. (in German).

Maibach, M. and Schneider, C. (2002) *External costs of corridors. A comparison between air, road and rail. Air Transport Action Group. Final Report, Zurich.* On WWW at http://www.atag.org/content/showpublications.asp?folderid=435&level1=4&level2=435&. Accessed 20.10.04.

Martin, G. (2005) Weather, climate and tourism. A geographical perspective. *Annals of Tourism Research* 32 (3), 571–591.

McCool, S.F., Brackett, B., Glover, G., Lewis, G., Martin, P., Traweek, D. and Turner, A. (1974) The energy shortage and vacation travel. *Tourism and Recreation Review* 3 (2), 1–5.

McDaniels, T., Axelrod, L.J. and Slovic, P. (1996) Perceived ecological risks of global change. *Global Environmental Change* 6 (2), 159–171.

McEvoy, D., Handley, J.F., Cavan, G., Aylen, J., Lindley, S., McMorrow, J. and Glynn, S. (2006) *Climate Change and the Visitor Economy: The Challenges and Opportunities for England's Northwest.* Oxford: Sustainability Northwest and UKCIP.

McKercher, B. (1999) A chaos approach to tourism. *Tourism Management* 20, 425–434.

McLaughlin, J.B., DePaola, A. and Bopp, C.A. (2005) Outbreak of *Vibrio parahaemolyticus* gastroenteritis associated with Alaskan oysters. *New England Journal of Medicine* 353 (14), 1463–1470.

McNicols J., Shone, M. and Horn, C. (2002) *Green Globe 21 Kaikoura Community Benchmarking Pilot Study.* Tourism Research and Education Centre (TREC), Lincoln University, Report No. 53.

Meehl, G.A. and Tebaldi, C. (2004) More intense, more frequent, and longer lasting heat waves in the 21st century. *Science* 305, 994–997.
Meijers, D. (2005) Tax flight. An investigation into the origins and developments of the exemption for various kinds of taxation of international aviation. International Centre for Integrative Studies (ICIS) Working paper: I05–E001.
Metz, B. (2000) International equity in climate change policy. *Integrated Assessment* (1), 111–126.
Miller, R.E. and Blair, P.D. (1985) *Input–Output Analysis: Foundations and Extensions*. Englewood Cliffs, NJ: Prentice-Hall.
Mills, E. (2003) The insurance and risk management industries: New players in the delivery of energy-efficient and renewable energy products and services. *Energy Policy* 31, 1257–1272.
Mills, B. and Andry, J. (2002) *The Potential Impacts of Climate Change on Transportation. In Climate Change and Transportation: Potential Interactions and Impacts. Workshop Summary and Discussion Papers*. The DOT Center for Climate Change and Environmental Forecasting, US Department of Transportation.
Mills, E., Roth, R. and Lecomte, E. (2005) *Availability and Affordability of Insurance Under Climate Change: A Growing Challenge for the US*, December 2005. On WWW at www.ceres.org. Accessed 25.11.05.
Ministry for the Environment (2005) Review of Climate Change Policies. 2 November, 2005. Wellington.
Moncrief, L.W., Mouser, T.W. and Pitrak, P. (1977) The influence of gasoline price availability upon recreation travel propensity. *Energy Communications* 3 (5), 431–468.
Morabito, M., Cecchi, L., Modesti, P.A., Crisci, A., Orlandini, S., Maracchi, G. and Gensini, G.F. (2004) The impact of hot weather conditions on tourism in Florence, Italy: The summers 2002–2003 experience. *2nd International Workshop on Climate, Tourism and Recreation.* Orthodox Academy of Crete, Kolimbari, Greece, 8–11 June 2004 (pp. 158–165).
Moscovici, S. (1981) On social representations. In J. Forgas (ed.) *Social Cognition* (pp. 181–209). London: Academic Press.
Motiva (ed) (1999) *Enduser's Energy Guidebook for Schools*. On WWW at www.motiva.fi. Accessed 02.08.01.
Müller, H.R. (1992) Ecological product declaration rather than 'green' symbol schemes. *Revue de Tourisme –The Tourist Review – Zeitschrift für Fremdenverkehr* 3, 7–10.
Müller, H.R. (1999) *ETH-Pilotprojekt '2000 Watt Gesellschaft': Arbeitsgruppe 'Freizeit und Energie'*. Bern.
Müller, H.R. and Mezzasalma, R. (1992) Transport Energiebilanz: Ein erster Schritt zu einer Öko-Bilanz für Reiseveranstalter. In *Jahrbuch der Schweizerischen Tourismuswirtschaft*. St. Gallen.
Munasinghe, M. (2001) Interactions between climate change and sustainable development – an introduction. *International Journal of Global Environmental Issues* 1 (2), 123–129.
Munich Re (1998) *World Map of Natural Hazards.* Munich, Germany.
Munich Re (2005) *Topics Geo. Annual Review: Natural Catastrophes 2003.* On WWW at http://www.munichre.com. Accessed 20.11.05.
National Drought Mitigation Center (1998) *Reported Effects of the 1997–98 El Niño*. Lincoln, Nebraska, USA: National Drought Mitigation Center, 11p.
National Ski Area Association (2004) *Keep Winter Cool Campaign.* On WWW at http://www.nsaa.org/nsaa/environment/climate_change/. Accessed 12.08.05.

Neuman, J.E., Yohe, G. and Nicholls, R. (2000) *Sea-level Rise and Global Climate Change: A Review of Impacts to US Coasts*. Pew Center on Global Climate Change.

New Scientist (2005) 11 August, 12.

Nicholls, R.J. and Klein, R.J.T. (2005) Climate change and coastal management on Europe's coast. In J.E. Vermaat, L. Ledoux, K. Turner, W. Salomons and L. Bouwer (eds) *Managing European Coasts: Past, Present and Future* (pp. 199–225). Berlin: Springer. Environmental Science Monograph Series.

Noble, I. and Scholes, R.J. (2001) Sinks and the Kyoto Protocol. *Climate Policy* 1, 5–25.

O'Connor, R.E., Bord, R.J. and Fisher, A. (1999) Risk perceptions, general environmental beliefs, and willingness to address climate change. *Risk Analysis* 19 (3), 461–471.

OECD (1998) *Maintaining Prosperity in an Ageing Society.* Policy Brief, June 1998. On WWW at www.oecd.org/dataoecd/21/10/2430300.pdf. Accessed 19.01.05.

OECD (2003) *Background Paper: Estimating Global Impacts from Climate Change. OECD Workshop on the Benefits of Climate Policy: Improving Information for Policy Makers.* On WWW at http://www.oecd.org/dataoecd/9/60/2482270.pdf. Accessed 12.12.05.

Olsthoorn, X. (2001) CO_2 emissions from international aviation: 1950–2050. *Journal of Air Transport Management* 7, 87–93.

Oouchi, K., Yoshimura, J., Yoshimura, H., Mizuta, R., Kusunoki, S. and Noda, A. (2006) Tropical cyclone climatology in a global-warming climate as simulated in a 20km-mesh global atmospheric model. *Journal of Meteorological Society of Japan* 84 (2), 259–267.

Organisation of Eastern Caribbean States (2005) Grenada: Macro-economic assessment of the damages caused by Hurricane Ivan. 7 September 2004. Castries, St Lucia: Organisation of Eastern Caribbean States.

Paavola, J. and Adger, N. (2006) Fair adaptation to climate change. *Ecological Economics* 56, 594–609.

Pal, J.S., Giorgi, F. and Bi, X. (2004) Consistency of recent European summer precipitation trends and extremes with future regional climate projections. *Geophysical Research Letters* 31, L13202.

Palutikof, J.P., Subak, S. and Agnew, M.D. (eds) (1997) *Economic Impacts of the Hot Summer and Unusually Warm Year of 1995.* Norwich: University of East Anglia.

Parry, M. (ed.) (2000) *Assessment of Potential Effects and Adaptations for Climate Change in Europe (ACACIA)*. Norwich, UK: University of East Anglia.

Patterson, M.G. (1996) What is energy efficiency? Concepts, indicators and methodological issues. *Energy Policy* 24 (5), 377–392.

Patterson, M.G. and McDonald, G. (2004) *How Green and Clean is New Zealand Tourism? Lifecycle and Future Environmental Impacts*. Lincoln: Manaki Whenua Press, Landcare Research.

Patz, J.A., Campbell-Lendrum, D., Holloway, T. and Foley, J.A. (2005) Impact of regional climate change on human health. *Nature* 438 (17), 310–317.

Paul, F., Kääb, A., Maisch, M., Kellenberger, T. and Haeberli, W. (2004) Rapid disintegration of Alpine glaciers observed with satellite data. *Geophysical Research Letters* 31, L21402.

Pearce, D. (1989) *Tourist development* (2nd edn). Harlow: Longman.

Pearce, B. and Pearce, D. (2000) Setting environmental taxes for aircraft: A case study of the UK. *CSERGE Working Paper GEC 2000–26*. On WWW at http://www.uea.ac.uk/env/cserge/pub/wp/gec/gec_2000_26.pdf. Accessed 7.11.05.

Peeters, P.M. (2000) Annex I: Designing aircraft for low emissions. Technical basis for the Escape Project. In: *Escape: Economic Screening of Aircraft Preventing Emissions-Background Report*. Delft: Centrum voor Energiebesparing en Schone Technologie.

Peeters, P., van Egmond, T. and Visser, N. (2004) *European Tourism, Transport and Environment*. Final Version. Breda: NHTV CSTT.

Peeters, P., Middel, J. and Hoolhorst, A. (2005) Fuel efficiency of commercial aircraft. An overview of historical and future trends. Amsterdam: National Aerospace Laboratory, NLR-CR-2005-669.

Peeters, P. and Schouten, F. (2006) Reducing the ecological footprint of inbound tourism and transport to Amsterdam. In S. Becken and B. Lane (eds) *Tourism and Transport – the Sustainability Dilemma. Special Issue: Journal of Sustainable Tourism* 14 (20), 157–171.

Penner, J., Lister, D., Griggs, D., Dokken, D. and McFarland, M. (eds) (1999) *Aviation and the Global Atmosphere.* A Special Report of IPCC Working Groups I and III. Published for the Intergovernmental Panel on Climate Change. Cambridge: Cambridge University Press.

Perry, A. (2004) Sports tourism and climate variability. In A. Matzarakis, C.R. de Freitas and D. Scott (eds) *Advances in Tourism Climatology* (pp. 174–179). Freiburg: Berichte des Meteorologischen Institutes der Universitaet Freiburg, Nr. 12.

Pickup, J. and de Dear, R. (2000) An outdoor thermal comfort index (OUT_SET*) – Part I – the model and its assumptions. In R.J. de Dear, J.D. Kalma, T.R. Oke and A. Auliciems (eds) *Biometeorology and Urban Climatology at the Turn of the Millennium* (pp. 279–283). WCASP 50: WMO/TD No. 1026. Geneva: World Meteorological Organization.

Poumadère, M., Mays, C., Pfeifle, G. and Vafeidis, A.T. (2005) *Worst Case Scenario and Stakeholder Group Decision: A 5-6 Meter Sea Level Rise in the Rhone Delta, France*. FNU-76. Hamburg: Hamburg University and Centre for Marine and Atmospheric Science. On WWW at http://www.uni-hamburg.de/Wiss/FB/15/Sustainability/Working_Papers.htm.

Powerlight (2005) *Case Study: Mauna Lani Resort, Hawaii.* On WWW at www.powerlight.com. Accessed 19.06.05.

Price, T. and Probert, D. (1995) Environmental impacts of air traffic. *Applied Energy* 50, 133–162.

Prideaux, B. (2000) Links between transport and tourism – past, present and future. In B. Faulkner, G. Moscardo and E. Laws (eds) *Tourism in the 21st Century. Lessons from Experience*. London, New York: Continuum.

Räisänen, J. (2005) Impact of increasing CO2 on monthly-to-annual precipitation extremes: Analysis of the CMIP2 experiments. *Climate Dynamics* 24 (2–3), 309–323.

Räisänen, J., Hansson, U., Ullerstig, A., Döscher, R., Graham, L.P., Jones, C., Meier, M., Samuelsson, P. and Willén, U. (2004) European climate in the late 21st century: Regional simulations with two driving global models and two forcing scenarios. *Climate Dynamics* 22 (1), 13–31.

Raksakulthai, V. (2003) *Climate Change Impacts and Adaptation for Tourism in Phuket, Thailand*. Pathumthani, Thailand: Asian Disaster Preparedness Centre.

Rayman-Bacchus, L. and Molina, A. (2001) Internet-based tourism services: Business issues and trends. *Futures* 33 (7), 589–605.

Regional Wood Energy Development Programme in Asia (RWEDP) (1997) *Energy and Environment Basics*. Compiled in Co-operation with Technology and

Development Group, University of Twente, Netherlands. RWEDP Report No. 29. Bangkok, Thailand.
Reiser, A. and Simmons, D. (2005) A quasi-experimental method for testing the effectiveness of ecolabel promotion. *Journal of Sustainable Tourism* 13 (6), 590–616.
Rignot, E., Casassa, G., Gogineni, P., Krabill, W., Rivera, A. and Thomas, R. (2004) Accelerated ice discharge from the Antarctic Peninsula following the collapse of Larsen B ice shelf. *Geophysical Research Letters* 31 (L18401), 1–4.
Ritchie, B. (1998) Bicycle tourism in the South Island of New Zealand: Planning and management issues. *Tourism Management* 19 (6), 567–582.
Rowell, D.P. (2005) A scenario of European climate change for the late twenty-first century: Seasonal means and interannual variability. *Climate Dynamics* 25, 837–849.
Royal Commission on Environmental Pollution (2002) *The Environmental Effects of Civil Aircraft in Flight*. Special Report. London: Royal Commission on Environmental Pollution.
Ruosteenoja, K., Carter, T.R., Jylha, K. and Tuomenvirta, H. (2003) Future climate in world regions: An intercomparison of model-based projections for the new IPCC emissions scenarios. *The Finnish Environment 644*. Finland: Finnish Environment Institute.
Russell, R. and Faulkner, B. (1999) Movers and shakers: Chaos makers in tourism development. *Tourism Management* 20 (4), 411–423.
Ryan, C. and Glendon, I. (1998) Application of leisure motivation scale to tourism. *Annals of Tourism Research* 25 (1), 169–184.
Sánchez, E., Gallardo, C., Gaertner, M.A. and Arribas, M. (2004) Future climate extreme events in the Mediterranean simulated by a regional climate model. *Global and Planetary Change* 4, 163–180.
Sausen, R., Isaksen, I., Grewe, V., Hauglustaine, D., Lee, D., Myhre, G., Koehler, M., Pitari, G., Schumann, U., Stordal, F. and Zerefos, C. (2005) Aviation radiative forcing in 2000: An update on IPCC (1999). *Meteorologische Zeitschrift* 14 (4), 555–561.
Schaeffer, M., Selten, F.M., Opsteegh, J.D. and Goosse, H. (2004) The influence of ocean convection patterns on high-latitude climate projections. *Journal of Climatology* 17, 4316–4329.
Schafer, A. and Victor, D.G. (1999) Global passenger travel: Implications for carbon dioxide emissions. *Energy* 24, 657–679.
Schär, C., Vidale, P.L., Lüthi, D., Frei, C., Haeberli, C., Liniger, M. and Appenzeller, C. (2004) The role of increasing temperature variability in European summer heat waves. *Nature* 427, 332–336.
Schellnhuber, H.J. (2001) Coping with earth system complexity and irregularity. *Challenges of a Changing Earth. Conference on Global Change*, 10–13 July, Amsterdam.
Schendler, A. (2003) Applying the principles of industrial ecology to the guest-service sector. *Journal of Industrial Ecology* 7 (1), 127–138.
Schiermeier, Q. (2005) Natural disasters: The chaos to come. *Nature* 438, 903–906.
Schneeberger, C., Blatter, H., Abe-Ouchi, A. and Wild, M. (2003) Modelling change in the mass balance of glaciers of the northern hemisphere for a transient 2xCO_2 scenario. *Journal of Hydrology* 274, 62–79.
Schröter, D., Polsky, C. and Patt, A.G. (2005) Assessing vulnerabilities to the effects of global change: An eight step approach. *Mitigation and Adaptation Strategies for Global Change* 10 (4), 573–595.

Schwarb, M. and Kundzewicz, Z.W. (2004) Alpine snow cover and winter tourism in the warming climate. *Papers on Global Change*, 11: 59–72.

Scott , D., Jones, B., Lemieux, C., McBoyle, G., Mills, B., Svenson, S., Wall, G. (2002) *The Vulnerability of Winter Recreation to Climate Change in Ontario's Lakelands Tourism Region*. Department of Geography Publication Series, Occasional Paper 18, University of Waterloo.

Scott, D. (2003) Climate change and tourism in the mountain regions of North America. *Proceedings of the First International Conference on Climate Change and Tourism*. 9–11 April, Tunisia. Madrid, Spain: World Tourism Organization.

Scott, D. and McBoyle, G. (2001) Using a 'tourism climate index' to examine the implications of climate change for climate as a natural resource for tourism. A. Matzarakis and C. de Freitas (eds) *Proceedings of the First International Workshop on Climate, Tourism and Recreation*. 5–10 October 2001, International Society of Biometeorology, Commission on Climate, Tourism and Recreation, Greece.

Scott, D. and Lemieux, C. (2005) Climate change and protected area policy and planning in Canada. *The Forestry Chronicle* 81 (5), 696–703.

Scott, D., Wall, G. and McBoyle, G. (2005) The evolution of the climate change issue in the tourism sector. In M. Hall and J. Higham (eds) *Tourism, Recreation and Climate Change* (pp. 44–60). Clevedon: Channel View Publications.

Selin, C. (2006) Trust and the illusive force of scenarios. *Futures* 38, 1–14.

Shank, G.V., Rauhe, B.R. and Schwegler, B.R. (2005) *Energy Efficient Design of Hong Kong Disneyland Infrastructure: Changing Business and Design Practices.* Walt Disney Imagineering Research & Development. On WWW at http://www.ctg-net.com/energetics/YourResources/WDIHKUSGBCPaper06nov021.htm. Accessed 21.06.05.

Shaw, G. and Williams, A.M. (2002) *Critical Issues in Tourism. A Geographical Perspective* (2nd edn). Oxford: Blackwell Publishers.

Shea, E.L., Dolcemascolo, G., Barnston, A., Hamnett, M. and Lewis, N. (2001) *Preparing for a Changing Climate: The Potential Consequences of Climate Variability and Change. Pacific Islands.* Honolulu, Hawaii: US Global Research Program. East-West Center, 102pp.

Simeonova, K. and Diaz-Bone, H. (2005) Integrated climate-change strategies of industrialised countries. *Energy* 30 (14), 2537–2557.

Simmons, C. and Lewis, K. (2001) *Take Only Memories, Leave Nothing but Footprints. An Ecological Footprint Analysis of Two Package Holidays*. Rough Draft Report. Oxford: Best Foot Forward Limited.

Simmons, D. and Becken, S. (2004) Ecotourism – the cost of getting there. In R. Buckley (ed.) *Case Studies in Ecotourism* (pp. 15–23). Wallingford: CAB International.

Simpson, M. (2003) Tourism, livelihoods, biodiversity, conservation and the climate change factor in Africa. *NATO Advanced Workshop on Climate Change and Tourism: Assessment and Doping Strategies*, Warsaw, Poland, November 2003.

Skinner, C.J. and de Dear, R.J. (2000) Climate and tourism – an Australian perspective. In R.J. de Dear, J.D. Kalma, T.R. Oke and A. Auliciems (eds) *Biometeorology and Urban Climatology at the Turn of the Millennium* (pp. 239–256). WCASP 50: WMO/TD No. 1026. Geneva: World Meteorological Organization.

Snyder, C. (1998) Zero-emissions aircraft? Scenarios for aviation's growth: Opportunities for advanced technology. In: *Nasa Environmental Compatibility Research Workshop* Iii. Monterey, CA: NASA.

Soden, B.J. and Held, I.M. (2006) An assessment of climate feedbacks in coupled ocean-atmosphere models. *Journal of Climatology* 19 (14), 3354–3360.

South Pacific Tourism Organization (2005) On WWW at http://www.spto.org/spto/cms/index.shtml. Accessed 20.01.06.

Spaargaren, G. (2003) Sustainable consumption: A theoretical and environmental policy perspective. *Society and Natural Resources* 16, 687–701.

Statistics on Tourism and Research (StarUK) (2003). UK Tourism Survey. Available at (19/03/2006) at: http://www.staruk.org.uk.

Sterk, W. and Bunse, M. (2004) Voluntary Compensation of Greenhouse Gas Emissions. *Policy Paper No. 3/2004*. Wuppertal: Wuppertal Institute for Climate, Environment and Energy.

Stern, N. (2006) *Stern Review on the Economics of Climate Change.* HM Treasury. Available at http://www.hm-treasury.gov.uk/independent_reviews/stern_review_economics_climate_change/sternreview_index.cfm. Accessed 14.11.06.

Stern, P., Dietz, T., Ruttan, V.W., Socolow, R.H. and Sweeney, J.L. (1997) *Environmentally Significant Consumption. National Research Council* (USA). Washington, D.C.: National Academy Press.

Stettler, J. (1997) *Sport und Verkehr. Sportmotiviertes Verkehrsverhalten der Schweizer Bevölkerung*. Bern: Berner Studien zu Freizeit und Tourismus, 36.

Stoll-Kleemann, S., O'Riordan, T. and Jaeger, C.C. (2001) The psychology of denial concerning climate mitigation measures: evidence from Swiss focus groups. *Global Environmental Change* 11, 107–117.

Subsidiary Body for Scientific and Technical Advice (1996) *Allocation and Control of Emissions from Bunker Fuels. Methodological Issues.* Comments from Parties and an international organization. Fourth session Geneva, 16–18 December 1996. On WWW at http://unfccc.int/cop4/misc5a01.htm. Accessed 7.11.05.

Swiss Agency for Environment, Forests and Landscape (2001) *Third National Communication of Switzerland, 2001.* Submitted to the United Nations Convention on Climate Change.

Tabatchnaia-Tamirisa, N., Loke, M.K., Leung, P. and Tucker, K.A. (1997) Energy and tourism in Hawaii. *Annals of Tourism Research* 24, 390–401.

Tatham, R.L. and Dornoff, R.J. (1971) Market segmentation for outdoor recreation. *Journal of Leisure Research* 13 (1), 5–16.

Tebaldi, C., Arblaster, J.M., Hayhoe, K. and Meehl, G.A. (2006) Going to the extremes: An intercomparison of model-simulated historical and future changes in extreme events. *Climate Change* 79 (3–4), 185–211.

Thaler, R. (2004) Ecotourism needs ecotourism needs ecomobility. Problems, strategies and good practices. *OECD Workshop Leisure Travel, Tourism Travel and the Environment*, 4–5 November, Berlin.

The Economist (2005) *All-Weather Wonderland. Snowmaking*. 26 November, 87.

Thiakoulis, T. and Kaldellis, J. (2001) Combined photovoltaic and wind energy opportunities for remote islands. International Conference: *Renewable Energies for Islands*. Chania, 14–16 June, Crete, Greece.

Timmerman, A., Oberhuber, J., Bacher, A., Esch, M., Latif, M. and Roeckner, E. (1999) Increased El Niño frequency in a climate model forced by future greenhouse warming. *Nature* 398, 649–696.

Timothy, D. (1999) Participatory planning. A view of tourism in Indonesia. *Annals of Tourism Research* 26 (2), 371–391.

Titus, J. (2002) *Does Sea Level Rise Matter to Transportation along the Atlantic Coast? The Potential Impacts of Climate Change on Transportation.* In: The Potential Impacts of Climate Change on Transportation: Workshop Summary and

Proceedings, Washington D.C., October 1–2, 2002. Available at (20/01/2006) at http://climate.volpe.dot.gov/workshop1002/titus.pdf.

Todd, G. (2003) The inter-relations between tourism and climate change. *Proceedings of the First International Conference on Climate Change and Tourism*. 9–11 April, Djerba, Tunisia. Madrid, Spain: World Tourism Organization.

Tol, R. (1998) Climate change and insurance: A critical appraisal. *Energy Policy* 26 (3), 257–262.

Tompkins, E., Nicholson-Cole, S., Hurlston, L., Boyd, E., Brooks Hodge, G., Clarke, J., Gray, G., Trotz, N. and Varlack, L. (2005) *Surviving Climate Change in Small Islands – A Guidebook.* October 2005. Norwich: Tyndall Centre for Climate Change Research.

Tour Operator Initiative (2002) Work in Progress. Activity Report 12 March 2000–12 March 2002. On WWW at www.toiinitiative.org. Accessed 20.03.05.

Transportation Association of Canada (1999) *Transportation and Climate Change: Options for Action.* Options Paper of the Transportation Climate Change Table, November 1999.

Travis, D.J., Carlton, A.M. and Lauriston, R.G. (2002) Contrails reduce daily temperature range. *Nature* 418, 601.

Tretheway, M. and Mak, D. (2006) Emerging tourism markets: Ageing and developing economies. *Journal of Air Transport Management* 12, 21–27.

Tribe, J. (2005) *The Economics of Recreation, Leisure and Tourism* (3rd edn). London: Butterworth Heinemann.

Trung, D.N. and Kumar, S. (2004) Resource use and waste management in Vietnam hotel industry. *Journal of Cleaner Production* 13, 109–116.

Twinshare (2005) *Tourism Accommodation and the Environment.* On WWW at http://twinshare.crctourism.com.au/aboutTwinshare.asp. Accessed 08.06.05.

Uchiyama, T., Mizuta, R., Kamiguchi, K., Kitoh, A. and Noda, A. (2005) Changes in temperature-based extremes indices due to global warming projected by a global 20 km mesh atmospheric model. *SOLA* 1, doi:10.2151/sola.

Uemura, Y., Kai, T., Natori, R., Takahashi, T., Hatate, Y. and Yoshida, M. (2003) Potential of renewable energy sources and its applications in Yakushima Island. *Renewable Energy* 29, 581–591.

UK CEED (1998) An Assessment of the Environmental Impacts of Tourism in St Lucia. Report 5/98. *British Airways and British Airways Holidays*. Cambridge, England.

United Nations (1993) *System of National Accounts 1993.* On WWW at http://unstats.un.org/unsd/sna1993. Accessed 07.05.

United Nations (2003) *Handbook of National Accounting. Integrated Environmental and Economic Accounting.* United Nations.

United Nations Development Programme (1998) *Handbook on Methods for Climate Change Impact Assessment and Adaptation Strategies*. Nairobi: UNEP.

United Nations Environment Programme (2002) *Sustainable Tourism.* On WWW at http://www.uneptie.org/pc/tourism/sust-tourism/home.htm. Accessed 14.12.04.

United Nations Environment Programme (2003) *Switched on. Renewable Energy Opportunities for the Tourism Industry.* Paris: UNEP.

United Nations Framework Convention on Climate Change (2005) *Compendium on Methods and Tools to Evaluate Impacts of, and Vulnerability and Adaptation to, Climate Change.* Bonn: UNFCC Secretariat, 155 pp.

United Nations World Tourism Organization (1999) *Tourism Satellite Account (TSA): The Conceptual Framework*. Madrid: UNWTO.

United Nations World Tourism Organization (2001) *Tourism 2020 Vision*. Madrid: World Tourism Organization.
United Nations World Tourism Organization (2003a) *WTO Tourism Market Trends* (2003 Edition). Madrid: WTO.
United Nations World Tourism Organization (2003b) *Chinese Outbound Tourism*. Madrid, Spain: United Nations.
United Nations World Tourism Organization (2003c) *Climate Change and Tourism. Proceedings of the 1st International Conference on Climate Change and Tourism*. Djerba, Tunisia, 9–11 April 2003. Madrid, Spain: United Nations.
United Nations World Tourism Organization (2004) *Sustainable Tourism*. On WWW at http://www.world-tourism.org/sustainable/top/concepts.htm. Accessed 30.11.04.
United Nations World Tourism Organization (2005) *WTO Tourism Market Trends* (2004 Edition). Madrid: UNWTO.
Urry, J. (1995) *Consuming Places*. London: Routledge.
US National Research Council (2002) *Abrupt Climate Change-Inevitable Surprises.* Washington, D.C.: National Academy Press.
US NAST (United States National Assessment Synthesis Team) (2001) *Climate Change Impacts on the United States: The Potential Consequences of Climate Variability and Change.* Report for the US Global Change Research Program. Cambridge: Cambridge University Press.
Uyarra M., Cote I., Gill, J., Tinch, R., Viner, D. and Watkinson, A. (2005) Island-specific preferences of tourists for environmental features: Implications of climate change for tourism-dependent states. *Environmental Conservation* 32 (1), 11–19.
Van den Brink, R.M. and Van Wee, B. (2001) Why has car-fleet specific fuel consumption not shown any decrease since 1990? Quantitative analysis of Dutch passenger car-fleet specific fuel consumption. *Transportation Research Part D* 6, 75–93.
Van den Broeke, A. and Korver, W. (2003) *Tourism by the Elderly in Europe*. Department of Traffic and Transport. On WWW at www.inro.tno.nl. Accessed 17.12.04.
Van den Vate, J.F. (1997) Comparison of energy sources in terms of their full energy chain emission factors of greenhouse gases. *Energy Policy* 25 (1), 1–6.
van Oldenborgh, G.J., Philip, S.Y. and Collins, M. (2005) El Niño in a changing climate: a multi-model study. *Ocean Science* 1, 81–95.
Vavrus, S.J., Walsh, J.E., Chapman, W.L. and Portis, D. (2006) The behavior of extreme cold air outbreaks under greenhouse warming. *International Journal of Climatology* 26 (9), 1133–1147.
Viner, D. and Agnew, M. (1999) *Climate Change and its Impact on Tourism*. Report prepared for WWF, United Kingdom, July 1999.
Vinnikov, K.Y., Robock, A., Stouffer, R.J., Walsh, J.E., Parkinson, C.L., Cavalieri, D.J., Mitchell, J.F.B., Garrett, D. and Zakharov, V.C. (1999) Global warming and northern hemisphere sea ice extent. *Science* 286, 1934–1937.
Wall, G. and Badke, C. (1994) Tourism and climate change: An international perspective. *Journal of Sustainable Tourism* 2 (4), 193–203.
Wall, G. (1998) Implications of global climate change for tourism and recreation in wetland areas. *Climatic Change* 40 (2), 371–389.
Wang, G. (2005) Agricultural drought in a future climate: Results from fifteen models participating in the IPCC 4th Assessment. *Climate Dynamics* 25 (7–8), 739–753.

Wang, X.L. and Swail, V.R. (2006) Climate change signal and uncertainty in projections of ocean wave heights. *Climate Dynamics* 26 (2–3), 109–126.

Warnken, J., Bradley, M. and Guilding, C. (2004) Exploring methods and practicalities of conducting sector-wide energy consumption accounting in the tourist accommodation industry. *Ecological Economics* 48, 125–141.

Warnken, J., Bradley, M. and Guilding, C. (2005) Eco-resorts vs. mainstream accommodation providers: An investigation of the viability of benchmarking environmental performance. *Tourism Management* 26, 367–379.

Weber, K. and Ladkin, A. (2003) The conventions industry in Australia and the UK: Key issues and competitive forces. *Journal of Travel Research* 42 (2), 12–132.

Weber, C. and Perrels, A. (2000) Modelling lifestyle effects on energy demand and related emissions. *Energy Policy* 28, 549–566.

Weisheimer, A. and Palmer, T.N. (2005) Changing frequency of occurrence of extreme seasonal temperatures under global warming. *Geophysical Research Letters* 32 (20), L09806.

Wen Pan, G.W. and Laws, E. (2001) Tourism marketing opportunities for Australia in China. *Journal of Vacation Marketing* 8 (1), 39–48.

Wheeler, B. (1993) Sustaining the ego. *Journal of Sustainable Tourism* 1 (2), 121–129.

Whitelegg, J. and Cambridge, H. (2004) *Aviation and Sustainability*. Stockholm Environment Institute. On WWW at www.sei.se. Accessed 20.01.05.

Wielke, L.-M., Laimberger, L. and Hantel, M. (2004) Snow cover duration in Switzerland compared to Austria. *Meteorologische Zeitschrift* 13, 13–17.

Wilbanks, T.J. (2003) Integrating climate change and sustainable development in a place-based context. *Climate Policy* 3S1, 147–154.

Williams, P.W., Burke, J.R. and Dalton, M.J. (1979) The potential impact of gasoline futures on 1979 vacation travel strategies. *Journal of Travel Research* 18 (1), 3–7.

Williams, V. and Noland, R.B. (2006) Potential applications of innovative air transport management tools and concepts to mitigate climate impacts of tourism. Presented at *Tourism and Climate Change Mitigation Workshop*. De Spreuweel, 11–14 June 2006, The Netherlands.

Williams, V., Noland, R.B. and Toumi, R. (2003) Air transport cruise altitude restrictions to minimise contrail formation. *Climate Policy* 3, 207–219.

Wing, S. (2004) *The Synthesis of Bottom-Up and Top-Down Approaches to Climate Policy Modeling: Electric Power Technology Detail in a Social Accounting Framework*. On WWW at http://people.bu.edu/isw/papers/top-down_bottom-up_sam.pdf. Accessed 25.11.04.

Wit, R.C.N., Dings, J.M.W., Mendes de Leon, P., Thwaites, L., Peeters, P., Greenwood, D. and Doganis, R. (2002) *Economic Incentives to Mitigate Green-House Gas Emissions from Air Transport in Europe*. Delft: CE.

Wit, R., Boon, B., van Velzen, A., Cames, M., Heuber, O. and Lee, D. (2005) *Giving Wings to Emission Trading. Inclusion of Aviation under the European Emission Trading System (ETS): Design and Impacts.* Summary of Draft final report. On WWW at www.ce.nl. Accessed 02.06.05.

World Bank Group (2006) *Managing Climate Risk: Integrating Adaptation into World Bank Group Operations*. Washington: World Bank Group Global Environment Facility Program, 42 pp.

World Travel and Tourism Council (2005) *Progress and Priorities 2005/06*. London: WTTC.

Xuejie, G., Zongci, Z. and Giorgi, F. (2004) Application of a regional climate model in climate change studies in China. *Climate Change Newsletter*. China IPCC

Office and Laboratory for Climate Studies, National Climate Center, China Meteorological Administration, May 2004, 38–40.

Yamaguchi, K. and Noda, A. (2006) Global warming patterns over the North Pacific: ENSO versus AO. *Journal of Meteorological Society of Japan* 84 (1), 221–241.

Yihui, D., Ying, X., Zongci, Z., Yong, L. and Xuejie, G. (2004) Climate change scenarios over East Asia and China in the future 100 years. *Climate Change Newsletter*. China IPCC Office and Laboratory for Climate Studies, National Climate Center, China Meteorological Administration, May 2004, 2–3.

Yin, J.H. (2005) A consistent poleward shift of the storm tracks in simulations of 21st century climate. *Geophysical Research Letters* 32, L18701.

Zehr, S. (2000) Public representations of scientific uncertainty about global climate change. *Understanding of Science* 9 (2), 85–103.

Index

Abrupt change 15, 20, 34, 62, 142
Adaptive management 230, 244, 245
Air conditioning 198, 212
Aircraft emissions 36, 72, 73, 75
Alpine tourism 36, 43
Antarctic 20, 121
Arctic 134, 142, 235
Attitudes 109, 198, 200, 295
Aviation (*see* International aviation)
Awareness raising 48, 51, 276, 296

Backpacker 57, 159, 164, 191, 204
Barriers 50, 51, 130, 151, 203, 227, 230, 261, 262, 270, 281, 283, 296, 304
Beaches 22, 63, 238, 252, 254, 257, 283
Behaviour (tourist) 106, 144, 151, 171
Benchmarking 114, 209
Best practice 13, 50, 211, 271
Biodiversity 49, 58, 110, 218, 221, 223, 237-239, 242, 245, 271, 279, 288, 301, 302
Biofuel 173, 303
Bottom-up models 151
Building materials 56-59, 186, 239

Calorific value 148, 156, 158, 185
Campervans 154-156, 187
Capacity building 245, 258
Capital costs 56, 178, 211, 214, 282
Carbon credit 97, 279, 285, 293
Carbon cycle 220
Carbon offsetting 1, 111, 217, 218
Carbon sinks 58, 174, 176, 218, 221, 285
Carbon tax 98, 283
Catastrophe 62, 63, 69
Cirrus clouds 71-75, 185
Clean Development Mechanism (CDM) 268, 269, 293
Climate change policy 112, 259, 268, 274, 287, 299
Climate scenarios 27, 104, 133, 142, 224, 247, 248, 255, 270
Climate–tourism hotspots 2, 7, 23, 26, 34, 35, 45, 67, 143, 230
Climatic index 233
Coastal tourism 27, 53, 229, 230, 23, 254, 298
Community (local) 4, 43, 50, 51, 66, 128, 200, 218, 227, 242, 258, 260, 271, 276, 283, 287, 290-295
Complex system 9, 14, 32, 36, 126, 133, 150

Contrails 72, 74, 75, 185
Coral bleaching 46, 48, 49, 122, 138, 139, 254, 255, 303
Cruise ship 106, 153, 157, 194, 195
Cycle tourism 202, 203, 280
Cyclone (*see* Tropical cyclone)

Decision making 127, 282
Deforestation 18, 55, 253
Demographic (change) 42, 98, 108
Dengue fever 137, 250
Desertification 239
Destination choice 6, 21, 22, 233, 248, 304
Developed countries 13, 76, 107, 109, 166, 262, 265, 268, 273-276, 294
Diesel generators 56, 206, 214, 216
Disaster management 4, 46, 84, 106, 223, 227

Early warning system 224, 230, 252
Ecolabel 13, 288
Ecological footprint 152, 181
Economic loss 48, 60, 63, 64, 138, 237
Ecosystem services 240, 243
Ecotourism 49, 57, 62, 99, 100, 109, 112, 113, 159, 279
Education (guests) 54, 57, 248
Embodied energy 144, 150, 193, 208
Emergency 53, 68, 105, 138
Emission factors (CO_2) 147, 148, 150
Emission factors (GHG) 149
Emission scenarios 3, 127, 128, 133, 141, 176
Emission trading 77
Energy audit 146, 210, 216
Energy conservation 9, 181, 211, 212
Equity 34, 35, 130, 227, 261, 262, 272-275, 301
Erosion (coastal) 30, 31, 46, 48, 117, 121, 138, 229, 254, 255
Ethanol 185, 196, 197, 303
European Alps 26, 34, 39, 43, 13
External costs 97, 178
Extinction (species) 238, 243, 245
Extreme events 7, 99, 45-46, 53, 66-68, 117, 121-123, 133, 140, 224, 226, 242, 260, 274-279, 283, 301

Feedback loops 14, 32, 142

Fire (*see* Forest fire)
Forecasts (tourism) 126, 143
Forest fire 31, 66, 218, 235
Freshwater lens, 240
Fuel cell 177, 184, 192, 196-198, 215-217

Global climate model 20, 116, 133, 140, 141, 246
Global tourist flows 15, 86, 88, 90, 97, 275

Health risks 7, 126, 137, 224, 250-253
Heat stress 21, 27, 51, 70, 234, 250-252
Heatwaves 24, 30, 117, 123, 138, 224, 250-252, 283, 305
Hurricane Katrina 65, 71, 123, 254, 257
Hybrid cars 196

Infrastructure (tourism) 7, 42, 97, 230, 238, 253, 279, 283, 298, 303
Input–Output models 166
Insurance claims 52, 62, 64-66
International agreements 4, 76, 285
International aviation 3, 35, 36, 71, 72, 78, 81, 98
Inventory (GHG) 150, 151, 160, 161

Land-use 7, 84, 133, 178, 285, 289
Long-haul flights 71-73, 78, 79, 88, 129, 131, 181, 275
Low-cost airline 78, 83, 101-104, 163

Malaria 70, 137, 224, 238, 250, 253
Mangrove forest 55, 250, 243, 301
Market-based 77, 188
Marketing strategies 42, 45, 53
Methane 16, 120
Modal shift 201
Monsoon 60, 131, 255, 257

Non-linearity 9, 15, 19, 21, 34, 152, 274, 302
No-regret measures 50, 226, 257, 275, 278, 284, 288
North Atlantic Oscillation 37, 122

Ocean currents 30, 48, 133
Offsetting (*see* Carbon offsetting)
Outdoor activities 43, 105, 106
Ozone depletion 116, 125, 126, 172

Planning (*see* Tourism planning)
Policy (*see* Climate change policy)
Polluter-pays principle 34, 261, 262, 272-274
Poverty 110, 262, 270, 291
Precautionary principle 34, 53, 245, 261, 263, 272, 278, 300
Proactive adaptation 7, 33, 230, 298
Protected areas 242-244
Public transport 45, 70, 173, 177, 193, 199-203, 283

Reforestation 69, 219, 243, 269, 284, 285
Risk management 2, 53, 59, 82, 223, 228, 229, 282, 302

Satisfaction (visitor) 85, 223, 231, 235, 241, 248, 252
Scenarios (*see* Emission scenarios, Climate scenarios or Tourism scenarios)
Seasonality 98, 104, 115, 158
Sequestration 218, 220, 221, 285, 301
Ski industry 112, 124, 211, 258
Small businesses 83, 261, 282, 289, 293, 299
Small islands 7, 46, 50, 55, 242, 265
Snowline 8, 255
'Soft mobility' 199, 201
Solar energy 6, 57, 170, 198, 210, 213, 216, 259
Storm surge 30, 46, 49, 54, 60, 117, 121, 234, 240, 254, 289, 299
Subtropics 117, 134, 213m 250
Supply chain 85, 166, 169, 203, 278, 280, 302
Surprises 3, 6, 19, 34, 62, 117, 140-143
Sustainable consumption 111, 303
Sustainable tourism development 4, 229, 274

Technological development 81, 101, 105
Technology transfer 287, 295, 296
Top-down models 22, 151, 152
Tour operators 45, 65, 97, 104, 112, 203, 209, 218, 219, 220, 257, 279
Tourism planning 50, 51, 224, 292, 298
Tourism scenarios 129, 251, 279
Tourist surveys 161-163
Traditional knowledge 243, 287
Trains 191-193, 199, 279
Transport (tourist) 72, 90, 150, 153, 173, 174, 176, 188, 190, 198, 200
Transport management 173, 199, 201
Tropics 49, 117, 134, 139
Tropical cyclone 9, 16, 32, 60, 117, 121, 122, 139, 140, 229, 239

Urban areas 173, 201, 241, 250

Voluntary initiatives 77, 110, 112, 219, 288, 299, 303

Water shortage 54, 240, 246, 274, 305
Weather forecast 104, 105, 257
Wildlife 47, 62, 63, 285
Wind energy 211, 214
Wind turbines (or farms) 44, 67, 146, 172, 176, 198, 211, 215-217, 259, 279